FIFTH MEXICAN SCHOOL OF PARTICLES AND FIELDS

AIP CONFERENCE PROCEEDINGS 317

FIFTH MEXICAN SCHOOL OF PARTICLES AND FIELDS

GUANAJUATO, MEXICO 1992

EDITORS: J. L. LUCIO M.
M. VARGAS
UNIVERSIDAD DE
GUANAJUATO

American Institute of Physics   New York

Authorization to photocopy items for internal or personal use, beyond the free copying permitted under the 1978 U.S. Copyright Law (see statement below), is granted by the American Institute of Physics for users registered with the Copyright Clearance Center (CCC) Transactional Reporting Service, provided that the base fee of $2.00 per copy is paid directly to CCC, 27 Congress St., Salem, MA 01970. For those organizations that have been granted a photocopy license by CCC, a separate system of payment has been arranged. The fee code for users of the Transactional Reporting Service is: 0094-243X/87 $2.00.

© 1994 American Institute of Physics.

Individual readers of this volume and nonprofit libraries, acting for them, are permitted to make fair use of the material in it, such as copying an article for use in teaching or research. Permission is granted to quote from this volume in scientific work with the customary acknowledgment of the source. To reprint a figure, table, or other excerpt requires the consent of one of the original authors and notification to AIP. Republication or systematic or multiple reproduction of any material in this volume is permitted only under license from AIP. Address inquiries to Series Editor, AIP Conference Proceedings, AIP Press, American Institute of Physics, 500 Sunnyside Boulevard, Woodbury, NY 11797-2999.

L.C. Catalog Card No. 94-72720
ISBN 1-56396-378-7
DOE CONF-9211305

Printed in the United States of America.

Contents

Preface	vii
Gauge Symmetries, Topology and Quantisation	1
A. P. Balachandran	
CP Violation in the Standard Model and Beyond	82
G. C. Branco	
Introduction to Chiral Perturbation Theory	95
A. Pich	
Knot Theory and Quantum Gravity in Loop Space: A Primer	141
J. Pullin	
Quantum Groups	191
M. Ruiz-Altaba	
New Results from Experiments at the HERA Storage Ring and from ARGUS	228
D. Wegener	
An Elementary Introduction to Conformal Field Theories	348
J. Weyers	
Quantum Mechanics and Decoherent Histories	388
J. Weyers	
List of Participants	399
Author Index	403

PREFACE

The Fifth Mexican School of Particles and Fields was held in Guanajuato, Mexico from November 29 to December 11, 1992 and was devoted to topics in the frontier of theoretical and experimental physics of the fundamental interactions.

It is a pleasure to thank Professors J. Appel, A. P. Balachandran, G. C. Branco, K. Eggert, J. F. Gunion, A. Pich, J. Pullin, M. Ruiz-Altaba, D. Wegener, and J. Weyers for their excellent lectures. Their lecture notes, with the exception of those of Professors Appel, Eggert, and Gunion, form the contents of these proceedings.

The school was made possible by generous support from the following institutions: Centro de Investigación y de Estudios Avanzados del Instituto Politécnico Nacional, Centro Latinoamericano de Física, Consejo Nacional de Ciencia y Tecnología, International Physics Groups of the American Physical Society, Universidad de Guanajuato, and the Universidad Nacional Autónoma de México. In addition, we would like to take this opportunity to express our appreciation and thanks to all those who helped to make this school possible. We would especially like to acknowledge the work of our school secretaries, Claudia Rodríguez and Soledad Teudosio.

J. L. Lucio M. and M. Vargas
León, Guanajuato 1993

GAUGE SYMMETRIES, TOPOLOGY AND QUANTISATION

A.P. BALACHANDRAN
Physics Department, Syracuse University
Syracuse, NY 13244-1130

Abstract

The following two loosely connected sets of topics are reviewed in these lecture notes: 1) Gauge invariance, its treatment in field theories and its implications for internal symmetries and edge states such as those in the quantum Hall effect. 2) Quantisation on multiply connected spaces and a topological proof the spin-statistics theorem which avoids quantum field theory and relativity. Under 1), after explaining the meaning of gauge invariance and the theory of constraints, we discuss boundary conditions on gauge transformations and the definition of internal symmetries in gauge field theories. We then show how the edge states in the quantum Hall effect can be derived from the Chern-Simons action using the preceding ideas. Under 2), after explaining the significance of fibre bundles for quantum physics, we review quantisation on multiply connected spaces in detail, explaining also mathematical ideas such as those of the universal covering space and the fundamental group. These ideas are then used to prove the aforementioned topological spin-statistics theorem.

Contents

1 INTRODUCTION

2 MEANING OF GAUGE INVARIANCE
  2.1 The Action
    2.1.1 Determinism
    2.1.2 Conservation Laws
  2.2 The Lagrangian
  2.3 The Hamiltonian

3 THE DIRAC-BERGMANN THEORY OF CONSTRAINTS
  3.1 Introduction
  3.2 Constraint Analysis
  3.3 Quantization Procedure

4 GAUGE CONSTRAINTS IN FIELD THEORIES
  4.1 Gauss Law Generates Asymptotically Trivial Gauge Transformations
  4.2 Internal Symmetries in Gauge Theories
  4.3 Nonabelian Examples

5 THE QUANTUM HALL EFFECT AND THE EDGE STATES OF CHERN-SIMONS THEORY
  5.1 Introduction
  5.2 Chern-Simons Field Theory and the Quantum Hall System
  5.3 Conformal Edge Currents
    5.3.1 The Canonical Formalism
    5.3.2 Quantization
  5.4 The Chern-Simons Source as a Conformal Family

6 QUANTIZATION AND MULTIPLY CONNECTED CONFIGURATION SPACES
  6.1 Configuration Space and Quantum Theory
  6.2 The Universal Covering Space and the Fundamental Group

6.3 Examples of Multiply Connected Configuration Spaces
6.4 Quantization on Multiply Connected Configuration Spaces
6.5 Nonabelian Fundamental Groups
6.6 The Case of the Asymmetric Rotor

7 TOPOLOGICAL SPIN-STATISTICS THEOREMS

Acknowledgements

References

Figures

1. INTRODUCTION

In recent years, there have been several important developments in low dimensional quantum physics such as those associated with conformal and Chern-Simons field theories, the quantum Hall effect and anyon physics. These lecture notes will address certain aspects of these developments, in particular those concerning gauge invariance and multiple connectivity and their consequences for low dimensional physics.

The material in these notes is organised as follows. In Chapter 2, we discuss the meaning of gauge symmetries and their distinction from conventional symmetries in general terms. The reason why gauge invariance leads to constrained Hamiltonian dynamics is also pointed out using qualitative arguments. An important tool for the quantisation of theories with constraints is the Dirac-Bergmann theory of constraints and that is briefly reviewed in Chapter 3.

Chapters 4 and 5 deal with important technical aspects regarding the treatment of constraints in gauge field theories and some of their physical consequences. The intimate and beautiful relationship between symmetries and gauge invariance is clarified and the general theory illustrated by examples from electrodynamics and the quantum Hall effect. The relation of the edge states and source excitations in the Hall system to gauge invariance is in particular explained in Chapter 5.

The remaining Chapters deal with quantisation of classical theories in multiply connected configuration spaces. As indicated previously, this topic has assumed importance in low dimensional physics. It has an especially crucial role in Hall effect and anyon physics where fractional statistics has a basic significance, statistics being a manifestation of configuration space connectivity. The notes conclude with a proof of the spin-statistics theorem in Chapter 7 using topological methods. This proof avoids the use of relativistic quantum fields and seems well adapted to condensed matter systems where such fields are not generally of any relevance.

2. MEANING OF GAUGE INVARIANCE

The subject matter of our first few Chapters is gauge invariance and its physical implications. We will introduce the topic of gauge transformations in this Chapter, discussing it in general conceptual terms [following ref. 1] and emphasizing its distinction from ordinary (global) symmetry transformations.

2.1. The Action

The action S is a functional of fields with values in a suitable range space. The domain of the fields is a suitable parameter space.

Thus for a nonrelativistic particle, the range space may be $\mathbf{R}^3$, a point of which denotes the position of the particle. The parameter space is $\mathbf{R}^1$, a point of which denotes an instant t of time. The fields q are functions from $\mathbf{R}^1$ to $\mathbf{R}^3$. Thus, if $F(\mathbf{R}^1, \mathbf{R}^3)$ is the collection of these fields,

$$F(\mathbf{R}^1, \mathbf{R}^3) = \{q\}, \quad q = (q_1, q_2, q_3), q(t) \in \mathbf{R}^3. \tag{2.1}$$

In other words, each field q assigns a point $q(t)$ in $\mathbf{R}^3$ to each instant of time t.

For a real scalar field theory in Minkowski space M^4, the parameter space is M^4, the range space is $\mathbf{R}^1$ and the set of fields $F(\mathbf{R}^4, \mathbf{R}^1)$ is the set of functions from $\mathbf{R}^4$ to $\mathbf{R}^1$.

Let us denote the parameter space by D, the range space by R and the set of fields by $F(D,R)$. Then the action S is a function on $F(D,R)$ with values in $\mathbf{R}^1$. It assigns a real number $S(f)$ to each $f \in F(D,R)$. For instance, in the nonrelativistic example cited above.

$$S(q) = \frac{m}{2} \int dt \frac{dq_i(t)}{dt} \frac{dq_i(t)}{dt}. \qquad (2.2)$$

[The action also depends on the limits of time integration. Since these limits are not important for us, they have been ignored here. If necessary, they can be introduced by restricting D suitably. In this case, for example, instead of $\mathbf{R}^1$, we can choose the interval $t_1 \leq t \leq t_2$ for D.]

The concept of a *global symmetry group* may be defined as follows: Suppose $G = \{g\}$ is a group with a specified action $r \to gr$ on $R \equiv \{r\}$. Then, G has a natural action $f \to gf$ on $F(D,R)$, where $(gf)(t) = gf(t)$. This group of transformations on $F(D,R)$ is the global group associated with G. We denote it by the same symbol G. We say further that G is a *global symmetry group* if

$$S(f) = S(gf) \qquad (2.3)$$

up to surface terms. For simplicity, we will assume hereafter that G is a connected Lie group.

As an example, consider the nonrelativistic free particle with $D = \{t \mid -\infty < t < \infty\}$, $R = \mathbf{R}^3$ and $G = SO(3)$. The rotation group has a standard action on $\mathbf{R}^3$. It can be "lifted" to the action $q \to gq$ on $F(\mathbf{R}^1, \mathbf{R}^3)$, where

$$[gq](t) = gq(t) \quad [\equiv (g_{ij}q_j(t))]. \qquad (2.4)$$

Thus in the usual language, g is a global rotation. Further, $SO(3)$ is a global symmetry group since for (2.4),

$$S(q) = S(gq). \qquad (2.5)$$

In contrast, the *gauge group* $\hat{\mathcal{G}}$ associated with a global group G is defined to be the set of all functions $F(D,G) = \{h\}$ from D to G [with a group composition law to be defined below]. An element h of $F(D,G)$ thus assigns an element $h(d)$ of G for each point d in D:

$$D \ni d \xrightarrow{h} h(d) \in G. \qquad (2.6)$$

[The hat for $\hat{\mathcal{G}}$ is put there to distinguish it from $\mathcal{G}$ which will occur later.] The group multiplication in $\hat{\mathcal{G}}$ is defined by $(hh')(d) = h(d)h'(d)$. This group as well has a natural action $f \to hf$ on $F(D,R)$ defined by $(hf)(d) = h(d)f(d)$. If S is invariant under $\hat{\mathcal{G}}$ (up to surface terms), that is, if $S(hf) = S(f)$+possible surface terms, then the gauge group is a *gauge symmetry group*.

It is possible that the sort of boundary conditions we impose on the set of functions in the gauge group can have serious consequences for the theory as we shall see in Chapter 4. See also ref. 2.

Let $\hat{\mathcal{G}}$ be a gauge symmetry group and let Γ be a global symmetry group where $\hat{\mathcal{G}}$ is not necessarily associated with Γ. Recall that the parameter space contains a coordinate which we identify as time t. *The profound difference between $\hat{\mathcal{G}}$ and Γ is due to the fact that $\hat{\mathcal{G}}$ contains time dependent transformations unlike Γ.* It affects the deterministic aspects of the theory and also has its impact on Noether's derivation of conservation laws. These twin aspects are manifested as constraints in the Hamiltonian framework. We can illustrate these remarks as follows:

2.1.1. Determinism

A trajectory, by which we mean a solution to the equations of motion, is a function $\bar{f} \in F(D,R)$ at which the action is an extremum. [The extremum is defined relative to a certain class of variations around $\bar{f}$. We will not discuss the details of these variations here.]

Suppose that $\bar{f}$ is a possible trajectory for a specified set of initial conditions $d^k \bar{f}/dt^k \,|_{t=0}, k = 0, 1, ..., n$. Since $\hat{\mathcal{G}}$ is a gauge symmetry group, $h\bar{f}$ is also a trajectory. Further, since the time dependence of h is at our disposal, we can *choose* h such that

$$\left. \frac{d^k(h\bar{f})(t)}{dt^k} \right|_{t=0} = \left. \frac{d^k \bar{f}(t)}{dt^k} \right|_{t=0}, k = 0, 1, ..., n. \qquad (2.7)$$

This does not constrain h to be trivial for *all* time [so that we can have $h\bar{f} \neq \bar{f}$]. The conclusion is that *there are several possible trajectories for specified initial conditions.* [We assume of course that $\hat{\mathcal{G}}$ acts nontrivially on fields.] In this sense, the theory does not determine the future from the present if the state of the system is given by the values of $\overline{f}$ and its derivatives at a given time.

In the customary formulation, determinism is restored by considering only those functions which are invariant under $\hat{\mathcal{G}}$. These gauge invariant functions and their derivatives at a given time are then *defined* to constitute the observables of the theory. (Such a definition of observables seems to have little direct bearing on whether they are accessible to experimental observation. It is a definition which is *internal* to the theory and dictated by requirements of determinism.)

In a Hamiltonian formulation with no constraints, the specification of Cauchy data (a point of phase space) allows us to uniquely specify the future state of the system (at least for sufficiently small times). The existence of a gauge symmetry group for the action S thus suggests that S should lead to a constrained Hamiltonian dynamics. This is in fact generally the case. An orderly way to treat constrained dynamics is due to Dirac and Bergmann. We will explain it briefly in the next Chapter.

2.1.2. Conservation Laws

The infinitesimal variation of S under a gauge transformation is characterized by arbitrary functions ϵ_α. If $\hat{\mathcal{G}}$ is a gauge symmetry, Noether's argument shows that there is a charge

$$Q = \int_{\overline{D}} \epsilon_\alpha Q_\alpha \qquad (2.8)$$

which is a constant of motion:

$$\frac{dQ}{dt} = 0 \ . \qquad (2.9)$$

Here $\overline{D}$ is a fixed time slice of D. Since the ϵ_α's are arbitrary functions, we can conclude that

$$Q_\alpha = 0 \ . \qquad (2.10)$$

Thus the generators of the gauge symmetry group vanish.

In electromagnetism, the analogues of (2.10) are Gauss' law

$$\vec{\nabla}\cdot\vec{E} + J_0 = 0 \qquad (2.11)$$

and the vanishing of the canonical momentum π^0 conjugate to A_0. The nonabelian generalizations of these equations are well known.

In the Hamiltonian framework, the equations $Q_\alpha = 0$ become first class constraints [cf. Chapter 3]. Quantization of the system often becomes highly nontrivial in their presence.

2.2. The Lagrangian

We will assume as previously that the theories we consider admit a choice of time. The configuration space in such a theory is usually identified with $F(\overline{D}, \mathbf{R}^1)$, where $\overline{D}$ is a fixed time slice of D. It is clear however that for precision, we should write $\overline{D}_t$ for the slice of D at time t. The customary hypothesis is that $\overline{D}_t$ for different t are diffeomorphic and that there is a natural identification of points of $\overline{D}_t$ for different times. Under these circumstances (which we assume), we are justified in writing $\overline{D}$.

As an example, consider a field theory on a four dimensional manifold with the topology of Minkowski space M^4. Slices at different times t give different three dimensional subspaces $\mathbf{R}_t^3$. Without further considerations, there is no natural identification of points of these spaces, that is, there is as yet no obvious meaning to the identity of spatial points for observations at different times. What is done in practice is as follows: On M^4, there is an action of the time translation group $\{U_\tau \mid -\infty < \tau < \infty\}$. The latter maps $\mathbf{R}_t^3$ to $\mathbf{R}_{t+\tau}^3$ in a smooth, invertible way. We then identify all points in $\mathbf{R}_t^3$ and $\mathbf{R}_{t+\tau}^3$ which are carried into each other by time translations $U_{\pm\tau}$. In the conventional coordinates $(\vec{x}, t)$,

$$U_\tau(\vec{x}, t) = (\vec{x}, t+\tau) \qquad (2.12)$$

and we think of $\vec{x}$ as referring to the *same* three dimensional point for all times.

A field $f \in F(D, R)$ restricted to a given time t is a function on $\overline{D}_t$. Since we have an identification of points of $\overline{D}_t$ for different t, the field f can be regarded as a one dimensional family of functions $f_t \in F(\overline{D}, R)$ parametrized by time. We have thus established a correspondence

$$F(D,R) \to F(\mathbf{R}^1, F(\overline{D}, R)) \tag{2.13}$$

between functions appropriate to the action principle and curves in the configuration space $F(\overline{D}, R)$.

The Lagrangian is a function of "coordinates and velocities." That is, it is a function of a point $\alpha \in F(\overline{D}, R)$ on the configuration space and of the tangent $\dot{\alpha}$ to this space at this point. This new space (a point of which is a point and a tangent at that point of the configuration space) is called the tangent bundle $T\,F(\overline{D}, R)$ on the configuration space.

When the action is reconstructed from the Lagrangian by the formula

$$S = \int dt \quad L(\alpha(t), \dot{\alpha}(t)), \tag{2.14}$$

we are integrating L along curves in the tangent bundle. This curve is not arbitrary since we require that $\dot{\alpha}(t) = d\alpha(t)/dt$. Such a curve in the tangent bundle is the "lift of a curve" from the configuration space. (It is defined by a "second order" vector field in the tangent bundle). With this restriction on curves, a curve in the tangent bundle is uniquely determined by a curve in $F(\overline{D}, R)$. Since such a curve in turn defines a function in $F(D, R)$, we recover the original interpretation of the action as a function on $F(D, R)$.

We need to investigate the action of the gauge group on the tangent bundle. It turns out that in its action on the tangent bundle, the gauge group, in its simplest version, is associated to the global group

$$\underline{G} \circledS G = \{(\ell, h) \mid \ell \in \underline{G}, \quad h \in G\} \tag{2.15}$$

where G is the global group appropriate for $\hat{\mathcal{G}}$, $\underline{G}$ is its Lie algebra and the group multiplication is

$$(\ell', h')(\ell, h) = (\ell' + Ad\ h'\ \ell, h'h) \tag{2.16}$$

[The sense in which the gauge group appropriate for the Lagrangian formalism can be thought of as associated with (2.15) will be explained below.] Here $Ad\ h'$ is the adjoint action of h' on $\underline{G}$. In the notation common in physics literature,

$$Ad\ h'\ \ell = h'\ell\ h'^{-1}. \tag{2.17}$$

Thus $\underline{G} \, \textcircled{s} \, G$ is the semi-direct product of $\underline{G}$ with G. This result has been discussed before by Sudarshan and Mukunda.

We denote the gauge group associated to $\hat{\mathcal{G}}$ at a given time by $\mathcal{G}$. It consists of functions $F(\overline{D}, G) = \{h\}$ with group multiplication defined by

$$(hh')(\bar{d}) = h(\bar{d})h'(\bar{d}), \quad \bar{d} \in \overline{D}. \tag{2.18}$$

The Lie algebra $\underline{G}$ is a group under addition and its associated gauge group $F(\overline{D}, \underline{G})$ at a given time will be denoted by $\underline{\mathcal{G}}$. Finally the gauge group associated to $\underline{G} \, \textcircled{s} \, G$ at a given time will be denoted by $\underline{\mathcal{G}} \, \textcircled{s} \, \mathcal{G}$.

In contrast to elements of $\mathcal{G}$, elements of the group $\hat{\mathcal{G}}$ introduced earlier had arbitrary time dependence. These two groups are to be carefully distinguished although both have been called gauge groups.

The group law (2.16) can be established by examining the way the action of the gauge group $\hat{\mathcal{G}}$ "projects down" to an action on coordinates and velocities. A function $f \in F(D, R)$ is transformed to hf. Thus the curve $\{\alpha(t) \in F(\overline{D}, \mathbf{R}^1)\}$ (t being time) is transformed into $\{(h\alpha)(t)\}$ where $h(t) \in \mathcal{G}$ is time dependent. Thus a point of the tangent bundle is transformed according to

$$(\alpha(t), \frac{d\alpha(t)}{dt} = \dot{\alpha}(t)) \to (h(t)\alpha(t), h(t)\frac{d\alpha(t)}{dt} + \ell(t)h(t)\alpha(t)) \tag{2.19}$$

where $l(t) \equiv \frac{dh(t)}{dt}h(t)^{-1} \in \underline{\mathcal{G}}$. In (2.19), all time dependences can henceforth be ignored since we are examining the action of the gauge group restricted to $TF(\overline{D}, R)$ at a given time. In writing (2.19), we have also assumed that the action of the gauge group is local in time, that is that

$$(h\alpha)(t) = h(t)\alpha(t) . \tag{2.20}$$

If $(h\alpha)(t)$ depends on $h(t)$ as well as (say) its derivatives $d^k h(t)/dt^k$. (2.19) will have to be modified. For Yang-Mills theories, this actually happens. (See below). We prefer to illustrate the idea without this complication. With this assumption, we can write

$$(\ell, h) \in \underline{\mathcal{G}} \, \textcircled{s} \, \mathcal{G}, \quad (\ell, h)(\alpha, \dot{\alpha}) = (h\alpha, h\dot{\alpha} + \ell(h\alpha)) . \tag{2.21}$$

The group multiplication (2.16) follows from

$$(\ell', h')(h\alpha, h\dot{\alpha} + \ell(h\alpha)) = (h'h\alpha, h'h\dot{\alpha} + (h'\ell h'^{-1})(h'h\alpha) + \ell'(h'h\alpha))$$
$$= (h'h\alpha, h'h\dot{\alpha} + (\ell' + Adh'\ell)(h'h\alpha)) =$$
$$(\ell' + Adh'\ell, h'h)(\alpha, \dot{\alpha})$$
(2.22)

The preceding considerations are easily illustrated by Yang-Mills theory where the vector potential A_μ has values in the Lie algebra $\underline{G}$ of the global group G and transforms as follows:

$$A_\mu \to h A_\mu h^{-1} + h \partial_\mu h^{-1}. \qquad (2.23)$$

Thus at a fixed time,

$$(\ell, h) A_i = h A_i h^{-1}, \qquad (2.24)$$

$$(\ell, h) A_0 = h A_0 h^{-1} - \ell \qquad (2.25)$$

where

$$\ell = \dot{h} h^{-1}. \qquad (2.26)$$

The group multiplication law (2.21) follows by considering the application of (ℓ', h') to the left hand sides of (2.24) and (2.25).

The transformation (2.25) on the configuration space variable A_0 is not local in time since (2.26) involves dh/dt. Nonetheless, the group multiplication (2.21) is unaffected.

The space on which the group is supposed to act however is not the space of A_μ, but of $(A_\mu, \dot{A}_\mu)$. If we consider the subspace $(A_i, \dot{A}_i)$, since (2.24) does not involve $\dot{h}$, we find the group $\underline{G} \circledS \mathcal{G}$. However, the argument has to be modified if $\dot{A}_0$ is considered since its transformation involves $\dot{\ell}$. An element of the gauge group is now a triple $(\ell, \dot{\ell}, h)$ with the action

$$(\ell, \dot{\ell}, h)(A_0, \dot{A}_0) = (h A_0 h^{-1} - \ell, h \dot{A}_0 h^{-1} + [\ell, h A_0 h^{-1}] - \dot{\ell}) \qquad (2.27)$$

and the multiplication law

$$\left(\ell_1, \dot{\ell}_1, h_1\right)\left(\ell_2, \dot{\ell}_2, h_2\right) = \left(\ell_1 + h_1\ell_2 h_1^{-1}, \dot{\ell}_1 + [\ell_1, h_1\ell_2 h_1^{-1}] + h_1\dot{\ell}_2 h_1^{-1}, h_1 h_2\right) \, .$$
(2.28)

The action of $(\ell, \dot{\ell}, h)$ on $(A_i, \dot{A}_i)$ is obtained from taking the derivative of (2.24). In this action, $\dot{\ell}$ is passive.

The general gauge group $\mathcal{G}_L$ at the Lagrangian level can thus in general involve $\ell, \dot{\ell}, \ddot{\ell}, \ldots$.

The group of *constant* functions from $\overline{D}$ to G is what is often called the global symmetry group. Since it is isomorphic to G, we can denote it by the same symbol G. It is a subgroup of $\mathcal{G}$ if all constant functions are allowed in $\mathcal{G}$. Thus, if the boundary conditions do not eliminate any such constant function, we can conclude the following: Since observables are gauge invariant or invariant under $\mathcal{G}$, they are invariant under the global group G. That is, all observables are globally neutral. But note however that there are as a rule conditions on the elements of $\mathcal{G}$ so that this conclusion is not always warranted.

2.3. The Hamiltonian

The Hamiltonian framework provides an algebraic formulation of the classical theory in terms of Poisson brackets (PB's). It is the essential step in the quantization of the classical theory.

In Chapter 3, we outline Dirac's procedure for setting up the canonical formalism in the presence of constraints. Certain subtle, but important aspects of this procedure involving the aforementioned boundary conditions will be explained in Chapter 4 and illustrated in Chapter 5.

3. THE DIRAC-BERGMANN THEORY OF CONSTRAINTS

3.1. Introduction

Constraints appear in the Hamiltonian formulation of all gauge theories we know of. We shall be applying the Dirac-Bergmann constraint theory for the treatment of these constraints. For readers unfamiliar with the subject, we give a very brief summary of this theory of constraints in the discussion which follows. [See refs. 3 and 2 for reviews and applications. They also contains further references on this subject.]

Let M be the space of "coordinates" appropriate to a Lagrangian L. It is the space Q on which equations of motion give trajectories if the Lagrangian is of the sort treated in elementary classical mechanics. More generally, it can be different from Q especially for gauge invariant systems. We denote the points of M by $m = (m_1, m_2, ...)$.

Now given any manifold M, it is possible to associate two spaces TM and T^*M to M. The space TM is called the tangent bundle over M. The coordinate of a point $(m, \dot m)[\dot m = (\dot m_1, \dot m_2, ...)]$ of TM can be interpreted as a position and a velocity. The Lagrangian is a function on TM. The space T^*M is called the cotangent bundle over M. The coordinate of a point $(m, p)[p = (p_1, p_2, ...)]$ of T^*M can be interpreted as a coordinate (or a "position") and a momentum so that in physicists' language, T^*M is the phase space. At each m, p belongs to the vector space dual to the vector space of velocities.

Poisson brackets (PB's) can be defined for any cotangent bundle T^*M. In the notation familiar to physicists, they read

$$\{m_i, m_j\} = \{p_i, p_j\} = 0 ,$$
$$\{m_i, p_j\} = \delta_{ij} .$$
(3.1)

Now given a Lagrangian L. there exists a map from TM to T^*M defined by

$$(m, \dot m) \rightarrow \left(m, \frac{\partial L}{\partial \dot m}(m, \dot m)\right).$$
(3.2)

If this map is globally one to one and onto, the image of TM is T^*M and we can express velocity as a function of position and momentum. This is the case in elementary mechanics and leads to the familiar rules for the passage from Lagrangian to Hamiltonian mechanics.

3.2. Constraint Analysis

It may happen, however, that the image of TM under the map (3.2) is not all of T^*M. Suppose for instance, that it is a submanifold of T^*M defined by the equations

$$P_j(m,p) = 0, \qquad j = 1, 2, \ldots . \tag{3.3}$$

Then we are dealing with a theory with constraints. The constraints P_j are said to be primary.

The functions P_j do not identically vanish on T^*M. Rather their zeros define a submanifold of T^*M. A reflection of the fact that P_j are not zero functions on T^*M is that there exist functions g on T^*M such that $\{g, P_j\}$ do not vanish on the surface $P_j = 0$. These functions g generate canonical transformations which take a point of the surface $P_j = 0$ out of this surface. It follows that it is incorrect to take PB's of arbitrary functions with both sides of the equations $P_j = 0$ and equate them. This fact is emphasized by rewriting (3.3), replacing the "strong" equality signs $=$ of these equations by "weak" equality signs $\approx$:

$$P_j(m,p) \approx 0. \tag{3.4}$$

When $P_j(m,p)$ are weakly zero, we can in general set $P_j(m,p)$ equal to zero only after evaluating all PB's.

In the presence of constraints, the Hamiltonian can be shown to be

$$\begin{aligned} H &= \dot{m}_j \frac{\partial L}{\partial \dot{m}_j}(m, \dot{m}) - L(m, \dot{m}) + v_j P_j(m, p) \\ &\equiv H_0(m, p) + v_j P_j(m, p) \end{aligned} \tag{3.5}$$

In obtaining H_0 from the first two terms of the first line, one can freely use the primary constraints. The functions v_j are as yet undetermined Lagrange multipliers. Some of them may get determined later in the analysis while the remaining ones will continue to be unknown functions with even their time dependence arbitrary.

Consistency of dynamics requires that the primary constraints are preserved in time. Thus we require that

$$\{P_m, H\} \approx 0. \tag{3.6}$$

These equations may determine some of the v_j or they may hold identically when the constraints $P_j \approx 0$ are imposed. Yet another possibility is that they lead to the "secondary constraints"

$$P'_m(q,p) \approx 0. \tag{3.7}$$

The requirement $\{P'_m, H\} \approx 0$ may determine more of the Lagrange multipliers, lead to tertiary constraints or be identically satisfied when (3.6) and (3.7) are imposed. We proceed in this fashion until no more new constraints are generated.

Let us denote all the constraints one obtains in this way by

$$C_k \approx 0. \tag{3.8}$$

Dirac divides these constraints into first class and second class constraints. First class constraints $F_\alpha \approx 0$ are those for which

$$\{F_\alpha, C_k\} \approx 0, \quad \forall k. \tag{3.9}$$

In other words, the Poisson brackets of F_α with C_k vanish on the surface defined by (3.8). The remaining constraints S_a are defined to be second class.

It can be shown that

$$\{F_\alpha, F_\beta\} = C^\gamma_{\alpha\beta} F_\gamma, \tag{3.10}$$

where $C^\gamma_{\alpha\beta}(= -C^\gamma_{\beta\alpha})$ are functions on T^*M. The proof is as follows: Eq.(3.9) implies that $\{F_\alpha, F_\beta\} = C^\gamma_{\alpha\beta} F_\gamma + D^a_{\alpha\beta} S_a$. But on using the Jacobi identity

$$\{F_\alpha, \{F_\beta, Sa\}\} + \{F_\beta\{Sa, F_\alpha\}\} + \{Sa, \{F_\alpha, F_\beta\}\} = 0,$$

we find,

$$0 \approx \{S_a, \{F_\alpha, F_\beta\}\} \approx D^b_{\alpha\beta}\{S_a, S_b\}.$$

In obtaining this result, we have used (3.9) which implies that $\{F_\alpha, S_a\}$ is of the form $\sum_k \xi^k_{\alpha a} C_k$. Now as regards S_a, we have the basic property

$$\det(\{S_a, S_b\}) \neq 0 \tag{3.11}$$

on the surface $C_k \approx 0$. Thus the matrix $(\{S_a, S_b\})$ is nonsingular on the surface $C_k \approx 0$. It then follows that $D^b_{\alpha\beta}$ weakly vanishes, proving (3.10).

Let $\mathcal{C}$ be the submanifold of T^*M defined by the constraints:

$$\mathcal{C} = \{(m,p) \mid C_k(m,p) = 0\}. \tag{3.12}$$

Then since the canonical transformations generated by F_α preserve the constraints, a point of $\mathcal{C}$ is mapped onto another point of $\mathcal{C}$ under the canonical transformations generated by F_α. Since the canonical transformations generated by S_a do not preserve the constraints, such is not the case for S_a.

Second class constraints can be eliminated by introducing the so-called Dirac brackets. They have the basic property that the Dirac bracket of S_a with any function on T^*M is weakly zero. We will not go into their details having no use for them in these lectures. Instead, we shall later follow the alternative route of finding all functions $\mathcal{F}$ with zero PB's with S_a. So long as we work with only such functions, we can use the constraints $S_a \approx 0$ as strong constraints $S_a = 0$ and eliminate variables using them even before taking PB's. Assuming that there are no first class constraints, the number N of functionally independent functions $\mathcal{F}$ is dimension of T^*M - number of S_a, $N = \dim(T^*M) - s$, s being the range of a. Thus s second class constraints eliminate s variables. Since $(\{S_a, S_b\})$ is nonsingular and antisymmetric, s is even. Since $\dim(T^*M)$ is even as well, N is even.

3.3. Quantization Procedure

Let us now imagine that there are only first class constraints and that $\mathcal{C}$ is defined exclusively by the zeros of F_α. (If there are second class constraints S_a as well, they can first be eliminated in the manner indicated above.) Dirac's prescription for the implementation of first class constraints in quantum theory is that they be imposed as conditions on the physically allowed states $|\cdot>$:

$$\hat{F}_\alpha |\cdot> = 0. \tag{3.13}$$

Here $\hat{F}_\alpha$ is the quantum operator corresponding to the classical function F_α.

The following may be observed in connection with (3.12). In writing it, there is the assumption that functions on T^*M have been realised (in some suitable sense) as operators on a vector space.

Since the PB's between F's involve only F's, this prescription is consistent (modulo factor ordering problems). That is, both sides of the equation

$$[\hat{F}_\alpha, \hat{F}_\beta] = iC_{\alpha\beta}^\gamma \hat{F}_\gamma \qquad (3.14)$$

annihilate the physical states. Here the commutator brackets $[\cdot,\cdot]$ are obtained from the PB's using the standard prescription of Dirac. [A similar argument shows that we cannot impose the conditions $\hat{S}_a \mid \cdot >= 0$ on physical states where $\hat{S}_a$ is the operator corresponding to the function S_a.]

An observable $\hat{\mathcal{O}}$ of the theory must preserve the condition (3.13) on the physical states. Requiring that $\hat{\mathcal{O}} \mid \cdot >$ is physical if $\mid \cdot >$ is, we find, for the set of quantum observables $\hat{\mathcal{A}}$, the condition

$$[\hat{\mathcal{O}}, \hat{F}_\alpha] = id_\alpha^\gamma(\hat{\mathcal{O}})\hat{F}_\gamma, \quad \hat{\mathcal{O}} \in \hat{\mathcal{A}}. \qquad (3.15)$$

For classical observables $\mathcal{O}$, this becomes

$$\{\mathcal{O}, F_\alpha\} = d_\alpha^\gamma(\mathcal{O})F_\gamma. \qquad (3.16)$$

Since the right hand side is zero on $\mathcal{C}$, we can regard $\mathcal{O}$ as a function on $\mathcal{C}$ which is constant on the orbits generated by F_α. If we regard these orbits as generating an equivalence relation $\sim$ between points of $\mathcal{C}$, then the classical observables are functions on the quotient of $\mathcal{C}$ by $\sim$. This quotient $\mathcal{C}/\sim$ may be regarded as the physical phase space. Note that if there are f first class constraints, then the dimension dim$[\mathcal{C}/\sim]$ of the physical phase space is dim$(T^*M) - 2f$, $\mathcal{C}$ having dimension dim$(T^*M) - f$ and each orbit in $\mathcal{C}$ having dimension f. [Here we assume that there is no nontrivial subgroup of the group of canonical transformations generated by F_α which leaves a point of this orbit invariant.]

An alternative method to deal with F_α consists in directly finding all the classical observables $\mathcal{O}$ and the corresponding classical PB algebra $\mathcal{A}$ of observables. This is the algebra of functions on $\mathcal{C}/\sim$. We then quantize it by replacing $\{.,.\}$ by $-i[.,.]$ and thus find $\hat{\mathcal{A}}$, and then look for a suitable representation of $\hat{\mathcal{A}}$ on a Hilbert space. In this approach, unlike in Dirac's approach, we do not first find a vector space V of vectors $\mid \cdot >$ with the property $\hat{F}_\alpha \mid \cdot >= 0$. Rather, we directly look for a representation of $\hat{\mathcal{A}}$.

In many examples, $C_{\alpha\beta}^\gamma$ are constants so that F_α generate a Lie algebra over reals and are associated with a group in a familiar manner. This group is in fact the Hamiltonian version of the group of gauge transformations for the action. Hence one says that first class constraints generate gauge transformations. An important fact one can prove is that the only undetermined Lagrange multipliers in H at the end of the constraint analysis multiply first class constraints. Since $\{\mathcal{O}, F_\alpha\} \approx 0$ for an observable, it follows that the time evolution of $\mathcal{O}$ does not depend on these arbitrary functions. Thus a well defined Cauchy problem can be posed on $\mathcal{A}$ and the time evolution of $\mathcal{O}$ can be determined uniquely from suitable initial data. The theory is therefore deterministic if we consider only $\mathcal{A}$. This ceases to be the case when nonobservables are also considered since their time evolution is influenced by the unknown Lagrange multipliers v_j. See Chapter 1 also in connection with these remarks.

Finally, we notice that there is an important symmetry structure associated with the first class constrained surfaces in phase space, the so-called BRST symmetry. [See ref. 2 for literature on this subject.] It is frequently used in the quantization of gauge theories, which are typically theories with first class constraints. We will not touch upon these considerations since we shall have no compelling reason for using the BRST approach to quantization.

4. GAUGE CONSTRAINTS IN FIELD THEORIES

4.1. Gauss Law Generates Asymptotically Trivial Gauge Transformations

In previous Chapters, we have outlined the physical reasons which lead to important distinctions between gauge invariance and invariance under time independent symmetry transformations. We have also sketched the classical theory of constraints and its extension to the quantum domain.

In this Chapter, we look more closely at gauge constraints in field theories. In field theories, even classical field theories, not all functions of fields and their conjugate momenta are admissible in the Hamiltonian formalism [3]. This is because not all functions generate well defined canonical transformations classically. Such functions, one presumes, are ill defined in quantum theory as well and are thus to be excluded. The restriction of allowed phase space functions using considerations along these lines has profound consequences for gauge field theories. It is this restriction which leads to the possibility of QCD θ-states and fractionally charged dyons, and to the edge states of Chern-Simons dynamics. The purpose of this Chapter is to explain this restriction and its physical implications.

It may be remarked that there are similar constraints on functions on the phase space $\mathcal{P}$ in classical mechanics as well. Thus in classical mechanics, we almost always deal with infinitely differentiable functions on $\mathcal{P}$ in order that all PB's and the finite canonical transformations obtained therefrom are well defined. The field theoretic conditions to be found below are conditions of this kind, and are therefore to be expected.

The sort of constraints we have in mind are best illustrated by a specific example. Let us consider the free electromagnetic field in 3+1 dimensional Minkowski space. Let the vector potential A_μ describe this field. The Lagrangian for this system contains no time derivative of A_0 so that the momentum field π_0 conjugate to A_0 vanishes weakly:

$$\pi_0 \approx 0 . \tag{4.1}$$

The momentum field π_i conjugate to A_i is the electric field and it has the equal time PB

$$\{A_i(x), E_j(y)\} = \delta_{ij}\delta^3(x-y) , \ x^0 = y^0 \tag{4.2}$$

with A_i. [All fields in this Section hereafter are at equal times and x for example is the same as $\vec{x}$.] The fields E_i are not all independent, but are also subject to the Gauss law constraint

$$\partial_i E_i \approx 0 . \tag{4.3}$$

The equations (4.1) and (4.3) constitute all the constraints in this system. They are first class, as their mutual PB is zero.

The constraint (4.1) is easy to deal with. Its PB with A_o is non-zero:

$$\{A_0(x), \pi_0(y)\} = \delta^3(x-y), x^0 = y^0. \tag{4.4}$$

It follows that A_0 is not an observable and that we can ignore it and π_0 as well hereafter and consider only functions of A_i and E_i. The latter have zero PB's with π_0 and are thus candidates for observables.

The constraint which merits delicacy of treatment is (4.3). Let us first rewrite it by smearing it with a 'test function' Λ^∞ :

$$\underline{g}^\infty(\Lambda^\infty) = \int d^3x \, \Lambda^\infty \partial_i E_i \approx 0 \,. \tag{4.5}$$

$\underline{g}^\infty(\Lambda^{(\infty)})$ is a generator of gauge transformations on A_i and E_i as shown by the PB's

$$\{A_i, \underline{g}^\infty(\Lambda^\infty)\} = -\partial_i \Lambda^\infty,$$
$$\{E_i, \underline{g}^\infty(\Lambda^\infty)\} = 0. \tag{4.6}$$

The underline on $\underline{g}^\infty$ has been put to indicate that it is associated with the Lie algebra of the gauge group rather than with the gauge group. The superscripts ∞ are to indicate certain boundary conditions at infinity which will emerge below.

The PB's of $\underline{g}^\infty$ with all quantities of interest are not well defined unless Λ^∞ is suitably restricted at spatial infinity. Such a restriction does not show up in (4.6) as it involves only the local fields A_i and E_i. Thus, consider for example the canonical expressions

$$J_i = \int d^3x \, E_j [(\vec{x} \times \vec{\nabla})_i \delta_{jk} + \theta(i)_{jk}] A_k \,,$$
$$\theta(i)_{jk} = \epsilon_{ijk} \tag{4.7}$$

for generators of rotations (components of angular momentum). The PB of J_i with $\underline{g}^\infty(\Lambda^\infty)$ can be computed by first evaluating it with $\partial_i E_i$ and then multiplying by Λ^∞ and integrating over x_i. Since

$$\{J_i, \partial \cdot E(x)\} = -\epsilon_{ijk} x_j \partial_k \, \partial \cdot E(x) \tag{4.8}$$

where $\partial \cdot E \equiv \partial_i E_i$, this method gives

$$\{J_i, \underline{g}^\infty(\Lambda^\infty)\} = \int d^3x \Lambda^\infty(x)\{J_i, \partial \cdot E\}(x)$$

$$= -\int d^3x \Lambda^\infty(x)(\vec{x} \times \vec{\nabla})_i \partial \cdot E(x)$$

$$= -\int_{|\vec{x}|\to\infty} d\Omega \mid \vec{x} \mid^2 \Lambda^\infty(x)(\vec{x} \times \tfrac{\vec{x}}{|\vec{x}|})_i \partial \cdot E(x) \quad (4.9)$$

$$+ \int d^3x [(\vec{x} \times \vec{\nabla})_i \Lambda^\infty(x)] \partial \cdot E(x)$$

$$= \underline{g}^\infty((\vec{x} \times \vec{\nabla})_i \Lambda^\infty)$$

where $d\Omega$ is the usual volume form on a two-sphere and $(\vec{x} \times \vec{\nabla})_i \Lambda^\infty$ is the function with value $(\vec{x} \times \vec{\nabla})_i \Lambda^\infty(x)$ at x.

We can also compute the PB $\{J_i, \underline{g}^\infty(\Lambda^\infty)\}$ by first evaluating the PB of $\underline{g}^\infty(\Lambda^\infty)$ with the integrand of J_i:

$$\{J_i, \underline{g}^\infty(\Lambda^\infty)\} = \int d^3x \; \{E_j[(\vec{x} \times \vec{\nabla})_i \delta_{jk} + \theta(i)_{jk}]A_k, \; \underline{g}^\infty(\Lambda^\infty)\}$$

$$= -\int d^3x \;\; E_j[(\vec{x} \times \vec{\nabla})_i \delta_{jk} + \theta(i)_{jk}]\partial_k \Lambda^\infty$$

$$= -\int d^3x \;\; E_j \partial_j (\vec{x} \times \vec{\nabla})_i \Lambda^\infty \quad (4.10)$$

$$= -\int_{|\vec{x}|\to\infty} d\Omega \mid \vec{x} \mid^2 \; \tfrac{\vec{x}\cdot\vec{E}}{|\vec{x}|}(\vec{x} \times \vec{\nabla})_i \Lambda^\infty + \underline{g}^\infty((\vec{x} \times \vec{\nabla})_i \Lambda^\infty) \;.$$

Thus the interchange of orders of integration in the evaluation of this PB changes its value unless conditions are imposed on Λ^∞. [See Chapter 5 (cf. Eq. (5.27)) or ref. 4 for another such example.] The simplest such condition is

$$\Lambda^\infty(x) \to 0 \text{ as } \mid \vec{x} \mid \to \infty \quad (4.11)$$

at some suitable rate. [We will not have to be more specific about this rate for the purposes of these notes.]

The condition (4.11) seems reasonable for our purposes. Besides J_i, there are also other functions such as momenta P_i and Lorentz boosts K_i which we must require to have well defined PB's with $\underline{g}^\infty(\Lambda^\infty)$, and they too can lead to boundary terms containing Λ^∞ like the one in (4.10). The condition (4.11) can serve to eliminate all these terms and to lead to well behaved PB's.

There is another way to look upon the boundary condition (4.11). Consider the variation of $g^\infty(\Lambda^\infty)$ under a variation δE_i of E_i:

$$\delta g^\infty(\Lambda^\infty) = \int_{|\vec{x}|\to\infty} d\Omega \,|\vec{x}|^2\, \Lambda^\infty\, \frac{\vec{x}\cdot\delta\vec{E}}{|\vec{x}|} - \int d^3x\, \partial_i\Lambda^\infty \delta E_i\,. \quad (4.12)$$

Now a function (or "functional") $\mathcal{F}$ of a collection of fields $\varphi^{(\alpha)}$ is said to be differentiable in $\varphi^{(\alpha)}$ if and only if we are able to write the variation $\delta\mathcal{F}$ of $\mathcal{F}$ under a variation $\delta\varphi^{(\alpha)}$ of $\varphi^{(\alpha)}$ in the form

$$\delta\mathcal{F} = \int d^3x\, \mathcal{F}_\alpha \delta\varphi^\alpha. \quad (4.13)$$

If (4.13) is possible, we then define the functional derivative $\delta\mathcal{F}/\delta\varphi^\alpha$ as $\mathcal{F}_\alpha$:

$$\frac{\delta\mathcal{F}}{\delta\varphi^{(\alpha)}(x)} = \mathcal{F}_\alpha[\varphi(x)], \varphi(x) = \varphi^1(x), \varphi^2(x), \ldots\,. \quad (4.14)$$

Differentiability of phase space functions in field theory is analogous to differentiability of phase space functions in classical mechanics and is among the simplest conditions we can impose to obtain well defined PB's. Comparison of (4.12) and (4.13) leads to the condition (4.11) when $g^\infty(\Lambda^\infty)$ is required to be differentiable.

Analogous considerations involving multiple PB's suggest that phase space functions may have to be infinitely differentiable while the requirement that they generate well defined canonical transformations can lead to more sophisticated conditions.

It is important to remark that if for some reason we exclude functions like J_i from consideration, then there is no reason to impose (4.11). Thus we really must examine the collection of all functionals of possible interest and their PB's before deciding on appropriate boundary conditions (BC's).

We will not study such difficult matters here, and will content ourselves with the BC (4.11). Let T^∞ denote the class of test functions Λ^∞ which fulfill the BC (4.11). Then the weak equality (4.5) is thus valid only if $\Lambda^\infty \in T^\infty$.

Using the same symbols for quantum and classical objects, it follows also that the quantum states $|\cdot>$ are annihilated only by such $\underline{g}^\infty(\Lambda^\infty)$:

$$\underline{g}^\infty(\Lambda^\infty)|\cdot>=0 \Leftrightarrow \Lambda^\infty \in T^\infty. \tag{4.15}$$

Furthermore, as we saw in Chapter 2, the observables commute with $\underline{g}^\infty(\Lambda^\infty)$.

The charge operator in electrodynamics is closely related to the Gauss law operator $\underline{g}^\infty(\Lambda^\infty)$. It is best discussed after first coupling the electromagnetic field to a charged field ψ with charge density J_0. The Gauss law and the physical state constraints (4.5) and (4.15) are then changed to

$$\underline{g}^\infty(\Lambda^\infty) = \int d^3x \Lambda^\infty [\partial_i E_i + J_0] \approx 0 \ , \quad \Lambda^\infty \in T^\infty,$$

$$\underline{g}^\infty(\Lambda^\infty)|\cdot>=0 \tag{4.16}$$

while the observables now commute with this $\underline{g}^\infty(\Lambda^\infty)$.

4.2. Internal Symmetries in Gauge Theories

It is convenient at this point to introduce some definitions. A general element $e^{i\Lambda}$ of the group $\mathcal{G}$ of gauge transformations (at a fixed time) in electrodynamics is a function (at a fixed time) on $\mathbf{R}^3$ with values in $U(1)$:

$$e^{i\Lambda} : \mathbf{R}^3 \to U(1),$$
$$x \to e^{i\Lambda(x)}. \tag{4.17}$$

It acts on A_i and ψ according to

$$A_i \to A_i + \partial_i \Lambda,$$
$$\psi \to e^{ie\Lambda} \psi \ . \tag{4.18}$$

We now wish to give names to several of its subgroups of particular interest, assuming as above that the spatial slice of spacetime is $\mathbf{R}^3$.

The group $\mathcal{G}^c$: The elements of $\mathcal{G}^c$ approach constant values as $|\vec{x}| \to \infty$. If $e^{i\Lambda^c} \in \mathcal{G}^c$, we thus have

$$\Lambda^c(x) \to \text{constant as } |\vec{x}| \to \infty \ . \tag{4.19}$$

The group $\mathcal{G}^\infty$: The elements of $\mathcal{G}^\infty$ approach 1 as $|\vec{x}| \to \infty$. Because of this boundary condition, we can identify $\mathcal{G}^\infty$ with the group of maps of the three-sphere S^3 to $U(1)$. This sphere is the one obtained by identifying all "points at ∞" of $\mathbf{R}^3$, that is by compactifying $\mathbf{R}^3$ to S^3 by adding a "point at ∞".

The group $\mathcal{G}_0^\infty$: This is the subgroup of $\mathcal{G}^\infty$ which is continuously connected to the identity. The generators of its Lie algebra are the Gauss law constraints $g^\infty(\Lambda^\infty)$ for all choices of $\Lambda^\infty \in T^\infty$.

The group G which is gauged in electrodynamics is $U(1)$, while it is $SU(3)$, in chromodynamics. All the preceding groups can be defined (in an obvious way) for the latter as well, and indeed for any choice of a Lie group G. In every case, it is easy to verify the important result that $\mathcal{G}^\infty$ is a normal subgroup of $\mathcal{G}$ (and hence of $\mathcal{G}^c$) :

$$\mathcal{G}^{(\infty)} \triangleleft \mathcal{G} . \tag{4.20}$$

And of course we have the standard result

$$\mathcal{G}_0^\infty \triangleleft \mathcal{G}^\infty . \tag{4.21}$$

But whereas $\mathcal{G}^\infty$ is the same as $\mathcal{G}_0^\infty$ when G is abelian, that is not the case when G is simple. When G is simple, we have instead the important result

$$\mathcal{G}^\infty / \mathcal{G}_0^\infty = \mathbf{Z} , \tag{4.22}$$

$\mathbf{Z}$ being the group of integers under addition. [We assume throughout this Chapter that the spatial manifold has the topology of $\mathbf{R}^3$.] For such a G, a typical element of $\mathcal{G}^\infty$ which is distinct from $\mathcal{G}_0^\infty$ is a winding number one transformation. Let us display such a transformation explicitly for $G = SU(2)$. If τ_α are Pauli matrices, then a winding number one element of $\mathcal{G}^\infty$ is $\hat{g}^\infty$ where

$$\begin{aligned} \hat{g}^\infty(x) &= e^{i\psi(r)\tau_\alpha \hat{x}_\alpha}, \\ r &= |\vec{x}|, \quad \hat{x}_\alpha = \frac{x_\alpha}{r}, \end{aligned} \tag{4.23}$$

and

$$\psi(0) = 0, \quad \psi(\infty) = 2\pi. \tag{4.24}$$

The group generated by $\hat{g}^\infty \mathcal{G}_0^\infty$ is the group Z.

Note that $\hat{g}^\infty$ is well defined at $r = 0$ because of the condition on $\psi(0)$ and becomes 1 at $r = \infty$, as it should being an element of $\mathcal{G}^\infty$.

The expression (4.23) is identical to Skyrme's ansatz in Skyrmion physics [2].

The generalization of (4.23) to simple Lie groups such as $G = SU(3)$ can be constructed by looking for example at its $SU(2)$ subgroups. Thus, if τ_α in (4.23) is replaced by λ_α ($1 \le \alpha \le 3$) where

$$\lambda_\alpha = \begin{bmatrix} \tau_\alpha & & 0 \\ & & 0 \\ 0 & 0 & 0 \end{bmatrix}, \tag{4.25}$$

then, for all x, $\hat{g}^\infty(x)$ is contained in a fixed $SU(2)$ subgroup of $SU(3)$ (realised as 3×3 unitary matrices of determinant 1) and $\hat{g}^\infty \mathcal{G}_0^\infty$ generates Z.

Now any connected Lie group is the quotient of the direct product of simple and abelian Lie groups by discrete abelian groups (which could be trivial). Using this fact, the preceding results can be generalized to arbitrary Lie groups.

We turn next to the examination of these groups in the canonical formalism and in quantum theory, limiting ourselves to $G = U(1)$ at this stage.

Closely associated to the Gauss law generator $\underline{g}^\infty(\Lambda^\infty)$ is another function obtained therefrom by partial integration and subsequent substitution of a new test function ξ for $\Lambda^{(\infty)}$. We thus consider

$$\underline{g}(\xi) = \int d^3x [-\partial_i \xi E_i + \xi J_0]. \tag{4.26}$$

It is clear that $\underline{g}(\xi)$ generates gauge transformations just as $\underline{g}^\infty(\Lambda^\infty)$ does:

$$\{A_i, \underline{g}(\xi)\} = -\partial_i \xi,$$

$$\{E_i, \underline{g}(\xi)\} = 0, \tag{4.27}$$

$$\{\psi, \underline{g}(\xi)\} = \xi \psi.$$

It furthermore appears to have no problems of differentiability in E_i regardless of boundary conditions on ξ, in contrast to what we found with g^∞. Thus we seem at first sight to have discovered the generators of $\mathcal{G}$.

But this conclusion is not quite correct. In electrodynamics, we encounter electric fields E_i which fall like $1/r^2$ as $r \equiv |\vec{x}| \to \infty$. If there is a charge distribution of compact support with total charge Q, its Coulomb field for example behaves as follows:

$$E_i = \frac{Q}{r^2}\frac{\hat{x}_i}{r} + 0\left(\frac{1}{r^3}\right) \text{ as } |\vec{x}| \to \infty, \hat{x}_i = \frac{x_i}{r}. \qquad (4.28)$$

This field in a moving frame has the behavior

$$E_i = \frac{Q}{r^2} v_i^{(-)}(\hat{x}) + 0\left(\frac{1}{r^3}\right) \text{ as } r \to \infty \qquad (4.29)$$

where $v_i^{(-)}$ is an odd function of its argument. The existence of these fields implies that $\underline{g}(\xi)$ will diverge unless we constrain ξ suitably. The simplest constraint for this purpose is

$$\xi = \Lambda^c. \qquad (4.30)$$

It is also what is universally assumed. It may be that there are more general permissible conditions on ξ compatible with the existence of $\underline{g}(\xi)$ and with Poincaré invariance. We will not however pursue this issue further here, but content ourselves with (4.29).

$\mathcal{G}^c$ is thus a canonically implementable group and presumably can be realised in quantum theory as well. As it acts on fields as a group of gauge transformations, it is also an invariance group of the Hamiltonian. In contrast, the full group $\mathcal{G}$ cannot be canonically implemented.

But $\mathcal{G}_0^\infty$ acts trivially on states and observables because of the Gauss law constraint. It is hence only the group

$$\mathcal{G}^c/\mathcal{G}_0^\infty \qquad (4.31)$$

which has a nontrivial action in the theory. As it is an invariance group of the Hamiltonian as well, we thus conclude that $\mathcal{G}^c/\mathcal{G}_0^\infty$ is the symmetry group of electrodynamics associated to $G = U(1)$. We will call it the internal symmetry group. [The full symmetry group is larger, containing for instance the Poincaré group.]

It is important to appreciate that the internal symmetry group in a gauge theory is a group like $\mathcal{G}^c/\mathcal{G}_0^\infty$. It is not necessarily G and may not even contain G. The examples below will illustrate these points.

But for $G = U(1)$ and when the spatial slice of spacetime is $\mathbf{R}^3$, it is not difficult to show that

$$\mathcal{G}^c/\mathcal{G}_0^\infty = U(1) \ . \tag{4.32}$$

The proof is as follows. As mentioned previously, $\mathcal{G}^\infty$ and $\mathcal{G}_0^\infty$ are identical for this case. Now if two elements $e^{i\Lambda_j^c}$ $(j = 1, 2)$ of $\mathcal{G}^c$ are both characterized by the same boundary values of Λ_j^c at spatial infinity, then

$$\left(e^{i\Lambda_1^c}\right)\left(e^{i\Lambda_2^c}\right)^{-1} = e^{i(\Lambda_1^c - \Lambda_2^c)} \tag{4.33}$$

approaches the value 1 at infinity and is hence an element of $\mathcal{G}_0^\infty$. Thus a coset $e^{i\Lambda^c}\mathcal{G}_0^\infty$ is entirely fixed by the value $e^{i\Lambda^c}\big|_\infty$ of any of its elements $e^{i\Lambda^c}$ at infinity, this value being independent of the choice of this element. Furthermore, the group multiplication law

$$\left(e^{i\Lambda_1^c}\mathcal{G}_0^\infty\right)\left(e^{i\Lambda_2^c}\mathcal{G}_0^\infty\right) = e^{i(\Lambda_1^c + \Lambda_2^c)}\mathcal{G}_0^\infty \tag{4.34}$$

in $\mathcal{G}^c/\mathcal{G}_0^\infty$ shows that the group multiplication law for the coset labels $e^{i\Lambda^c}\big|_\infty$ is the standard multiplication of phases. We have thus the result (4.32).

Let $\underline{g}^c(\Lambda^c)$ denote the generators of $\mathcal{G}^c$. Our discussion shows that in quantum theory, when acting on states subject to the Gauss law constraint, all that matters is the asymptotic value $\Lambda^c|_\infty = \lambda$ of Λ^c. So we may as well take the function with the constant value λ for Λ^c as the test function in $\underline{g}^c(\Lambda^c)$ and call it the generator $Q(\lambda)$ of $\mathcal{G}^c/\mathcal{G}_0^\infty$. From (4.26), we see that

$$Q(\lambda) = \lambda \int d^3x J_o \ . \tag{4.35}$$

Hence $Q(1)$ is what is called the charge Q in electrodynamics.

4.3. Nonabelian Examples

We will here limit ourselves to brief sketches about the structure of the symmetry group when we stray away from electrodynamics.

In chromodynamics, with $\mathbf{R}^3$ as the spatial slice, the discussion of test functions like Λ^∞ and Λ^c is similar to their discussion in electrodynamics. For example, the Gauss law generator in chromodynamics which generalizes $g^\infty(\Lambda^\infty)$ is

$$g^\infty(\Lambda^\infty) = \int d^3x \, Tr\Lambda^\infty D_i E_i \,, \quad \Lambda^\infty \to 0 \text{ as } r \to \infty \,. \tag{4.36}$$

Here we have not changed the notation for the generator or Λ^∞, $D_i E_i = \partial_i E_i + [A_i, E_i]$ and Λ^∞, A_i and E_i are valued in the Lie algebra of $SU(3)$. For example, $\Lambda^\infty = \Lambda^\infty_\alpha \lambda_\alpha$ where λ_α are the Gell-Mann matrices.

The symmetry group as before is $\mathcal{G}^c/\mathcal{G}^\infty_0$. But in this case, $\mathcal{G}^\infty/\mathcal{G}^\infty_0$ is Z instead of being trivial. Now one can easily show, as for electrodynamics, that $\mathcal{G}^c/\mathcal{G}^\infty$ is G. It is also easy to show that $\mathcal{G}^c/\mathcal{G}^\infty$ is $(\mathcal{G}^c/\mathcal{G}^\infty_0)/(\mathcal{G}^\infty/\mathcal{G}^\infty_0)$. In other words, the symmetry group $\mathcal{G}^c/\mathcal{G}^\infty_0$ is an extension of $G = SU(3)$ by $\mathbf{Z} = \mathcal{G}^\infty/\mathcal{G}^\infty_0$. As elements of this Z are readily seen to commute with elements of $\mathcal{G}^c/\mathcal{G}^\infty_0$, the extension is central. The symmetry group is thus a central extension of $SU(3)$ by Z.

This extension is actually trivial:

$$\mathcal{G}^c/\mathcal{G}^\infty_0 = SU(3) \times \mathbf{Z} \,. \tag{4.37}$$

The generators of $SU(3)$ in (4.36) can be obtained from the nonabelian analogue of (4.26) with the help of constant test functions [valued in the Lie algebra of $SU(3)$].

The states in quantum theory can be associated with the unitary representations of the symmetry group (4.37).

The group Z has unitary irreducible representations ρ_θ which are in one-to-one correspondence with the points of the circle S^1. The image of $n \in \mathbf{Z}$ in the UIR ρ_θ is $e^{in\theta}$:

$$\rho_\theta : n \to e^{in\theta} \,. \tag{4.38}$$

The angle θ here is the famous QCD θ parameter.

The UIR's of $SU(3)$ in (4.37) account for colour in QCD.

A more complicated and interesting example is the 't Hooft-Polyakov model for monopoles [5]. It is a model of an $SO(3)$ or $SU(2)$ gauge theory which in its simplest version contains a real Higgs field $\varphi = (\varphi_1, \varphi_2, \varphi_3)$ transforming like an $SU(2)$ triplet. The vacuum value $<\varphi>$ of φ is a constant nonzero vector which we may take to be $(0,0,v), v \neq 0$. The $U(1)$ or $SO(2)$ group of rotations in the 1-2 plane leaves this $<\varphi>$ invariant so that $SO(3)$ is said to be spontaneously broken to $U(1)$ in this model.

It was shown by 't Hooft and Polyakov that the model admits finite energy configurations of φ_i and the gauge potential A_i with the asymptotic conditions

$$\varphi_i(x) \to \varphi_i^\infty(x) = v\,\hat{x}_i \,,$$
$$A_i \equiv A_i^\alpha \tau_\alpha \to \tfrac{1}{2}\,\vec{\tau}\cdot\hat{x}\,\partial_i\,\vec{\tau}\cdot\hat{x}$$
(4.39)

for large r. It was also established that these configurations provide a field theoretic version of Dirac's magnetic monopole.

A general element of $\mathcal{G}$ for the 't Hooft-Polyakov model is a map

$$g: R^3 \to SU(2) \,,$$
$$x \to g(x) \,,$$
(4.40)

while, for the boundary conditions (4.39), the analogue of $\mathcal{G}^c$ is a certain group of gauge transformations which leave φ^∞ invariant at ∞. Let $\mathcal{G}^c$ still denote this group. It is defined as follows. Let $g^c \in \mathcal{G}^c$. Then

$$g^c: \mathbf{R}^3 \to SU(2) \,,$$
$$x \to g^c(x) \,,$$
(4.41)

$$g^c(x) \underset{r\to\infty}{\to} e^{i\alpha_\infty \vec{\tau}\cdot\hat{x}} \,,$$
(4.42)

α_∞ being a constant independent of $\hat{x}$. Such a g_c clearly leaves φ^∞ invariant.

Next suppose that g_j^c ($j = 1, 2$) have both the same asymptotic limit as $r \to \infty$. Then

$$g_1^c(x)^{-1} g_2^c(x) \underset{r\to\infty}{\to} 1$$
(4.43)

so that

$$g_1^{c-1} g_2^c \in \mathcal{G}^\infty \tag{4.44}$$

where the elements of $\mathcal{G}^\infty$ as before go to identity at infinity. A generic g^c with the asymptotic behaviour (4.42) is therefore given by

$$g^c = g_0^c\, g^\infty, \quad g^\infty \in \mathcal{G}^\infty, \tag{4.45}$$

g_0^c being a particular solution of the condition (4.42).

One such particular solution is

$$g_0^c(x) = e^{i\alpha(r)\vec{\tau}\cdot\hat{x}}, \tag{4.46}$$

where for $\alpha(r)$ we insert any one function with the properties

$$\alpha(0) = 0,$$
$$0 \leq \alpha(\infty) \equiv \alpha_\infty < 2\pi, \tag{4.47}$$

the last condition here eliminating the ambiguity in the determination of α_∞ from the asymptotic limit of g_0^c.

The symmetry group is $\mathcal{G}^c/\mathcal{G}_0^\infty$. An element of this group is $g_0^c\, g^\infty \mathcal{G}_0^\infty$. Now if h^∞ and k^∞ are two elements of $\mathcal{G}^\infty$ with the same winding number, then $h^\infty = k^\infty g_0^\infty$ for some $g_0^\infty \in \mathcal{G}_0^\infty$. Hence $h^\infty \mathcal{G}_0^\infty = k^\infty \mathcal{G}_0^\infty$, so that we can choose any one typical winding number n map for g^∞. One such choice is specified by

$$g^\infty(x) = (\hat{g}^\infty(x))^n = e^{in\psi(r)\vec{\tau}\cdot\hat{x}}. \tag{4.48}$$

We may thus choose $g_0^c\, g^\infty$ according to

$$g_0^c(x) g^\infty(x) = e^{i\beta(r)\vec{\tau}\cdot\hat{x}},$$
$$\beta(0) = 0 \tag{4.49}$$

for insertion into the expression $g_0^c\, g^\infty\, \mathcal{G}_0^\infty$. In contrast to (4.47), we here do not restrict $\beta(\infty)$. Further, as two $\beta(r)$ with the same $\beta(\infty)$ give the same element of the symmetry group, it suffices to consider one $\beta(r)$ for each $\beta(\infty)$.

For each $\beta(\infty)$, we have thus an element γ $\mathcal{G}_0^\infty$ of the symmetry group with $\gamma(x) = e^{i\beta(r)\hat{\tau}\cdot\hat{x}}$, this correspondence being onto the group. It is also one-to-one. For suppose that the images $\gamma_j \mathcal{G}_0^\infty$ ($j = 1, 2$) of $\beta_j(\infty)$ are equal, γ_j being defined by

$$\gamma_j(x) = e^{i\beta_j(r)\hat{\tau}\cdot\hat{x}} \qquad (4.50)$$

$[\beta_j(0)$ being of course zero.] Then

$$\gamma_1 \gamma_2^{-1} \in \mathcal{G}_0^\infty . \qquad (4.51)$$

Since $\gamma_1\gamma_2^{-1}(x) = e^{i[\beta_1(r)-\beta_2(r)]\hat{\tau}\cdot\hat{x}}$, it follows that

$$\beta_1(\infty) = \beta_2(\infty) . \qquad (4.52)$$

Thus the elements of $\mathcal{G}^c/\mathcal{G}_0^\infty$ are all uniquely labelled by a real number $\beta(\infty)$. A formula similar to (4.34) also shows that the group composition in $\mathcal{G}^c/\mathcal{G}_0^\infty$ induces addition as the group composition on $\beta(\infty)$.

We have thus proved the remarkable result due to Witten [5] that the symmetry group $\mathcal{G}^c/\mathcal{G}_0^\infty$ is the additive group $\mathbf{R}^1$ of real numbers:

$$\mathcal{G}^c/\mathcal{G}_0^\infty = \mathbf{R}^1 . \qquad (4.53)$$

This result is to be contrasted with (4.32), (4.37). In analogy to those expressions, we might have anticipated the symmetry group here to be $U(1) \times \mathbf{Z}$. But it is not, it is rather the nontrivial central extension $\mathbf{R}^1$ of $U(1)$ by $\mathbf{Z}$. The critical fact which leads to this result is that the map γ defined above becomes a winding number one transformation when $\beta(\infty) = 2\pi$. Had it instead been an element of $\mathcal{G}_0^\infty$, it would have acted trivially on states. In such a case, there would be periodicity of the elements of the symmetry group in $\beta(\infty)$ and this group would contain $U(1)$.

There are striking physical consequences of (4.53). The charges associated with $U(1)$ are quantized whereas those associated with $\mathbf{R}^1$ are not. Therefore, there is the possibility of fractionally charged excitations (dyons) of the 't Hooft-Polyakov monopole as first established by Witten [5].

It is to be noted that the symmetry group $\mathbf{R}^1$ of the monopole sector does not contain $U(1)$ as a subgroup even though φ was supposed to spontaneously break $SU(2)$ to $U(1)$.

The result (4.53) is valid in the monopole sector. In the vacuum sector, the symmetry group is $U(1)$ as one can readily show.

In Chapter 5, we will illustrate the application of some of the ideas developed here to Chern-Simons theories.

5. THE QUANTUM HALL EFFECT AND THE EDGE STATES OF CHERN-SIMONS THEORY

5.1. Introduction

In this Chapter, we will review certain results due to Friedman, Sokoloff, Widom and Srivastava, Fröhlich and Kerler, and Fröhlich and Zee [cf. ref. 6 and citations therein] who show that the quantum Hall (QH) system is related to the pure Chern Simons (CS) gauge theory. We then consider CS theory on a disk, and using the methods of Chapter 4, show that there are chiral currents of a conformal field theory at the edge of the disk. This result is originally due to Witten. For the QH system, the existence of these currents has been demonstrated from microscopic considerations by Halperin.

As we set the speed of light c equal to 1, the magnetic flux can be measured in units of h/e.

5.2. Chern-Simons Field Theory and the Quantum Hall System

Let us begin our discussion by examining a QH system characterized by zero longitudinal resistance. The conductivity tensor σ can then be written as

$$\sigma = \begin{pmatrix} 0 & \sigma_H \\ -\sigma_H & 0 \end{pmatrix} \tag{5.1}$$

In QH systems, σ_H is quantized and is a rational multiple of e^2/h. The idea which emerges from the works mentioned above is that this fact may have a universal explanation emerging from rational conformal field theories.

As the longitudinal conductivity σ_L is zero for a two-dimensional system with σ given by Eq. (5.1), the current density j induced by an electric field E is given by

$$j^a(\vec{x}, t) = \sigma_H \epsilon^{ab} E_b(\vec{x}, t); \quad a, b = 1, 2; \quad \epsilon^{ab} = -\epsilon^{ba}, \epsilon^{12} = 1. \quad (5.2)$$

Here $E_a = -F_{0a}$, $F_{\mu\nu}$ being the electromagnetic field strength tensor.

Now if j^0 is the charge density, then we have the continuity equation

$$\frac{\partial j^0}{\partial x^0} + \vec{\nabla} \cdot \vec{j} = 0, \quad x^0 = t. \quad (5.3)$$

Also B and E are related by the Maxwell's equation

$$\frac{\partial B}{\partial x^0} = -\epsilon^{ab} \partial_a E_b, \quad (5.4)$$

where $B = F_{12}$. Equations (5.2), (5.3) and (5.4) give

$$\sigma_H \frac{\partial B}{\partial x^0} = \frac{\partial}{\partial x^0} j^0. \quad (5.5)$$

We thus obtain

$$j^0 = \sigma_H (B + B_c). \quad (5.6)$$

Here B_c is an integration constant representing a time independent background magnetic field.

Let us assume that the three-dimensional manifold M has the topology of $R^1 \times D$ with D characterizing the two-dimensional space of the sample, and R^1 describing time. Furthermore, let $\eta = (\eta_{\mu\nu})$ be any metric of Euclidean or Lorentzian signature on M. Then Eqs. (5.2) and (5.6) can be extended to a generally covariant form valid for arbitrary metrics as well as follows.

Let

$$J_{\alpha\beta}(x) = |\operatorname{Det} \eta(x)|^{-1/2} \epsilon_{\alpha\beta\gamma} j^\gamma(x), \quad x = \vec{x}, t \quad (5.7)$$

and

$$j^\alpha(x) = \frac{1}{2} |\operatorname{Det} \eta(x)|^{1/2} \sigma_H \epsilon^{\alpha\beta\gamma} F_{\beta\gamma}(x). \quad (5.8)$$

Here, $\alpha, \beta, \gamma = 0, 1, 2$, $\epsilon_{\alpha\beta\gamma}$ is the totally antisymmetric symbol with $\epsilon_{012} = 1$ and $t = x^0$ is time. By (5.7) and (5.8),

$$J_{\alpha\beta}(x) = \sigma_H F_{\alpha\beta}(x) . \tag{5.9}$$

(5.9) reduces to (5.2) and (5.6) for a flat metric.

Using the language of differential forms, we can write Eqs. (5.9) and (5.7) as

$$J = \sigma_H F,$$
$$J = *j \tag{5.10}$$

where $J = \frac{1}{2} J_{\alpha\beta} \, dx^\alpha \wedge dx^\beta$ and $*$ is the Hodge dual. The one form $j(x)$ is defined as

$$j(x) = \sum_\alpha \left(\sum_\beta \eta_{\alpha\beta}(x) j^\beta(x) \right) dx^\alpha . \tag{5.11}$$

The continuity equation (5.3) can be written as

$$dJ = 0 \tag{5.12}$$

where d is the exterior derivative.

We shall assume that σ_H is a constant. Equation (5.10) then gives the Maxwell equations

$$dF = 0 . \tag{5.13}$$

Here, we can write $F = dA'$, $A' = A + A_c$ where A_c is the vector potential corresponding to a constant magnetic field B_c (see Eq. 5.6), A represents the vector potential of a fluctuation field due to localized sources and A' the total vector potential.

Now, Eq. (5.12) implies that

$$J = da \tag{5.14}$$

where a is a one form. Equation (5.10) can then be written in terms of the one forms a and A' as

$$da = \sigma_H dA' . \tag{5.15}$$

We now note that this last equation can be obtained from an action principle with the action S_{CS} given by

$$S_{CS} = \frac{1}{2\sigma_H} \int_M (a - \sigma_H A') \wedge d(a - \sigma_H A') \tag{5.16}$$

or in terms of components,

$$S_{CS} = \frac{1}{2\sigma_H} \int_M \epsilon^{\alpha\beta\gamma}(a_\alpha - \sigma_H A'_\alpha)\partial_\beta(a_\gamma - \sigma_H A'_\gamma) \, d^3x. \tag{5.17}$$

The overall normalization of S_{CS} is here fixed by the requirement that the coupling of A'_μ to j^μ is by the term $-j^\mu.A'_\mu$ in the Lagrangian density.

The action S_{CS} is the Chern-Simons action for the gauge field $a - \sigma_H A'$.

It is important to note at this step that the derivation of Eq. (5.16) from the QH effect is valid only in the scaling limit when both length and 1/frequency scales are large. This is because although the continuity equation (5.12) is exact, Eq. (5.2) is experimentally observed to be valid only at large distance and time scales.

The action S_{CS} can be naturally generalized to the case where there are several independently conserved electric current densities $j^{(i)}, i = 1, ...m$. For example, for m filled Landau levels, if one neglects mixing of levels (which is a good approximation due to the large gaps between Landau levels), each level can be treated as dynamically independent with electric currents in each level being separately conserved. We will not however pursue such generalizations here.

We will continue in the next Section with the exploration of the relationship between the Quantum Hall system and the Chern-Simons theory and we will demonstrate how the edge currents in a Quantum Hall system arise naturally from the Chern-Simons theory. This result is first due to Witten. We follow the approach of Balachandran et al. [4] [see also ref. 6] who derive further results in Chern-Simons theory using this approach.

5.3. Conformal Edge Currents

The Lagrangians considered in this Section follow from (5.16) by setting

$$\bar{a} = (a - \sigma_H A')[2\pi/\mid k\sigma_H \mid]^{1/2} \tag{5.18}$$

and calling $\bar{a}$ again as a, k being $\mid k \mid (\frac{\mid\sigma_H\mid}{\sigma_H})$. We do so in order to be consistent with the form of the Chern-Simons Lagrangian most frequently encountered in the literature.

In this Section, we will use natural units where $\hbar = c = 1$.

5.3.1. The Canonical Formalism

Let us start with a $U(1)$ Chern-Simons (CS) theory on the solid cylinder $D \times R^1$ (D being a disk) with action given by

$$S = \frac{k}{4\pi} \int_{D \times R^1} ada, \quad a = a_\mu dx^\mu, \quad ada \equiv a \wedge da \tag{5.19}$$

where a_μ is a real field.

The action S is invariant under diffeos of the solid cylinder and does not permit a natural choice of a time function. As the notion of time is all the same indispensable in the canonical approach, we arbitrarily choose a time function denoted henceforth by x^0. Any constant x^0 slice of the solid cylinder is then the disc D with coordinates x^1, x^2.

It is well known that the phase space of the action S is described by the equal time Poisson brackets (PB's)

$$\{a_i(x), a_j(y)\} = \epsilon_{ij} \frac{2\pi}{k} \delta^2(x-y) \text{ for } i,j = 1,2, \; \epsilon_{12} = -\epsilon_{21} = 1 \tag{5.20}$$

and the constraint

$$\partial_i a_j(x) - \partial_j a_i(x) \equiv f_{ij}(x) \approx 0 \tag{5.21}$$

where $\approx$ denotes weak equality in the sense of Dirac. [Cf. Chapter 3.] All fields are evaluated at the same time x^0 in these equations, and this will continue to be the case when dealing with the canonical formalism or quantum operators in the remainder of the paper. The connection a_0 does not occur as a coordinate of this phase space. This is because, just as in electrodynamics, its conjugate momentum is weakly zero and first class and hence eliminates a_0 as an observable.

The constraint (5.21) is somewhat loosely stated. As emphasized in Chapter 4, it is important to formulate it more accurately by first smearing it with a suitable class of "test" functions $\Lambda^{(0)}$. Thus we write, instead of (5.21),

$$g(\Lambda^{(0)}) := \frac{k}{2\pi} \int_D \Lambda^{(0)}(x) da(x) \approx 0. \tag{5.22}$$

It remains to state the space $T^{(0)}$ of test functions $\Lambda^{(0)}$. For this purpose, we recall from Chapter 4 that a functional on phase space can be relied on to generate well defined canonical transformations only if it is differentiable. The meaning and implications of this remark can be illustrated here by varying $g(\Lambda^{(0)})$ with respect to a_μ,

$$\delta g(\Lambda^{(0)}) = \frac{k}{2\pi} \left(\int_{\partial D} \Lambda^{(0)} \delta a - \int_D d\Lambda^{(0)} \delta a \right), \tag{5.23}$$

∂D being the boundary of D. By definition, $g(\Lambda^{(0)})$ is differentiable in a only if the boundary term - the first term - in (5.23) is zero. We do not wish to constrain the phase space by legislating δa itself to be zero on ∂D to achieve this goal. This is because we have a vital interest in regarding fluctuations of a on ∂D as dynamical and hence allowing canonical transformations which change boundary values of a. We are thus led to the following condition on functions $\Lambda^{(0)}$ in $T^{(0)}$:

$$\Lambda^{(0)} |_{\partial D} = 0 . \tag{5.24}$$

It is useful to illustrate the sort of troubles we will encounter if (5.24) is dropped. Consider

$$q(\Lambda) = \frac{k}{2\pi} \int_D d\Lambda a . \tag{5.25}$$

It is perfectly differentiable in a even if the function Λ is nonzero on ∂D. It creates fluctuations

$$\delta a |_{\partial D} = d\Lambda |_{\partial D} \tag{5.26}$$

of a on ∂D by canonical transformations. It is a function we wish to admit in our canonical approach. Now consider its PB with $g(\Lambda^{(0)})$:

$$\{g(\Lambda^0), q(\Lambda)\} = \frac{k}{2\pi} \int d^2x d^2y \Lambda^{(0)}(x) \epsilon^{ij} [\partial_j \Lambda(y)] \left[\frac{\partial}{\partial x^i} \delta^2(x-y)\right] \quad (5.27)$$

where $\epsilon^{ij} = \epsilon_{ij}$. This expression is quite ill defined if

$$\Lambda^{(0)}|_{\partial D} \neq 0 \, . \quad (5.28)$$

Thus integration on y gives zero for (5.27). But if we integrate on x first, treating derivatives of distributions by usual rules, one finds instead,

$$-\int_D d\Lambda^0 d\Lambda = -\int_{\partial D} \Lambda^0 d\Lambda \, . \quad (5.29)$$

Thus consistency requires the condition (5.24).

The constraints $g(\Lambda^{(0)})$ are first class since

$$\{g(\Lambda_1^{(0)}), g(\Lambda_2^{(0)})\} = \tfrac{k}{2\pi} \int_D d\Lambda_1^{(0)} d\Lambda_2^{(0)} \quad (5.30)$$

$$= \tfrac{k}{2\pi} \int_{\partial D} \Lambda_1^{(0)} d\Lambda_2^{(0)}$$

$$= 0 \text{ for } \Lambda_1^{(0)}, \Lambda_2^{(0)} \in \mathcal{T}^{(0)}.$$

$g(\Lambda^{(0)})$ generates the gauge transformation $a \to a + d\Lambda^{(0)}$ of a.

Next consider $q(\Lambda)$ where $\Lambda|_{\partial D}$ is not necessarily zero. Since

$$\{q(\Lambda), g(\Lambda^{(0)})\} = -\tfrac{k}{2\pi} \int_D d\Lambda d\Lambda^{(0)}$$

$$= \tfrac{k}{2\pi} \int_{\partial D} \Lambda^{(0)} d\Lambda = 0 \text{ for } \Lambda^{(0)} \in \mathcal{T}^{(0)} \, , \quad (5.31)$$

they are first class or the observables of the theory. More precisely, observables are obtained after identifying $q(\Lambda_1)$ with $q(\Lambda_2)$ if $(\Lambda_1 - \Lambda_2) \in \mathcal{T}^{(0)}$. For then,

$$q(\Lambda_1) - q(\Lambda_2) = -g(\Lambda_1 - \Lambda_2) \approx 0 \, . \quad (5.32)$$

The functions $q(\Lambda)$ generate gauge transformations $a \to a + d\Lambda$ involving Λ's which do not necessarily vanish on ∂D.

It may be remarked that the expression for $q(\Lambda)$ is obtained from $g(\Lambda^{(0)})$ after a partial integration and a subsequent substitution of Λ for $\Lambda^{(0)}$. It too generates gauge transformations like $g(\Lambda^{(0)})$, but the test function spaces for the two are different. The pair $q(\Lambda), g(\Lambda^{(0)})$ thus resemble the pair $\underline{g}^c(\Lambda^c), \underline{g}^\infty(\Lambda^\infty)$ of electrodynamics discussed in Chapter 4. The resemblance suggests that we think of $q(\Lambda)$ as akin to the generator of a global symmetry transformation. It is natural to do so for another reason as well : the Hamiltonian is a constraint for a first order Lagrangian such as the one we have here, and for this Hamiltonian, $q(\Lambda)$ is a constant of motion.

In quantum gravity, for asymptotically flat spatial slices, it is often the practice to include a surface term in the Hamiltonian which would otherwise have been a constraint and led to trivial evolution. However, we know of no natural choice of such a surface term, except zero, for the CS theory.

The PB's of $q(\Lambda)$'s are easy to compute :

$$\{q(\Lambda_1), q(\Lambda_2)\} = \frac{k}{2\pi} \int_D d\Lambda_1 d\Lambda_2 = \frac{k}{2\pi} \int_{\partial D} \Lambda_1 d\Lambda_2 . \qquad (5.33)$$

Remembering that the observables are characterized by boundary values of test functions, (5.33) shows that the observables generate a $U(1)$ Kac-Moody algebra localized on ∂D. [Literature must be consulted for information on Kac-Moody algebras. Knowledge of these algebras is not important for understanding this Chapter.] Note that it is a Kac-Moody algebra for "zero momentum" or "charge". For if $\Lambda \mid_{\partial D}$ is a constant, it can be extended as a constant function to all of D and then $q(\Lambda) = 0$. The central charges (given by the right hand side of (5.33)) and hence the representation of (5.33) are different for $k > 0$ and $k < 0$, a fact which reflects parity violation by the action S.

Let $\theta \ (mod \ 2\pi)$ be the coordinate on ∂D and ϕ a free massless scalar field moving with speed v on ∂D and obeying the equal time PB's

$$\{\phi(\theta), \dot{\phi}(\theta')\} = \delta(\theta - \theta') . \qquad (5.34)$$

If μ_i are test functions on ∂D and $\partial_\pm = \partial_{x^0} \pm v \partial_\theta$, then

$$\left\{\frac{1}{v} \int \mu_1(\theta) \partial_\pm \phi(\theta), \frac{1}{v} \int \mu_2(\theta) \partial_\pm \phi(\theta)\right\} = \pm 2 \int \mu_1(\theta) d\mu_2(\theta), \qquad (5.35)$$

the remaining PB's involving $\partial_\pm \varphi$ being zero. Also $\partial_\pm \partial_\pm \phi = 0$. Thus the algebra of observables is isomorphic to that generated by the left moving $\partial_+ \phi$ or the right moving $\partial_- \phi$.

5.3.2. Quantization

Our strategy for quantization relies on the observation that if

$$\Lambda \mid_{\partial D} (\theta) = e^{iN\theta}, \tag{5.36}$$

then the PB's (5.33) become those of creation and annihilation operators. These latter can be identified with the similar operators of the chiral fields $\partial_\pm \phi$.

Thus let Λ_N be any function on D with boundary value $e^{iN\theta}$:

$$\Lambda_N \mid_{\partial D} (\theta) = e^{iN\theta}, N \in \mathbf{Z} - \{0\}. \tag{5.37}$$

[$N = 0$ is excluded here in view of a remark above, $\Lambda_0 \mid_{\partial D}$ being a constant.] These Λ_N's exist. All $q(\Lambda_N)$ with the same $\Lambda_N \mid_{\partial D}$ are weakly equal and define the same observable. Let $\langle q(\Lambda_N)\rangle$ be this equivalence class of weakly equal $q(\Lambda_N)$ and q_N any member thereof. [q_N can also be regarded as the equivalence class itself.] Their PB's follow from (5.33):

$$\{q_N, q_M\} = -iNk\, \delta_{N+M,0}. \tag{5.38}$$

The q_N's are the CS constructions of the Fourier modes of a massless chiral scalar field on the circle S^1.

We can now proceed to quantum field theory. Let $\mathcal{G}(\Lambda^{(0)}), Q(\Lambda_N)$ and Q_N denote the quantum operators for $g(\Lambda^{(0)}), q(\Lambda_N)$ and q_N. We then impose the constraints

$$\mathcal{G}(\Lambda^{(0)}) \mid \cdot \rangle = 0 \tag{5.39}$$

on all quantum states. It is an expression of their gauge invariance. Because of this equation, $Q(\Lambda_N)$ and $Q(\Lambda'_N)$ have the same action on the states if Λ_N and Λ'_N have the same boundary values. We can hence write

$$Q_N \mid \cdot \rangle = Q(\Lambda_N) \mid \cdot \rangle. \tag{5.40}$$

Here, in view of (5.38), the commutator brackets of Q_N are

$$[Q_N, Q_M] = Nk\delta_{N+M,0}. \quad (5.41)$$

Thus if $k > 0$ ($k < 0$), Q_N for $N > 0$ ($N < 0$) are annihilation operators (up to a normalization) and Q_{-N} creation operators. The "vacuum" $|0>$ can therefore be defined by

$$Q_N |0> = 0 \text{ if } Nk > 0. \quad (5.42)$$

The excitations are obtained by applying Q_{-N} to the vacuum.

When the spatial slice is a disc, the observables are all given by Q_N and our quantization is complete. When it is not simply connected, however, there are further observables associated with the holonomies of the connection a and they affect quantization. We will not examine quantization for nonsimply connected spatial slices here.

The CS interaction does not fix the speed v of the scalar field in (5.34) and so its Hamiltonian, a point previously emphasized by Fröhlich and Kerler, and Fröhlich and Zee. This is but reasonable. For if we could fix v, the Hamiltonian H for ϕ could naturally be taken to be the one for a free massless chiral scalar field moving with speed v. It could then be used to evolve the CS observables using the correspondence between this field and the latter. But we have seen that no natural nonzero Hamiltonian exists for the CS system. It is thus satisfying that we cannot fix v and hence a nonzero H.

5.4. The Chern-Simons Source as a Conformal Family

From the physical point of view, it is of great interest to study Chern-Simons dynamics in the presence of point sources. It is known that the statistics and spin of particles are changed by interaction with the CS field and that they acquire fractional statistics and spin for suitable choices of the coupling strength. [Cf. ref 4 and references therein.]. For this reason, Chern-Simons dynamics with sources can provide a useful means to describe anyons. Also we saw in Section 5.2 that the abelian CS field theory furnishes a description of the QH effect. The sources of the CS field can therefore be thought of as quasiparticle excitations in the QH system, giving us another reason to study these sources. One can also argue that there are sound mathematical reasons for studying these sources, since their spacetime history are connected to Wilson lines and Wilson lines are important for knot theory.

As mentioned above, when a point source is immersed in the CS field, its statistics is affected thereby. As interaction renormalizes statistics, it must renormalize spin as well if, as some of us may conservatively desire, CS dynamics incorporates the canonical spin-statistics connection. One purpose of this Section is to discuss this spin renormalization using a generalization of the canonical approach to source free quantum CS dynamics developed in the last Section. We will see that the specific mechanism for spin renormalization is a novel one: the configuration space of particle mechanics is enlarged by a circle S^1. A point of S^1 can be regarded as parametrizing a tangent direction or an orthonormal frame (although not canonically). A spinless source thus ends up acquiring a configuration space which is that of a two-dimensional rotor with translations added on. What occurs in CS theory is a massless chiral (conformal) quantum field on this S^1 (and time) with ability to change its location in space and with precisely the right spin to maintain the spin-statistics connection. The necessity for framing the particle has been emphasized in the literature before. The qualitative reason for the emergence of this frame is regularization, which surrounds the particle with a tiny hole H which is eventually shrunk to a point. The CS action is then no longer for a disc D, but for $D\backslash H$, which is a disc with a hole. In contrast to D, the latter has an additional boundary ∂H, which is the circle S^1 mentioned above. Just as ∂D, this boundary as well can be associated with a massless chiral scalar field. The internal states of a CS anyon for a fixed location on D thus form an infinite dimensional family of quantum states and are not described by just a single ray. This remark was first stated by Witten and applies with equal force to the Quantum Hall quasiparticle if described in the Chern-Simons framework. It is also noteworthy that the CS source is not a first quantized framed particle, but is better regarded as a "particle" with a first quantized position and a second quantized frame. One intention of this Section is to explain these striking results with hopefully transparent arguments.

Suppose that a spinless point source with coordinate z is coupled to A_μ with coupling $eA_\mu(z(x^0))\dot{z}^\mu$, $z^0 = x^0$. The field equation $\partial_1 A_2 - \partial_2 A_1 = 0$ is thereby changed to

$$\partial_1 A_2 - \partial_2 A_1 = -\frac{2\pi e}{k}\delta^2(x-z)\,. \tag{5.43}$$

If $\mathcal{C}$ is a contour enclosing z with positive orientation, then by (5.43),

$$\oint_{\mathcal{C}} A = -\frac{2\pi e}{k}\,. \tag{5.44}$$

On letting $\mathcal{C}$ shrink to a point, it now follows that $A(x) = A_j(x)dx^j$ has no definite limit when x approaches z. This singularity of A demands regularization. A good way to regularize is to punch a hole H containing z, and eventually to shrink the hole to a point.

Once this hole is made, the action is no longer for a disc D, but for $D\backslash H$, a disc with a hole. $D\backslash H$ has a new boundary ∂H and it must be treated exactly like ∂D. The Gauss law must accordingly be changed to

$$g(\Lambda^{(1)}) \approx 0 \tag{5.45}$$

where the new test function space $\mathcal{T}^{(1)}$ for $\Lambda^{(1)}$ is defined by

$$\Lambda^{(1)}|_{\partial D} = \Lambda^{(1)}|_{\partial H} = 0\,. \tag{5.46}$$

The quantum operator $\mathcal{G}(\Lambda^{(1)})$ for $g(\Lambda^{(1)})$ annihilates all the physical states.

There are now two KM algebras of the type (5.33), one each for ∂D and ∂H. The former is defined by observables $q(\xi^{(0)})$ with test functions $\xi^{(0)}$ which vanish on ∂H, the latter by observables $q(\xi^{(1)})$ with test functions $\xi^{(1)}$ which vanish on ∂D. Let us now define the KM generators for the outer and inner boundaries as

$$\begin{aligned}q^{(0)}_N &\equiv q(\xi^{(0)}_N),\quad \xi^{(0)}_N(\theta)|_{\partial D}= e^{iN\theta},\quad \xi^{(0)}_N|_{\partial H}= 0\,;\\ q^{(1)}_N &\equiv q(\xi^{(1)}_N),\quad \xi^{(1)}_N(\theta)|_{\partial H}= e^{-iN\theta},\quad \xi^{(1)}_N|_{\partial D}= 0\,,\end{aligned} \tag{5.47}$$

θ (mod 2π) being an angular coordinate on ∂H. [The coordinates θ on both ∂D and ∂H increase, say, in the anticlockwise sense.] The corresponding quantum operators will be denoted by $Q^{(0)}_N$ and $Q^{(1)}_N$. Note that the boundary conditions exclude the choice ξ^α = a constant nonzero function on $D\backslash H$. Hence we may not exclude $N = 0$ now.

An interpretation of the observables localized on ∂H is as follows. Let θ (mod 2π) be an angular coordinate on D which reduces to the θ coordinates we have fixed on ∂D and ∂H. A typical A compatible with (5.44) has a blip $-\frac{2\pi e}{k}\delta(\theta - \theta_0)d\theta$ localized on ∂H at θ_0. The behaviour of a general A on ∂H can be duplicated by an appropriate superposition of these blips. The observable $q(\xi^{(1)})$ has zero PB with the left side of (5.44) and hence preserves the flux enclosed by $\mathcal{C}$. In fact, the finite canonical transformation generated by $q(\xi^{(1)})$ changes A to $A + d\xi^{(1)}$ where the fluctuation $d\xi^{(1)}$ creates zero net flux through $\mathcal{C}$. All A compatible with (5.44) can be generated from any one A, such as an A with a blip, by these transformations. Thus the KM algebra of observables $Q_N^{(1)}$ on ∂H generates all connections on ∂H with a fixed flux from any one of these connections.

We have now reproduced Witten's observation that the CS anyon or the CS version of the quantum Hall quasiparticle is a conformal family.

A point of ∂H can be regarded as a frame (alluded to previously) attached to the particle. The restriction (pull back) of a connection A to ∂H can be regarded as a field on these frames. It follows from this remark that the observables localised at ∂H can be regarded as describing spin excitations.

We refer the reader to the original papers [4] for further developments of the approach outlined here.

6. QUANTIZATION AND MULTIPLY CONNECTED CONFIGURATION SPACES

It has been mentioned in Chapter 5 that the sources coupled to the CS field have their statistical properties changed in a way compatible with their spin renormalisation and the spin-statistics theorem (although for reasons of length of the article, and time available for the lectures, we have not gone into the details of this statistics renormalisation). It is thus natural at this point to examine the theoretical foundations of statistics [7] and the spin-statistics theorem [8,9]. The remaining Chapters of this review will be devoted to this task.

It has been known for some time that the statistics of identical particles in two or more dimensions can be understood in terms of the topology of their configuration space Q, their connectivity playing a particularly significant role. In this Chapter, after having first explained why topology, and in particular connectivity, is important for quantisation, we will systematically develop a method of quantisation on multiply connected spaces, providing the necessary mathematical background along the way. The Chapter concludes with several examples of physical systems for which multiple connectivity is significant. Chapter 7 will be our final Chapter. There we outline a purely topological proof of the spin-statistics theorem which completely avoids relativistic quantum field theory (RQFT) and is entirely based on the topology of the configuration space. There are several interesting physical systems governed by nonrelativistic dynamics such as those of holes in a Fermi sea or excitations above that sea. The topological proof discussed here is applicable to many of these systems whereas a RQFT proof looks at best contrived.

Further discussion of the material of this and subsequent Chapter and pertinent references can be found in refs. 2 and 7 to 10.

6.1. Configuration Space and Quantum Theory

The dynamics of a system in classical mechanics can be described by equations of motion on a configuration space Q. These equations are generally of second order in time. Thus if the position $q(t_0)$ of the system in Q and its velocity $\dot{q}(t_0)$ are known at some time t_0, then the equations of motion uniquely determine the trajectory $q(t)$ for all time t.

When the classical system is quantized, the state of a system at time t_0 is not specified by a position in Q and a velocity. Rather, it is described by a wave function ψ which in elementary quantum mechanics is a (normalized) function on Q. The correspondence between the quantum states and wave functions however is not one to one since two wave functions which differ by a phase describe the same state. The quantum state of a system is thus an equivalence class $\{e^{i\alpha}\psi \mid \alpha \text{ real}\}$ of normalized wave functions. The physical reason for this circumstance is that experimental observables correspond to functions like $\psi^*\psi$ which are insensitive to this phase.

In discussing the transformation properties of wave functions, it is often convenient to enlarge the domain of definition of wave functions in elementary quantum mechanics in such a way as to naturally describe all the wave functions of an equivalence class. Thus instead of considering wave functions as functions on Q, we can regard them as functions on a larger space

$\hat{Q} = Q \times S^1 \equiv \{(q, e^{i\alpha})\}$. The space $\hat{Q}$ is obtained by associating circles S^1 to each point of Q and is said to be a $U(1)$ bundle on Q. Wave functions on $\hat{Q}$ are not completely general functions on $\hat{Q}$, rather they are functions with the property $\psi(q, e^{i(\alpha+\theta)}) = \psi(q, e^{i\alpha})e^{i\theta}$. [Here we can also replace $e^{i\theta}$ by $e^{in\theta}$ where n is a fixed integer]. Because of this property, experimental observables like $\psi^*\psi$ are independent of the extra phase and are functions on Q as they should be. The standard elementary treatment which deals with functions on Q is recovered by restricting the wave functions to a surface $\{(q, e^{i\alpha_0}) \mid q \in Q\}$ in Q where α_0 has a fixed value. Such a choice α_0 of α corresponds to a phase convention in the elementary approach.

When the topology of Q is nontrivial, it is often possible to associate circles S^1 to each point of Q so that the resultant space $\hat{Q} = \{\hat{q}\}$ is not $Q \times S^1$, although there is still an action of $U(1)$ on $\hat{Q}$. We shall indicate this action by $\hat{q} \to \hat{q}e^{i\theta}$. It is the analogue of the transformation $(q, e^{i\alpha}) \to (q, e^{i\alpha}e^{i\theta})$ we encountered earlier. We shall require this action to be free, which means that $\hat{q}e^{i\theta} = \hat{q}$ if and only if $e^{i\theta}$ is the identity of $U(1)$. When $\hat{Q} \neq Q \times S^1$, the $U(1)$ bundle $\hat{Q}$ over Q is said to be twisted. It is possible to contemplate wave functions which are functions on $\hat{Q}$ even when this bundle is twisted provided they satisfy the constraint $\psi(\hat{q}e^{i\theta}) = \psi(\hat{q})e^{in\theta}$ for some fixed integer n. If this constraint is satisfied, experimental observables being invariant under the $U(1)$ action are functions on Q as we require. However, when the bundle is twisted, it does not admit globally valid coordinates of the form $(q, e^{i\alpha})$ so that it is not possible (modulo certain technical qualifications) to make a global phase choice, as we did earlier. In other words, it is not possible to regard wave functions as functions on Q when $\hat{Q}$ is twisted.

The classical Lagrangian L often contains complete information on the nature of the bundle $\hat{Q}$. We can regard the classical Lagrangian as a function on the tangent bundle $T\hat{Q}$ of $\hat{Q}$. The space $T\hat{Q}$ is the space of positions in $\hat{Q}$ and the associated velocities. When $\hat{Q}$ is trivial, it is possible to reduce any such Lagrangian to a Lagrangian on the space TQ of positions and velocities associated with Q thereby obtaining the familiar description. On the other hand, when $\hat{Q}$ is twisted, such a reduction is in general impossible. Since the equations of motion deal with trajectories on Q and not on $\hat{Q}$, it is necessary that there is some principle which renders the additional $U(1)$ degree of freedom in such a Lagrangian nondynamical. This principle is the principle of gauge invariance for the gauged group $U(1)$. Thus under the gauge transformation $\hat{q}(t) \to \hat{q}(t)e^{i\theta(t)}$, these Lagrangians change by constant times $d\theta/dt$, where t is time. Since the equations of motion therefore involve only gauge invariant quantities which can be regarded as functions of positions and velocities associated with Q, these equations describe dynamics on Q. The Lagrangians we often deal with split into two terms L_0 and L_{WZ}, where L_0 is gauge invariant while L_{WZ} changes as indicated above. This term L_{WZ} has a geometrical interpretation. It is the one which is associated with the nature of the bundle $\hat{Q}$.

In particle physics, such a topological term was first discovered by Wess and Zumino in their investigation of nonabelian anomalies in gauge theories. The importance and remarkable properties of such "Wess-Zumino terms" have been forcefully brought to the attention of particle physicists in recent years because of the realization that they play a critical role in creating fermionic states in a theory with bosonic fields and in determining the anomaly structure of effective field theories.

In point particle mechanics, the existence and significance of Wess-Zumino terms have long been understood. For example, such terms play an essential role in the program of geometric quantization and related investigations which study the Hamiltonian or Lagrangian description of particles of fixed spin. A similar term occurs in the description of the charge-monopole system and has also been discussed in the literature. Such terms have been found in dual string models as well.

The Wess-Zumino term affects the equations of motion and has significant dynamical consequences already at the classical level. Its impact however is most dramatic in quantum theory where, as was indicated above, it affects the structure of the state space. For example, in the $SU(3)$ chiral model, it is this term which is responsible for the fermionic nature of the Skyrmion.

The preceding remarks on the nature of wave functions in quantum theory can be generalized by replacing the group $U(1)$ by more general abelian or nonabelian groups. A particularly important class of physical systems where such groups are discrete are those with multiply connected configuration spaces.

Multiply connected configuration spaces play an important role in many branches of physics. Examples are molecular physics, condensed matter and quantum field theories, and quantum gravity. Exotic statistics, which has recently assumed an important physical role in condensed matter theory for example, can be understood in terms of the multiple connectivity of the configuration space. In this Chapter, we will also give a few examples of such physical systems.

As a prelude to the discussion of multiply connected configuration spaces, we shall first generalize the preceding remarks on the nature of wave functions.

The arguments above which led to the consideration of $U(1)$ bundles on Q were based on the observation that since only observables like $\psi^*\psi$ are required to be functions on Q, it is permissible to consider wave functions ψ which are functions on a $U(1)$ bundle $\hat{Q}$ over Q provided all wave functions fulfill the property $\psi(\hat{q}e^{i\theta}) = \psi(\hat{q})e^{in\theta}$. We shall now show that we can meet this requirements on observables even with vector valued wave functions $\psi = (\psi_1, ..., \psi_K)$ which are functions on an H bundle $\bar{Q}$ over Q, the group H not being necessarily $U(1)$.

The general definition of an H bundle $\bar{Q}$ over Q is as follows. In an H bundle $\bar{Q} = \{\bar{q}\}$ over Q, there is an action $\bar{q} \to \bar{q}h$ of the group $H = \{h\}$ on $\bar{Q}$ with the property

$$\bar{q} = \bar{q}h \text{ if and only if } h = \text{ identity } e. \tag{6.1}$$

As indicated earlier, such an action of a group H is said to be free. Furthermore, in an H bundle, when all points of $\bar{Q}$ connected by this H action are identified, we get back the space Q. The space Q is thus the quotient of $\bar{Q}$ by the H action:

$$Q = \bar{Q}/H. \tag{6.2}$$

A point of Q can be thought of the set of all points $\{\bar{q}h \mid h \in H\} \equiv \bar{q}H$ connected to $\bar{q}$ by the H action.

If the action of H on $\bar{Q}$ is written in the form $\bar{q} \to \rho(h^{-1})\bar{q} \equiv \bar{q}h$, then $\rho(h_1)\rho(h_2) = \rho(h_1 h_2)$. Hence the map $\rho : h \to \rho(h)$ from H to these transformations on $\bar{Q}$ is a homomorphism. It is in fact an isomorphism in view of (6.1). Note that the image of h under ρ acts on $\bar{q}$ according to $\bar{q} \to \bar{q}h^{-1}$ and not according to $\bar{q} \to \bar{q}h$. Nevertheless, following the convention in the mathematical literature, we shall often regard the action of h on $\bar{Q}$ as being given by $\bar{q} \to \bar{q}h$.

An example of $\bar{Q}$ is the trivial H bundle $\bar{Q} = Q \times H = \{(q,s) \mid s \in H\}$. It carries the free H action $(q,s) \to (q,s)h \equiv (q,sh)$. The quotient of $\bar{Q}$ by this action is Q. A point of Q is q which can be identified with $(q,s)H$ (for any s).

In the mathematical literature, the space $\bar{Q}$ is known as the *bundle space* and H is known as the *structure group*. The map

$$\pi : \bar{Q} \to Q,$$
$$\bar{q} \to \bar{q}H$$
(6.3)

is known as the *projection map*. The set of points in $\bar{Q}$ which project to the same point q of Q under π is known as the fibre over q. The entire structure $(\bar{Q}, \pi, Q, H)$ is known as a *principal fibre bundle*. We shall however call $\bar{Q}$ itself as a principal fibre bundle (or as an H bundle).

It follows from the relation (6.2) between $\bar{Q}$ and Q that any function σ on $\bar{Q}$ which is invariant under the H action $[\sigma(\bar{q}h) = \sigma(\bar{q})$ for all $h \in H]$ can be regarded as a function on Q. Let $h \to D(h)$ define a representation Γ of $H = \{h\}$ by $K \times K$ unitary matrices. Let us demand of our wave functions that they transform by Γ under the action of H:

$$\psi_i(\bar{q}h) = \psi_j(\bar{q})D_{ji}(h).$$
(6.4)

Then for any two wave functions ψ and ψ', the expression

$$<\psi, \psi'>(\bar{q}) \equiv \psi_i^*(\bar{q})\psi_i'(\bar{q})$$
(6.5)

is invariant under H and $<\psi, \psi'>$ may be thought of as a function on Q. If we define the scalar product (ψ, ψ') on wave functions by appropriately integrating $<\psi, \psi'>$ over Q, then it is clear that there is no obvious conceptual problem in working with wave functions of this sort.

We shall see that such vector valued wave functions with $N \geq 2$ will occur in the general theory of multiply connected configuration spaces if H is nonabelian. When that happens, as Sorkin has proved, the space of wave functions we have described above is too large when the dimension of Γ exceeds 1, even when Γ is irreducible. The reduction of this space to its proper size will also be described following Sorkin and will be seen to lead to interesting consequences.

A result of particular importance we shall see later and which merits emphasis is that the quantum theory of systems with multiply connected configuration spaces is ambiguous, there being as many inequivalent ways of quantizing the system as there are distinct unitary irreducible representations (UIR's) of $\pi_1(Q)$. The angle θ which labels the vacua in QCD, for example, can be thought as the label of the distinct UIR's of $\mathbf{Z}$, $\mathbf{Z}$ being $\pi_1(Q)$ for such a theory. As is well-known, the quantum theories associated with different $e^{i\theta}$ are inequivalent.

6.2. The Universal Covering Space and the Fundamental Group

Given any manifold such as a configuration space Q, it is possible to associate another manifold $\bar{Q}$ to Q which is simply connected. The space $\bar{Q}$ is known as the universal covering space of Q. The group $\pi_1(Q) = H$ acts freely on $\bar{Q}$ and the quotient of $\bar{Q}$ by this action is Q. Thus $\bar{Q}$ is a principal fibre bundle over Q with structure group H. The space $\bar{Q}$ plays an important role in the construction of possible quantum theories associated with Q. In this Section, we shall describe the construction of $\bar{Q}$. We shall also explain the concept of the fundamental group $\pi_1(Q)$ of Q and its action on $\bar{Q}$.

We shall assume in what follows that Q is path-connected, that is that if q_0, q_1 are any two points of Q, we can find a continuous curve $q(t) \in Q$ with $q(0) = q_0, q(1) = q_1$.

The first step in the construction of $\bar{Q}$ is the construction of the path space $\mathcal{P}Q$ associated with Q. Let q_0 be any point of Q which once chosen is held fixed in all subsequent considerations. Then $\mathcal{P}Q$ is the collection of all paths which start at q_0 and end at any point q of Q. We shall denote the paths ending at q by $\Gamma_q, \tilde{\Gamma}_q, \Gamma'_q$ etc. It is to be noted that these paths Γ_q are *oriented* and *unparametrized*. The former means that they are to be regarded as *starting* at the base point q_0 and *ending* at q. Each of these paths has thus an arrow attached pointing from q_0 to q. The implication of the statement that Γ_q is "unparametrized" is that (besides its orientation) only its geographical location in Q matters. If we introduce a parameter s to label points of Γ_q and write the associated parametrized path as

$$\gamma_q = \{\gamma_q(s) \mid \gamma_q(0) = q_0,\ \gamma_q(1) = q\}, \tag{6.6}$$

then Γ_q is the equivalence class of all such parametrized paths (with parameters compatible with the orientation of Γ_q) with the same location in Q.

We next introduce an equivalence relation $\sim$ on the paths known as homotopy equivalence. We say that two paths Γ_q and $\tilde{\Gamma}_q$ with the same end point q are *homotopic* and write

$$\Gamma_q \sim \tilde{\Gamma}_q \tag{6.7}$$

if Γ_q can be continuously deformed to $\tilde{\Gamma}_q$ while holding q (and of course q_0) fixed.

A more formal definition of homotopy equivalence is the following: If there exists a continuous family of paths $\Gamma_q(t)$ $[0 \leq t \leq 1]$ in Q (all from q_0 to q) such that

$$\Gamma_q(0) = \Gamma_q, \ \Gamma_q(1) = \tilde{\Gamma}_q, \tag{6.8}$$

then $\Gamma_q \sim \tilde{\Gamma}_q$.

Let $[\Gamma_q]$ denote the equivalence class of all paths ending at q which are homotopic to Γ_q. The *universal covering space* $\bar{Q}$ of Q is just the collection of all these equivalence classes:

$$\bar{Q} = \{[\Gamma_q]\}. \tag{6.9}$$

It can be shown that $\bar{Q}$ is simply connected.

Of particular interest to us are the equivalence classes $[\Gamma_{q_0}]$ of all loops Γ_{q_0} starting and ending at q_0. These equivalence classes have a natural group structure. The group product is defined by

$$[\Gamma_{q_0}][\tilde{\Gamma}_{q_0}] = [\Gamma_{q_0} \cup \tilde{\Gamma}_{q_0}], \tag{6.10}$$

where in the loop $\Gamma_{q_0} \cup \tilde{\Gamma}_{q_0}$, we first traverse Γ_{q_0} and then traverse $\tilde{\Gamma}_{q_0}$. The inverse is defined by

$$[\Gamma_{q_0}]^{-1} = [\Gamma_{q_0}^{-1}], \tag{6.11}$$

where the loop $\Gamma_{q_0}^{-1}$ has the same geographical location in Q as Γ_{q_0}, but has the opposite orientation. The identity e is the equivalence class of the loop consisting of the single point q_0. It is clear that

$$[\Gamma_{q_0}][\Gamma_{q_0}^{-1}] = [\Gamma_{q_0}^{-1}][\Gamma_{q_0}] = e . \qquad (6.12)$$

The group $\pi_1(Q)$ with elements $[\Gamma_{q_0}]$ and the group structure defined above is known as the *fundamental group* of Q. If $\pi_1(Q)$ is nontrivial $[\pi_1(Q) \neq \{e\}]$, the space Q is said to be multiply connected. We shall see examples of multiply connected spaces in Section 6.3. They will show in particular that $\pi_1(Q)$ can be abelian or nonabelian. In any case, it is always discrete.

The group $\pi_1(Q)$ has a free action on $\bar{Q}$. It is defined by

$$[\Gamma_{q_0}] : [\Gamma_q] \to [\Gamma_{q_0}][\Gamma_q] \equiv [\Gamma_{q_0} \cup \Gamma_q], \qquad (6.13)$$

where in $\Gamma_{q_0} \cup \Gamma_q$, we first traverse Γ_{q_0} and then traverse Γ_q. It is a simple exercise to show that this action is free.

We now claim that the quotient of $\bar{Q}$ by this action is Q, the associated projection map $\pi : \bar{Q} \to Q$ being defined by

$$\pi : [\Gamma_q] \to \pi([\Gamma_q]) = q. \qquad (6.14)$$

This means the following: a) All the points $[\Gamma_q], [\tilde{\Gamma}_q],\ldots$ with the same image q under π are related by $\pi_1(Q)$ action, and b) these are the only points related by $\pi_1(Q)$ action. To show a), let $\tilde{\Gamma}_q \cup \Gamma_q^{-1}$ be the loop based at q_0 where we first go along $\tilde{\Gamma}_q$ from q_0 to q and then return to q_0 along Γ_q (in a sense opposite to the orientation of Γ_q). It is clear that

$$[\tilde{\Gamma}_q] = [\tilde{\Gamma}_q \cup \Gamma_q^{-1}][\Gamma_q] \ , \ [\tilde{\Gamma}_q \cup \Gamma_q^{-1}] \in \pi_1(Q). \qquad (6.15)$$

This proves a). As regards b), elements of $\pi_1(Q)$ act by attaching loops at the starting point q_0 of Γ_q and hence map $[\Gamma_q]$ to some $[\tilde{\Gamma}_q]$. Both $[\Gamma_q]$ and $[\tilde{\Gamma}_q]$ project under π to the same point q of Q. This proves b).

We have now proved that $\bar{Q}$ is a principal fibre bundle over Q with structure group $\pi_1(Q)$.

6.3. Examples of Multiply Connected Configuration Spaces

It is appropriate at this point to give some examples of multiply connected spaces. We will avoid examples from gauge and gravity theories for reasons of simplicity. There are several such relevant examples and we shall pick three.

1. Let $x_1, x_2, ..., x_N$ be N distinct points in the plane $\mathbf{R}^2$ and let Q be the complement of the set $\{x_1, x_2, ..., x_N\}$ in $\mathbf{R}^2$:

$$Q = \mathbf{R}^2 \backslash \{x_1, x_2, ..., x_N\}. \tag{6.16}$$

Thus Q is the plane with N holes $x_1, x_2, ..., x_N$. The fundamental group $\pi_1(Q)$ of this Q is of infinite order. It is nonabelian for $N \geq 2$. The generators of this group are constructed as follows: Let q_0 be any fixed point of Q and let C_M be any closed curve from q_0 to q_0 which encloses x_M and none of the remaining holes. It is understood that C_M winds around x_M exactly once with a particular orientation. Let C_M^{-1} be the curve with orientation opposite to C_M, but otherwise the same as C_M. Let $[C_M]$ and $[C_M^{-1}] = [C_M]^{-1}$ be the homotopy classes of C_M and C_M^{-1}. Then $\pi_1(Q)$ consists of all possible products like $[C_M][C_{M'}][C_{M''}]^{-1}...$ and is the free group with generators $[C_M]$. The products of homotopy classes are defined here as in the last Section. For example, $[C_M][C_{M'}] = [C_M \cup C_{M'}]$ where $C_M \cup C_{M'}$ is the curve where we first trace C_M and then trace $C_{M'}$. For $N = 1$, the group $\pi_1(Q)$ has one generator and is $\mathbf{Z}$. The relevance of this Q for the treatment of the Aharonov-Bohm effect should be evident.

2. In the collective model of nuclei, one considers nuclei with asymmetric shapes with three distinct moments of inertia I_i along the three principal axes. There are also polyatomic molecules such as the ethylene molecule C_2H_4 which can be described as such asymmetric rotors. The configuration space Q in these cases is the space of orientations of the nucleus or the molecule. These orientations can be described by a real symmetric 3×3 matrix T (the moment of inertia tensor) with three distinct but fixed eigenvalues I_i. We now show that this Q has a nonabelian fundamental group.

Any $T \in Q$ can be written in the form

$$T \equiv \mathcal{R} T_0 \mathcal{R}^{-1},$$

$$T_0 = \begin{bmatrix} I_1 & & 0 \\ & I_2 & \\ 0 & & I_3 \end{bmatrix}, \tag{6.17}$$

where $\mathcal{R}$ being in $SO(3)$ is regarded as a real orthogonal matrix of determinant 1. Hence Q is the orbit of T_0 under the action of $SO(3)$ given by (6.17). If $\mathcal{R}_i(\pi)$ is the rotation by π around the ith axis,

$$\mathcal{R}_1(\pi) = \begin{bmatrix} 1 & & 0 \\ & -1 & \\ 0 & & -1 \end{bmatrix}, \quad \mathcal{R}_2(\pi) = \begin{bmatrix} -1 & & 0 \\ & 1 & \\ 0 & & -1 \end{bmatrix},$$

$$\mathcal{R}_3(\pi) = \begin{bmatrix} -1 & & 0 \\ & -1 & \\ 0 & & 1 \end{bmatrix},$$

(6.18)

then T is invariant under the substitution $\mathcal{R} \to \mathcal{R}\mathcal{R}_i(\pi)$. So Q is the space of cosets of $SO(3)$ with respect to the four element subgroup $\{1, \mathcal{R}_1(\pi), \mathcal{R}_2(\pi), \mathcal{R}_3(\pi)\}$.

It is convenient to view this coset space as the coset space $SU(2)/H$ of $SU(2)$ with regard to an appropriate subgroup H. For this purpose let us introduce the standard homomorphism $R : SU(2) \to SO(3)$. The definition of R is

$$s\tau_i s^{-1} = \tau_j R_{ji}(s) , \quad s \in SU(2),$$

(6.19)

τ_i being Pauli matrices. [Here we think of $SU(2)$ concretely as the group of 2×2 unitary matrices of determinant 1.] Then we can write any T in the form

$$T = R(s)T_0 R(s^{-1})$$

(6.20)

and hence view Q as the orbit of T_0 under $SU(2)$. Since by (6.19),

$$R(-s) = R(s),$$

$$R(\pm s i \tau_i) = R(\pm s e^{i\pi\tau_i/2}) = R(s)\mathcal{R}_i(\pi),$$

(6.21)

the stability group H of T_0 is the quaternion (or binary dihedral) group D_8^* :

$$H = D_8^* = \{\pm 1, \pm i\tau_1, \pm i\tau_2, \pm i\tau_3\}.$$

(6.22)

Thus
$$Q = SU(2)/D_3^*. \tag{6.23}$$

It is well known that $SU(2)$ is simply connected $[\pi_1(SU(2)) = \{e\}]$. A consequence of this fact [which will not be proved here] is that

$$\pi_1(Q) = D_3^*. \tag{6.24}$$

The loops in Q associated with the elements of D_3^* can be constructed as follows. Consider a curve $\{s(t)\}$ in $SU(2)$ from identity to $h \in D_3^*$:

$$s(t) \in SU(2) , \ s(0) = 1 , \ s(1) = h. \tag{6.25}$$

The image of this curve in Q is $\{T(t)\}$ where

$$T(t) = R[s(t)]T_0 \ R[s(t)^{-1}] . \tag{6.26}$$

Since $T(0) = T(1) = T_0$, this is a loop in Q based at T_0. Two loops $T(t)$ and $T'(t)$ with different $s(1) \in D_3^*$ are not homotopic, whereas all loops $T(t)$ and $T'(t)$ with the same $s(1) \in D_3^*$ are homotopic and form a homotopy class. Such homotopy classes can be thought of as the elements h of $\pi_1(Q)$.

The relation (6.23) shows that Q is the quotient of $SU(2)$ by the free action

$$s \to sh , \ s \in SU(2) , \ h \in D_3^* \tag{6.27}$$

of D_3^*. Furthermore $\pi_1(Q) = D_3^*$. Therefore in this example, $SU(2)$ as a manifold is the universal covering space of Q.

There are molecules with configuration spaces Q such that $\pi_1(Q)$ is any one of the finite discrete subgroups of $SU(2)$, the binary dihedral group being just one of these possibilities. Reference [10] can be consulted for further discussion of this fact and for citations to the literature.

3. The last example we shall give is relevant for discussing possible statistics of particles in k spatial dimensions. Consider N identical spinless particles in $\mathbf{R}^k$ [for $N \geq 2$] and assume first that $k \geq 3$. A configuration of these particles is given by the unordered set $[x_1, x_2, ..., x_N]$ where $x_j \in \mathbf{R}^k$. The set must be regarded as unordered (so that for example $[x_1, x_2, ..., x_N] = [x_2, x_1, ..., x_N]$) because of the assumed indistinguishability

of the particles. Let us also assume that no two particles can occupy the same position so that $x_i \neq x_j$ if $i \neq j$. The resultant space of these sets can be regarded as the configuration space Q of this system. It can be shown that $\pi_1(Q)$ is identical to the permutation group S_N. The closed curves in Q associated with the transpositions $s_{ij} \in S_N$ of two particles can be constructed as follows. Choose the base point q_0 to be $[x_1^0, x_2^0, ..., x_N^0]$. Let $\{\gamma_{ij}(t); 0 \leq t \leq 1\}$ be the loop in Q defined by

$$\gamma_{ij}(t) = [x_1^0, x_2^0, ..., x_{i-1}^0, x_i(t), x_{i+1}^0, ..., x_{j-1}^0, x_j(t), x_{j+1}^0, ..., x_N^0] ,$$

$$x_i(0) = x_i^0 , \quad x_i(1) = x_j^0 , \qquad (6.28)$$

$$x_j(0) = x_j^0 , \quad x_j(1) = x_i^0 .$$

$\{\gamma_{ij}(t)\}$ is a loop since the set $[x_1^0, x_2^0, ..., x_N^0]$ is unordered. The homotopy class of this loop can be identified with s_{ij}.

The distinct quantum theories of this system are labelled by the UIR's of S_N and are associated with parastatistics. Special cases of these theories describe bosons and fermions.

We can describe the configuration space of N identical particles for $k = 2$ as well in a similar way. The fundamental group $\pi_1(Q)$ for $k = 2$ however is not S_N, but a very different (infinite) group known as the braid group B_N. It is because $\pi_1(Q) = B_N$ for $k = 2$ that remarkable possibilities for statistics (such as fractional statistics) arise in two spatial dimensions.

For $N = 2$, it is simple to illustrate the difference between B_2 and S_2. The discussion also shows why fractional statistics is possible in two dimensions. Thus consider the square s_{12}^2 of the transposition for two particles. It is easy to see that it is the homotopy class of the curve where x_1^0 is held fixed, say at the origin, and x_2 goes around it from x_2^0 to x_2^0. For $k = 2$, that is in a plane, this curve is a loop with x_1^0 at its middle. It can not be shrunk to a point since $x_i \neq x_j$ for points of Q. Thus $s_{12}^2 \neq$ identity e for $k = 2$. A similar argument shows that no power of s_{12} is e. The group B_2 is abelian and is generated by s_{12}. Its UIR's are given by $s_{12} \to e^{i\theta}$ where θ is real. All real θ are allowed since we have argued above that no power of s_{12} is e. We therefore have the possibility of fractional statistics which describe neither bosons (for which $s_{12} \to 1$) nor fermions (for which $s_{12} \to -1$) for $k = 2$.

[The next two Sections describe how to realise quantum theories for distinct UIR's of $\pi_1(Q)$.]

Now for $k > 2$, s_{12}^2 is still the homotopy class of a loop like the one described above. But this loop can be shrunk to a point for $k > 2$. For example, it can be taken to be in a plane not enclosing x_1^0, if necessary after first deforming it. It can then be shrunk to a point on this plane. Thus $s_{12}^2 =$ identity e and the corresponding $\pi_1(Q)$ is $S_2 = \mathbf{Z}_2$. There are only two UIR's of S_2 and they are given by $s_{12} \to 1$ and $s_{12} \to -1$. They describe bosons and fermions respectively.

6.4. Quantization on Multiply Connected Configuration Spaces

We shall now describe the general approach to quantization when the configuration space Q is multiply connected.

As indicated previously, this quantization can be carried out by introducing a Hilbert space $\mathcal{H}$ of complex functions on $\bar{Q}$ with a suitable scalar product and realizing the classical observables as quantum operators on this space. Since the classical configuration space is Q and not $\bar{Q}$, classical observables are functions of $q \in Q$ and of their conjugate momenta. Let us concentrate on functions of q. Let $\alpha(q)$ define a function of q and let $\hat{\alpha}$ be the corresponding quantum operator. The definition of $\hat{\alpha}$ consists in specifying the transformed function $\hat{\alpha}f$ for a generic function $f \in \mathcal{H}$. Thus given the function f, we have to specify the value of $\hat{\alpha}f$ at every $\bar{q}$. This is done by the rule

$$(\hat{\alpha}f)(\bar{q}) = \alpha[\pi(\bar{q})]f(\bar{q}). \qquad (6.29)$$

The group $\pi_1(Q)$ acts on $\mathcal{H}$. Let t denote a generic element of $\pi_1(Q)$. If $\hat{t}$ is the operator which represents t on $\mathcal{H}$, and $\hat{t}f$ is the transform of a function $f \in \mathcal{H}$ by $\hat{t}$, $\hat{t}$ is defined by specifying the function $\hat{t}f$ as follows:

$$(\hat{t}f)(\bar{q}) \equiv f(\bar{q}t). \qquad (6.30)$$

Now $\hat{\alpha}$ commutes with $\hat{t}$:

$$\begin{aligned}
(\hat{\alpha}\hat{t}f)(\bar{q}) &= \alpha[\pi(\bar{q})](\hat{t}f)(\bar{q}) \\
&= \alpha[\pi(\bar{q})]f(\bar{q}t)\,, \\
(\hat{t}\hat{\alpha}f)(\bar{q}) &= (\hat{\alpha}f)(\bar{q}t) \\
&= \alpha[\pi(\bar{q}t)]f(\bar{q}t) \\
&= \alpha[\pi(\bar{q})]f(\bar{q}t) \\
&= (\hat{\alpha}\hat{t}f)(\bar{q}).
\end{aligned} \qquad (6.31)$$

Here we have used the fact that $\pi(\bar{q}t) = \pi(\bar{q})$. [See (6.14) and the remarks which follow.]

Since the operators $\hat{t}$ are not all multiples of the identity operator, Schur's lemma tells us that this representation of the observables $\hat{\alpha}$ on $\mathcal{H}$ is not irreducible. We can proceed in the following way to reduce it to its irreducible components. Let $\Gamma_1, \Gamma_2, \ldots$ denote the distinct irreducible representations of $\pi_1(Q)$. Let $\mathcal{H}_\beta^\ell$ ($\beta = 1, 2, \ldots$) be the subspaces of $\mathcal{H}$ which transform by Γ_ℓ, β being an index to account for multiple occurrences of Γ_ℓ in the reduction. Let us also define

$$\mathcal{H}^{(\ell)} = \bigoplus_\beta \mathcal{H}_\beta^{(\ell)}. \qquad (6.32)$$

Then

$$\mathcal{H} = \bigoplus_\ell \mathcal{H}^{(\ell)}. \qquad (6.33)$$

Since $\hat{\alpha}$ commutes with $\hat{t}$, it can not map a vector transforming Γ_ℓ to one transforming by Γ_m ($m \neq \ell$) since Γ_ℓ and Γ_m are inequivalent. Thus

$$\hat{\alpha}\mathcal{H}^{(\ell)} \subset \mathcal{H}^{(\ell)}. \qquad (6.34)$$

In other words, we can realize our observables on any one subspace $\mathcal{H}^{(\ell)}$ and ignore the remaining subspaces. Quantization on the subspaces $\mathcal{H}^{(\ell)}$ and $\mathcal{H}^{(m)}$ are known to be inequivalent when $\ell \neq m$. Thus there are at least as many distinct ways to quantize the system as the number of inequivalent irreducible representations of $\pi_1(Q)$. It may also be shown that the representation of the algebra of observables on any one $\mathcal{H}^{(\ell)}$ is irreducible if $\pi_1(Q)$ is abelian, while some additional reduction is possible if it is nonabelian as shown by Sorkin and as we shall see below.

Here we have not discussed how the momentum variables conjugate to the coordinates are realized on $\mathcal{H}^{(\ell)}$. It can be shown that for the problems at hand, these momentum variables can also be consistently realized.

6.5. Nonabelian Fundamental Groups

Let us now consider nonabelian $\pi_1(Q)$ in more detail. Let γ_ℓ ($\ell = 1, 2, ...$) denote its distinct one dimensional UIR's and let $\bar{\gamma}_\alpha$ ($\alpha = 1, ...$) denote its distinct UIR's of dimension greater than 1. [For simplicity, we assume here that the indexing sets for both abelian and nonabelian UIR's are countable.] The subspaces of $\mathcal{H}$ which carry γ_ℓ will be called $h_k^{(\ell)}$ and the subspaces which carry $\bar{\gamma}_\alpha$ will be called $\bar{h}_\sigma^{(\alpha)}$, k and σ being indices to account for multiple occurrences of a given UIR in the reduction of $\mathcal{H}$. If we set

$$h^{(\ell)} = \bigoplus_k h_k^{(\ell)}, \qquad (6.35)$$

then as in the abelian case the algebra of observables is represented irreducibly on $h^{(\ell)}$, and the representations on different $h^{(\ell)}$ are inequivalent. The novelty is associated with the representations on

$$\bar{h}^{(\alpha)} = \bigoplus_\sigma \bar{h}_\sigma^{(\alpha)}. \qquad (6.36)$$

They are inequivalent for different α, but they are not irreducible. We now show this fact.

Let $e_\sigma(j)$ ($j = 1, 2, ..., n > 1$) be a basis for $\bar{h}_\sigma^{(\alpha)}$ chosen so that they transform in the same way under $\pi_1(Q)$ for different σ:

$$\hat{t} e_\sigma(j) = e_\sigma(k) D(t)_{kj}. \qquad (6.37)$$

Here $t \to D(t)$ defines the representation $\bar{\gamma}_\alpha$. [Since α can be held fixed in the ensuing discussion, an index α has not been put on the vectors $e_\sigma(j)$ or on the matrices $D(t)$.]

Now if $\hat{L}$ is any linear operator such that $\hat{L} e_\sigma(j)$ transforms in the same way as $e_\sigma(j)$,

$$\hat{t} \hat{L} e_\sigma(j) = [\hat{L} e_\sigma(k)] D_{kj}(t), \qquad (6.38)$$

that is if $[\hat{L}, \hat{t}] = 0$, then by Schur's lemma $\hat{L}$ acts only on the index σ:

$$\hat{L}e_\sigma(j) = e_\lambda(j)\mathcal{D}_{\lambda\sigma}(\hat{L}). \tag{6.39}$$

Furthermore, again by Schur's lemma, $\mathcal{D}(\hat{L})$ is independent of j. Since $\hat{\alpha}$ in (6.29) shares the preceding property of $\hat{L}$, it follows that

$$\hat{\alpha}e_\sigma(j) = e_\lambda(j)\mathcal{D}_{\lambda\sigma}(\hat{\alpha}). \tag{6.40}$$

It can be shown that there is a similar formula for momentum observables as well.

Thus the subspace spanned by the vectors $e_\sigma(j)$ [$\sigma = 1, 2, ...$] for any fixed j is invariant under the action of observables. Also, since $\mathcal{D}(\hat{\alpha})$ is independent of j, the representation of the algebra of observables on the subspaces associated with different j are equivalent. It is thus sufficient to retain just one such subspace, the remaining ones may be discarded. When we do so, we also obtain an irreducible representation of the algebra of observables.

Further insight into the nature of this representation is gained by working with a "basis" for $\mathcal{H}$ consisting of states localized at points of Q. These are analogous to the states $|\vec{x}>$ which are localized at positions $\vec{x}$ in the standard nonrelativistic quantum mechanics of spinless particles. But while there is only one such linearly independent state for a given $\vec{x}$, we have $\dim \pi_1(Q)$ [$\equiv$ dimension of $\pi_1(Q)$] worth of such linearly independent states $\{|\bar{q}t>\}$ localized at q, because under π, $\bar{q}t$ projects to q independently of t. [Here $\bar{q}$ is any conveniently chosen point of $\bar{Q}$ with $\pi(\bar{q}) = q$.] The group $\pi_1(Q)$ acts on these states according to

$$\hat{s}|\bar{q}>=|\bar{q}s^{-1}>, \quad s \in \pi_1(Q). \tag{6.41}$$

Clearly this representation of $\pi_1(Q)$ on the subspace spanned by the vectors $\{|\bar{q}t>\} = \{|\bar{q}s^{-1}>\}$ (for fixed $\bar{q}$) is isomorphic to the regular representation of $\pi_1(Q)$. As is well known, when this representation is fully reduced, each UIR occurs as often as its dimension. Thus each γ_ℓ occurs once and is carried by a one dimensional vector space with basis $F^{(\ell)}$ say, while each $\bar{\gamma}_\alpha$ occurs $\dim \bar{\gamma}_\alpha$ times and is carried by a vector space with basis $E_\sigma^{(\alpha)}(j)[j, \sigma = 1, 2..., \dim \bar{\gamma}_\alpha]$ say. The transformation law of $E_\sigma^{(\alpha)}(j)$ under $\pi_1(Q)$ is

$$\hat{t}E_\sigma^{(\alpha)}(j) = E_\sigma^{(\alpha)}(k)D_{kj}(t). \tag{6.42}$$

According to our previous argument, the reduction of the representation of the algebra of observables is achieved by retaining only the subspace $V_j(q)$ spanned by the vectors $E_\sigma^{(\alpha)}(j)$ for a fixed j [and a fixed α].

Now every nonzero vector in $V_j(q)$ is localized at q. Thus even after this reduction, there are $\dim \bar\gamma_\alpha$ linearly independent vectors localized at q. In nonrelativistic quantum mechanics, if the system has internal symmetry (or quantum numbers like intrinsic spin), the linearly independent states localized at $\vec x$ are of the form $|\vec x, m>$ ($m = 1, 2, ..., k$) where the index m carries the representation of internal symmetry. In this case, there are k linearly independent vectors localized at $\vec x$. The situation we are finding when $\pi_1(q)$ is nonabelian has points of resemblance to this familiar quantum mechanical situation in the sense that here as well there are many states localized at q.

It is of interest to know the physical observables $\hat O$ which mix the indices σ of the basis $E_\sigma^{(\alpha)}(j)$. That is, it is of interest to find the observables $\hat O$ with the property

$$\hat O E_\sigma^{(\alpha)}(j) = E_\lambda^{(\alpha)}(j) \mathcal{D}_{\lambda\sigma}(\hat O) \qquad (6.43)$$

such that their representation on $V_j(q)$ is irreducible. There is an elegant, but local, geometrical construction for a family of such operators which we now describe. Consider loops from q to q, they can be divided into homotopy classes $[C_t(q)][t \in \pi_1(Q)]$ labelled by elements of $\pi_1(Q)$. The class $[C_t(q)]$ consists of closed loops which are homotopic to each other. The labels can be so chosen that $[C_s(q)][C_t(q)] = [C_{st}(q)]$ where the multiplication of homotopy classes has been described in Section 6.2. [Note however that the loops $C_t(q)$ are based at q and not at the base point q_0 of Section 6.2.] Pick one closed curve $C_t(q)$ from $[C_t(q)]$ and consider the operator which parallel transports wave functions around $C_t(q)$. It can be shown that the change of a wave function as a result of parallel transporting it around a loop in $C_t(q) \in [C_t(q)]$ is independent of the choice of the loop in the class $[C_t(q)]$. Thus the parallel transport operator depends only on the homotopy class $[C_t(q)]$ and not on the choice of the closed curve in $[C_t(q)]$. It can hence be denoted by $\hat O_t$. These operators $\hat O_t$ can serve as the observables we are seeking.

The above description of the operators $\hat O_t$ is rather loose however since $\hat O_t$ is defined only if the transform $\hat O_t \psi$ of a wave function ψ is defined and this involves specifying $(\hat O_t \psi)(\bar q)$ for all $\bar q$. Hence we must associate a homotopy class $[C_t(q)]$ to each $t \in \pi_1(Q)$ and all q. This association must be smooth in q and fulfill the property $[C_s(q)][C_t(q)] = [C_{st}(q)]$. Consider what happens if we smoothly change $[C_t(q)]$ as q is taken around a closed loop in the homotopy class $[C_s(q)]$, $s \in \pi_1(Q)$. It is then easy to convince oneself that $[C_t(q)]$ evolves into the homotopy class $[C_{sts^{-1}}(q)]$. When $\pi_1(Q)$ is nonabelian, $[C_t(q)]$ will not be equal to $[C_{sts^{-1}}(q)]$ for all t and s. A consequence is that the operators $\hat{O}_t$ are not all well defined when the UIR of $\pi_1(Q)$ defining the quantum theory is nonabelian. [Nonetheless, the representation of the algebra of observables we have described can be shown to be irreducible.] The obstruction in defining all the operators $\hat{O}_t$ here is similar to the obstruction in defining the colour group in the presence of nonabelian monopoles or the helicity group for massless particles in higher dimensions.[See ref. 2 for references on these topics.]

It is remarkable that when $\pi_1(Q)$ is nonabelian, quantization can lead to a multiplicity of states all localized at the same point. The consequences of this multiplicity have not yet been sufficiently explored in the literature.

6.6. The Case of the Asymmetric Rotor

We shall now briefly illustrate these ideas by the example of the asymmetric rotor described in Section 6.3. The treatment given here is equivalent for example to the standard treatment molecules with D_8^* as the symmetry group [that is, $\pi_1(Q)$] or of nuclei with three distinct moments of inertia in the collective model approach to nuclei. See ref. 10 in particular in this connection.

Let $\bar{Q}$ be the manifold of the group $SU(2)$ and let s denote a point of $\bar{Q}$. We regard s as a 2×2 unitary matrix of determinant 1. Let D_8^* be the quaternion subgroup of $SU(2)$:

$$D_8^* = \{\pm 1, \pm i\tau_i \ (i = 1, 2, 3)\}. \qquad (6.44)$$

It has the free action

$$s \to sh, \ h \in D_8^* \equiv H \qquad (6.45)$$

on $\bar{Q}$. If we identify all the eight points which are taken into each other by this action, we get a space Q which as we saw in Section 6.3 is the configuration space of the asymptotic rotor.

The group D_3^* has five inequivalent UIR's. Four of these are abelian and may be described as follows. In one, the trivial one, all elements of D_3^* are represented by the unit operator. In one of the remaining three, ± 1 and $\pm i\tau_1$ are represented by $+1$ while $\pm i\tau_2$ and $\pm i\tau_3$ are represented by -1. The two remaining one dimensional UIR's are constructed similarly, ± 1 and $\pm i\tau_2$ being represented by $+1$ in one and ± 1 and $\pm i\tau_3$ being represented by $+1$ in the other. As regards the two dimensional UIR, it is the defining representation (6.44) involving Pauli matrices.

There are thus five ways of quantizing this system. We now concentrate on the quantization method involving the two dimensional nonabelian UIR of D_3^*.

A basis for all functions on $SU(2)$ are the matrix elements $D^j_{\rho\sigma}(s)[s \in SU(2)]$ of the rotation matrices. The group $D_3^* = \{h\}$ acts by operators $\hat{h}$ on these functions according to the rule

$$(\hat{h}D^j_{\rho\sigma})(s) = D^j_{\rho\sigma}(sh). \qquad (6.46)$$

Since

$$D^j_{\rho\sigma}(sh) = D^j_{\rho\lambda}(s)D^j_{\lambda\sigma}(h) \qquad (6.47)$$

and since for integer j, $h \to D^j(h)$ for $h \in D_3^*$ defines an abelian representation of D_3^*, we can and shall restrict j to half odd integer values.

The next step is to reduce the representation $h \to D^j(h)$ into its irreducible components. It then splits into a direct sum of the two dimensional UIR's (6.44). [Only the two dimensional UIR's occur in this reduction. This is because the image of $(i\tau_i)^2$ being a 2π rotation is represented by -1, j being half an odd integer.] The basis vectors for the vector spaces which carry such UIR's are of the form $e^j_{\rho,m,a}$, $m = 1, 2, ..., N$; $a = 1, 2$ where $2N$ equals $2j + 1$. Under the transformations $s \to sh$, their behavior is given by

$$e^j_{\rho,m,a}(sh) = e^j_{\rho,m,b}(s)h_{ba}. \qquad (6.48)$$

The vector space which carries the algebra of observables irreducibly is spanned by e^j_{ρ,m,a_0} with one fixed value a_0 and with j, ρ, and m taking on all allowed values. The vectors $e^j_{\rho,m,a'}$ with the remaining values a' for a are to be discarded.

When the asymmetric rotor model is used to describe nuclei, m can be interpreted in terms of the third component of angular momentum in the body fixed frame.

We have not discussed a scalar product for this vector space. A suitable scalar product may be

$$(\alpha, \beta) = \int_{SU(2)} d\mu(s) \alpha^*(s)\beta(s). \tag{6.49}$$

Here we have regarded the elements of our vector space as functions on $SU(2)$ and $d\mu(s)$ is the invariant measure on $SU(2)$.

In the preceding discussion, we have not referred to a Lagrangian or a Hamiltonian. They are of course important from a dynamical point of view. They do not however play a critical role in the construction of the vector space for wave functions that we have outlined because this construction is valid for a large class of Lagrangians and Hamiltonians.

7. TOPOLOGICAL SPIN-STATISTICS THEOREMS

In nonrelativistic quantum mechanics or relativistic quantum field theory (RQFT) in three or more (spatial) dimensions, one encounters two sorts of particles or localized solitonic excitations. One of these is characterized by tensorial states, which are invariant under 2π rotation, and the other by spinorial states, which change sign under this rotation. If we limit ourselves to Bose and Fermi systems, the spin-statistics correlation in three or more dimensions amounts to the assertion that the former are bosons and the latter are fermions. Thus according to this assertion, the change in the phase of a state under the exchange of two identical systems of spin S is $\exp[i2\pi S]$. In two dimensions, there are more general possibilities for spin and statistics such as fractional spin and fractional statistics. But here as well, the above correlation asserts that the exchange operation is associated with the phase $\exp[i2\pi S]$ for a spin S "anyon" subject to fractional statistics. [It may be emphasized here however that the notions of spin and statistics are more fragile in two dimensions. There the assignment of a well-defined statistics ceases to make sense when generic, velocity-dependent forces ("magnetic fields") are present; and spin is subject to a similar loss of meaning. In such situations, the spin-statistics correlation is vacuous and our discussion will not apply.]

There are different sorts of proofs of this correlation currently available in the literature. One class of proofs typically uses RQFT in one of its formulations such as the one initiated by Wightman, or the algebraic formulation of quantum field theory. In the Wightman framework, for example, it is shown that tensorial fields commute and spinorial ones anticommute for space like separations, and this result is interpreted as a proof of the spin-statistics connection. A second approach to the spin-statistics theorem due to Finkelstein and Rubinstein applies to solitons or "kinks". It is a "topological" proof which does not use the heavy machinery of RQFT. It examines the fundamental group $\pi_1(Q)$ of the configuration space Q appropriate for solitons and shows that 2π rotation of a soliton and exchange of two identical solitons are the same element of $\pi_1(Q)$. This proof in particular does not use relativistic invariance, but does use the facts that solitons are continuous structures in field theories and that each soliton has its antisoliton.

The spin-statistics theorem is pertinent in disciplines such as atomic physics where relativity or field theory does not play a significant role. It is therefore desirable to prove it for point particles in a topological manner that would dispense with these assumptions. We may also hope that such a proof would make sense for topological geons in quantum gravity, where again the assumptions of flat space quantum field theory are too restrictive. Indeed there are reasons to hope that, once we see how such a derivation would go, we will have an important clue to the dynamical rules governing the change of spacetime topology. A derivation of this sort will be outlined in this Section.

References 8 and 9 can be consulted for citations to the literature on topological spin-statistics theorems, including those discussed here.

The existence of an antiparticle is an indispensable ingredient in the topological proofs for solitons, and will be so here as well. The concept of antiparticle in this context can be associated with any state which on suitable pairing with a particle state acquires the quantum numbers of the ground state. The proof below is thus applicable to condensed matter systems with particle-hole excitations. There are however many situations in low energy physics where even such antiparticles are not available. Electron pair production energies being several orders of magnitude larger than typical energies in atomic physics for example, the spin-statistics connection hence seems to provide us an example where a high energy result has a profound influence on low energy physics.

For purposes of simplicity, we shall assume here that the particle and antiparticle are distinct when they have spin, although this assumption can be dispensed with. We will not use such an assumption here when the particles are spinless.

We may at this juncture point out an important implication of the topological spin-statistics theorems for particles moving in $\mathbf{R}^d$ ($d \geq 2$): They exclude "nonabelian" statistics. Thus according to these theorems, paraparticles of order 2 and more for $d \geq 3$, and particles associated with nonabelian braid group representations for $d = 2$, could not exist in nature.

Let us first outline the proof for spinless particles with distinct antiparticles. As discussed in Section 6.3, in one conventional approach to statistics in particle mechanics, the configuration space Q_M for M identical spinless particles in $\mathbf{R}^d$ ($d \geq 2$) is

$$Q_M = \left\{ [x^{(1)}, x^{(2)}, ..., x^{(M)}] \,\middle|\, \begin{array}{l} x^{(i)} \in \mathbf{R}^d; x^{(i)} \neq x^{(j)} \text{ if } i \neq j; \\ [x^{(1)}, ..., x^{(i)}, ..., x^{(j)}, ..., x^{(M)}] \\ = [x^{(1)}, ..., x^{(j)}, ..., x^{(i)}, ..., x^{(M)}] \end{array} \right\}. \quad (7.1)$$

The configuration space $\bar{Q}_N$ for N spinless antiparticles is obtained from (7.1) by replacing M by N and $x^{(i)}$'s by $\bar{x}^{(i)}$'s. Next consider the Cartesian product

$$Q_{M,N} = Q_M \times \bar{Q}_N, \quad M, N \geq 1. \quad (7.2)$$

Define also

$$Q_{M,0} = Q_M, \quad M \geq 1.$$
$$Q_{0,N} = \bar{Q}_N, \quad N \geq 1, \quad (7.3)$$

and introduce the vacuum ("VAC") by setting

$$Q_{0,0} = \{VAC\}. \quad (7.4)$$

The final configuration space C_K is obtained by imposing an equivalence relation $\sim$ on the disjoint union

$$\bigsqcup_{\substack{K \text{ fixed} \\ N+K\geq 0 \\ N\geq 0}} Q_{N+K,N} \tag{7.5}$$

which makes creation and annihilation processes possible. According to this relation, a particle and an antiparticle at the same location "annihilate" to VAC, and conversely they emerge from VAC by separating from an identical location. This is illustrated in Fig. 1 and can also be expressed in equations as follows:

$$([x];[\bar{x}]) \sim \text{VAC if } x = \bar{x},$$

$$([x^{(1)},...,x^{(i)},...,x^{(N+K)}];[\bar{x}^{(1)},...,\bar{x}^{(j)},...,\bar{x}^{(N)}])$$
$$\sim ([x^{(1)},...,\underline{x^{(i)}},...,x^{(N+K)}];[\bar{x}^{(1)},...,\underline{\bar{x}^{(j)}},...,\bar{x}^{(N)}]) \tag{7.6}$$

$$\text{if } x^{(i)} = \bar{x}^{(j)}.$$

Here the underlined entries are to be deleted and equations such as $([x^{(1)}, x^{(2)}, \underline{x^{(3)}}];[\underline{\bar{x}}]) = [x^{(1)}, x^{(2)}]$ are to be understood. C_K is the quotient of (7.5) by this equivalence relation. Elements of C_K which are equivalence classes containing points such as $([x^{(1)},...,x^{(N+K)}];[\bar{x}^{(1)},...,\bar{x}^{(N)}])$ will be denoted by $[x^{(1)},...,x^{(N+K)};\bar{x}^{(1)},...,\bar{x}^{(N)}]$. They fulfill identities which follow from (7.6). The significance of K is that the particle number (= number of particles-number of antiparticles) for a point of C_K is K. Note that C_K is infinite dimensional and not a manifold.

The spin-statistics connection for spinless particles reduces to the statement that the particles are bosons. To establish this we will show that the exchange operation is associated with the trivial element of $\pi_1(C_K)$. That this topological triviality of exchange does in fact entail Bose statistics in the ordinary sense is not something we will prove here, plausible though it is. Chapter 6 can be consulted regarding this point.

The result that particle interchange is trivial in $\pi_1(C_K)$ will be shown in C_2 adopting the following conventions, the proof for any C_K being similar.

The homotopy parameter t will increase upwards in the figures. their horizontal sections being $\mathbf{R}^d$. Following Feynman, a "particle travelling backward in t" will be used to represent an antiparticle. We will sometimes refer to t as time. The base point for homotopy will correspond to two particles located say on the 1-axis.

The curve for exchange is Fig. 2(a) whereas the trivial curve describing static particles is Fig. 2(g). Figures (a-g) show how to deform the first to the last of these figures thereby demonstrating the theorem. Exchanges being the identity of $\pi_1(C_K)$, nonabelian statistics are also excluded. Note that p_1 and p_2 [q_1 and q_2] are VAC, and superposing them as in the passage from Fig. 2(b) to 2(c) [2(e) to 2(f)] is a legitimate activity.

If the particle and its antiparticle are not distinct, the configuration space is

$$D_K = \bigcup_{M=K \bmod 2} \left\{ [x^{(1)}, x^{(2)}, ..., x^{(M)}] \left| \begin{array}{l} [x^{(1)}, ..., x^{(i)}, ..., x^{(j)}, ..., x^{(M)}] \\ = [x^{(1)}, ..., x^{(j)}, ..., x^{(i)}, ..., x^{(M)}] ; \\ [x^{(1)}, ..., x^{(i)}, ..., x^{(j)}, ..., x^{(M)}] \\ = [x^{(1)}, ..., \underline{x^{(i)}}, ..., \underline{x^{(j)}}, ..., x^{(M)}] \\ \text{if } x^{(i)} = x^{(j)} . \end{array} \right. \right\}$$
(7.7)

Here K is either 0 or 1, underlined entries are as usual to be deleted and we employed the convention that $[x^{(1)}, x^{(2)}, ..., x^{(M)}] :=$ VAC when $M = 0$. The spin-statistics connection and the exclusion of nonabelian statistics can be proved for D_K exactly as before.

We now turn to particles with spin. We will in a well-known way account for spin by attaching a frame to each particle. [The physical origin of these frames is documented further in the second paper of ref. 8.] Let $\mathcal{F}^d$ be the set of all frames, orthonormal with respect to the Euclidean metric on $\mathbf{R}^d$ and with a fixed orientation. The generalization Q_M^{SPIN} of Q_M to spinning particles is then

$$Q_M^{SPIN} = \{[(x^{(1)}, F^{(1)}), ..., (x^{(M)}, F^{(M)})]\} \tag{7.8}$$

where $x^{(i)} \in \mathbf{R}^d, F^{(i)} \in \mathcal{F}^d$, the elements of Q_M^{SPIN} are invariant under permutations of the $(x^{(i)}, F^{(i)})$ and we require $x^{(i)} \neq x^{(j)}$ if $i \neq j$. The antiparticle space $\bar{Q}_N^{SPIN}$ which generalizes $\bar{Q}_N$ is similarly obtained. Its elements are denoted by $[(\bar{x}^{(1)}, \bar{F}^{(1)}), ..., (\bar{x}^{(N)}, \bar{F}^{(N)})], \bar{F}^{(i)} \in \bar{\mathcal{F}}^d$ where now $\bar{\mathcal{F}}^d$ is the set of orthonormal frames oppositely oriented to elements of $\mathcal{F}^d$. Such an orientation reversal is suggested by the fact that the CP or CPT transform of a left handed particle is a right handed antiparticle. It is also suggested by the Finkelstein-Rubinstein work. The particle and antiparticle are distinct since $\mathcal{F}^d \neq \bar{\mathcal{F}}^d$.

Our final spinning particle configuration space C_K^{SPIN} for particle number K is obtained from the disjoint union

$$\bigsqcup_{\substack{K \text{ Fixed} \\ N+K \geq 0 \\ N \geq 0}} Q_{N+K,N}^{SPIN} \tag{7.9}$$

by specifying a condition which makes annihilation and creation possible. [The definition of $Q_{M,N}^{SPIN}$ is essentially analogous to the definition of $Q_{M,N}$. See the second or third paper of ref. 8 for a more precise treatment.] For this purpose, consider for simplicity a particle i and an antiparticle j moving towards each other along a straight line L and colliding at ξ. Let $\mathcal{P}$ be the plane through ξ normal to L. Our central assumption is that i and j annihilate at ξ if and only if the antiparticle frame $\bar{F}^{(j)}$ approaches the reflection of the particle frame $F^{(i)}$ in $\mathcal{P}$. There is a similar rule for pair production. These assumptions are shown in Fig. 3 for $d = 2$. [The axes of the particle (antiparticle) frames are drawn in figures with single (double) lines]. They imply equations such as

$$\lim_{x^{(i)}, \bar{x}^{(j)} \to \xi} [(x^{(i)}, F^{(i)}); (\bar{x}^{(j)}, \bar{F}^{(j)})] = VAC;$$

$$\lim_{x^{(i)}, \bar{x}^{(j)} \to \xi} [(x^{(1)}, F^{(1)}), ..., (x^{(i)}, F^{(i)}), ...; (\bar{x}^{(1)}, \bar{F}^{(1)}), ..., (\bar{x}^{(j)}, \bar{F}^{(j)}), ...]$$

$$= [(x^{(1)}, F^{(1)}), ..., \underline{(x^{(i)}, F^{(i)})}, ...; (\bar{x}^{(1)}, \bar{F}^{(1)}), ..., \underline{(\bar{x}^{(j)}, \bar{F}^{(j)})}, ...] \tag{7.10}$$

where the limit is taken with $x^{(i)}$ and $\bar{x}^{(j)}$ approaching ξ along L and the antiparticle frame approaching the appropriate reflection of the particle frame (explained above) in the limit. The rest of the new notation follows the earlier one.

The exchange diagram Fig. 2(a) is as before homotopic to Fig. 2(e) where now an appropriate frame is supposed to be attached to each point of these figures. We now show that the left hand side Fig. 4(a) of Fig. 2(e) is homotopic to Fig. 4(b,c) where the frame of the outgoing particle undergoes 2π rotation as t evolves, thereby showing the theorem.

The homotopy of Fig. 4(b) to Fig. 4(c) is obtained by coalescing C and D. We must thus prove the homotopy of Fig. 4(a) and Fig. 4(b). For this purpose, it is convenient to assume that the particles and antiparticle in these pictures are moving along the 1-axis except within the dashed circle when the particle created by pair production takes a little excursion in the 1-2 plane and then returns to the 1-axis.

The process in Fig. 4(a) is redrawn in Fig. 5, which shows only the first two axes of the frames. At times $t < t_1$, a particle, call it 1, is moving to the right on 1-axis. A pair is produced at $t = t_1$, with the particle 2 of the pair to the left of antiparticle $\bar{2}$. As t evolves, 1 and $\bar{2}$ annihilate at $t = t_2$ while 2 moves to the left on 1-axis, makes a detour in the 1-2 plane and then returns to the 1-axis. Fig. 4(b) is likewise redrawn in Fig. 6. Note that the alignment of the frames in Figs. 5 and 6 is consistent with (7.10).

A comparison of these figures shows that the left-right order of the $2 - \bar{2}$ pair at the moment of production is reversed in going from Fig. 5 to Fig. 6. The homotopy of Fig. 5 to Fig. 6 thus involves gradually changing the production angle θ of 2 from π as in Fig. 5 to zero as in Fig. 6. [We assume that 2 is produced in the 1-2 plane in the successive stages of the homotopy.] Fig. 7 shows the frame of 2 as θ is so changed, the $\bar{2}$ frame being held fixed. Clearly, because of the mirror rule involved in (7.10), the frame of 2 rotates by $2(\theta_1 - \pi)$ when θ decreases from π to θ_1. This means that when Fig. 5 is deformed so that θ becomes θ_1, the frame of 2 will rotate by $2(\pi - \theta_1)$ before 2 reaches its final destination. This is shown in Fig. 8. This rotation being 2π for $\theta = 0$, the homotopy of Figs. 4(a,b) is thus established.

Nonabelian statistics can be shown to be excluded by a simple extension of the preceding arguments. Thus consider M particles in C_K^{SPIN} say. By the above, the exchange σ_{ij} of particles i and j is equal to 2π rotation $R_{2\pi}^{(i)}$

of the frame i. Repeating the argument, we have further $R_{2\pi}^{(i)} = \sigma_{1i} = R_{2\pi}^{(1)}$ whence all exchanges and all rotations are homotopic to each other. This shows that all exchanges commute thereby establishing the result.

In a more complete treatment, we must define suitable topologies for C_K and C_K^{SPIN} and derive equations like (7.10) as consequences of these topologies. This task is carried out in the second paper of ref. 8.

There are several physical systems of interest other than point particles in $\mathbf{R}^d$ to which the techniques outlined here can be extended. It has been shown elsewhere for example [2] that there exist exotic possibilities for the statistics of strings in $\mathbf{R}^3$ if antistrings are ignored. [These strings can be vortex rings in He^4 or strings produced in GUT's during phase transitions.] A spin-statistics theorem for these strings as well has been proved in ref. 9 by including antistrings and creation-annihilation processes, and it will rule out these exotic statistics.

Acknowledgements

This work was supported by the U.S. Department of Energy under Contract Number DE-FG02-85ER40231. I am grateful to Garry Trahern for very helpful suggestions about the manuscript.

References

The references which follow are not exhaustive. They may be consulted for further citations to the literature. The material in Chapter 4 uses unpublished results obtained by A.P. Balachandran, G. Marmo, N. Mukunda, J.S. Nilsson and E.C.G. Sudarshan.

[1] A.P. Balachandran, G. Marmo, B.-S. Skagerstam, and A. Stern, "Gauge Symmetries and Fibre Bundles - Applications to Particle Dynamics", Springer-Verlag Lecture Notes in Physics 188 [Springer-Verlag, 1983].

[2] A.P. Balachandran, G. Marmo, B.-S. Skagerstam, and A. Stern, "Classical Topology and Quantum States" [World Scientific, 1991].

[3] A.J. Hanson, T. Regge and C. Teitelboim, "Constrained Hamiltonian Systems" (Accademia Nazionale dei Lincei, Roma, 1976). The work reviewed here on boundary conditions for the diffeomorphism group is originally due to T. Regge and C. Teitelboim, Ann. Phys. $\underline{88}$ (1974) 286. For other reviews, see E.C.G. Sudarshan and N. Mukunda, "Classical Mechanics in a Modern Perspective" [Wiley, 1978]; K. Sundermeyer, "Constrained Dynamics", Springer-Verlag Lecture Notes in Physics 169 [Springer-Verlag, 1982]. For further work on such conditions, especially in gauge theories, see J.L. Gervais, B. Sakita and S. Wadia, Phys. Letters $\underline{B63}$ (1976) 55; S. Wadia, Phys. Rev. $\underline{D15}$ (1977) 3615; J.L. Friedman and R.D. Sorkin, Commun. Math. Phys. $\underline{89}$ (1983) 483 and 501; refs. 2, 4 and references therein.

[4] A.P. Balachandran, G. Bimonte, K.S. Gupta and A. Stern, Int. J. Mod. Phys. $\underline{A7}$ (1992) 4655 and Syracuse University preprint SU-4228-491 (1992) [to be published in the International Journal of Modern Physics].

[5] E. Witten, Phys. Lett. $\underline{86B}$ (1979) 283.

[6] M.H. Friedman, J.B. Sokoloff, A. Widom and Y.N. Srivastava, Phys. Rev. Lett $\underline{52}$ (1984) 1587; Y. Srivastava and A. Widom, Phys. Reports $\underline{148}$ (1987) 1; A.P. Balachandran and A.M. Srivastava, Minnesota preprint TPI-MINN-91/38-T [SU-4228-492] [to be published by World Scientific in a volume edited by L.C Gupta and M.S. Multani].

[7] A.P. Balachandran, Int. J. Mod. Phys. $\underline{B5}$ (1991) 2585.

[8] A.P. Balachandran, A. Daughton, Z.-C. Gu, G. Marmo, A.M. Srivastava and R.D. Sorkin, Mod. Phys. Lett. $\underline{A5}$ (1991) 1575; A.P. Balachandran, A. Daughton, Z.-C. Gu, G. Marmo, A.M. Srivastava and R.D. Sorkin,

Syracuse University preprint SU-4228-433 (1992); A.P. Balachandran, W.D. McGlinn, L. O'Raifeartaigh, S. Sen and R.D. Sorkin, Syracuse University preprint 4228-496 (1991) and Int.J. Mod. Phys. A (in press).

[9] A.P. Balachandran, W.D. McGlinn, L. O'Raifeartaigh, S. Sen, R.D. Sorkin and A.M. Srivastava, Mod. Phys. Lett. A7 (1992) 1427.

[10] A.P. Balachandran, A. Simoni and D.M. Witt, Int. J. Mod. Phys. A7 (1992) 2087.

74 Gauge Symmetries, Topology

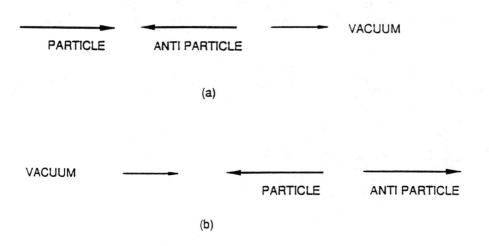

ANNIHILATION AND CREATION PROCESSES FOR
SPINLESS PARTICLES

FIG. 1

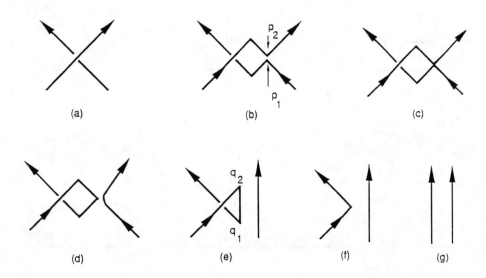

THESE FIGURES SHOW THAT THE EXCHANGE CURVE (a) IS HOMOTOPIC TO THE TRIVIAL CURVE (g) FOR SPINLESS PARTICLES.

FIG. 2

Gauge Symmetries, Topology

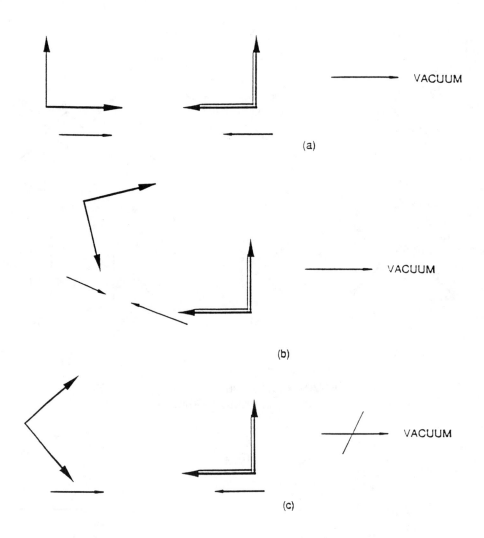

(a) AND (b) SHOW TWO ANNIHILATION PROCESSES AND (c) A SEQUENCE WHICH DOES NOT ANNIHILATE FOR d = 2.

FIG. 3

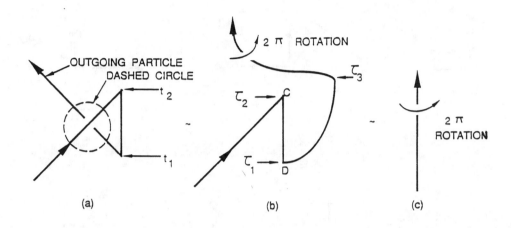

THESE CURVES ARE HOMOTOPIC. "2π ROTATION" INDICATES THAT THE FRAME UNDERGOES A 2π ROTATION AS WE GO ALONG THE LINE. ALL PARTICLES AND THE ANTIPARTICLE ARE ON THE 1-AXIS EXCEPT THE OUTGOING PARTICLE WITHIN THE DASHED CIRCLE WHEN IT IS IN THE 1-2 PLANE.

FIG. 4

78 Gauge Symmetries, Topology

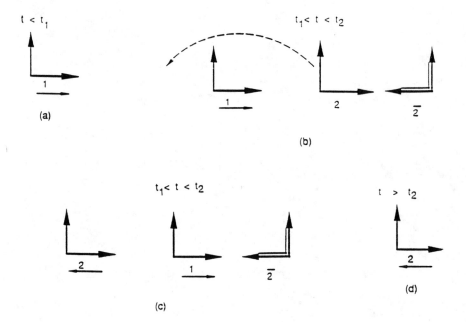

THIS IS A REDRAWING OF FIG. 4(a) AS A PROCESS EVOLVING IN t.

FIG. 5

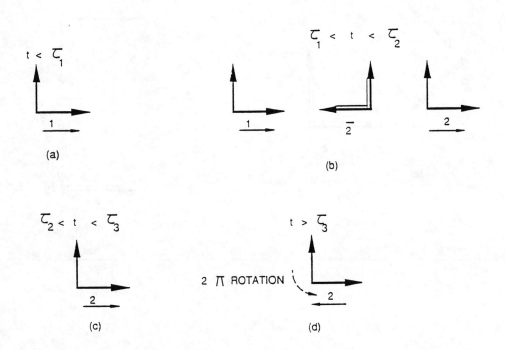

THIS IS A REDRAWING OF FIG. 4(b) AS A PROCESS EVOLVING IN t.

FIG. 6

80 Gauge Symmetries, Topology

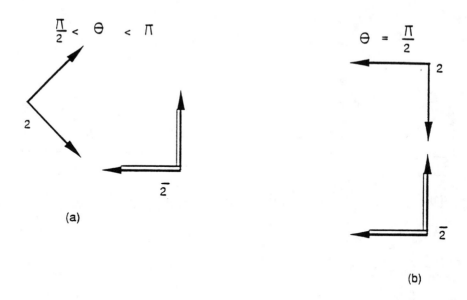

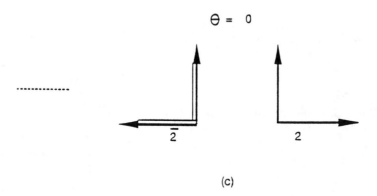

THE FRAME OF 2 FOR VARIOUS PRODUCTION ANGLES WITH THE FRAME OF $\bar{2}$ HELD FIXED.

FIG. 7

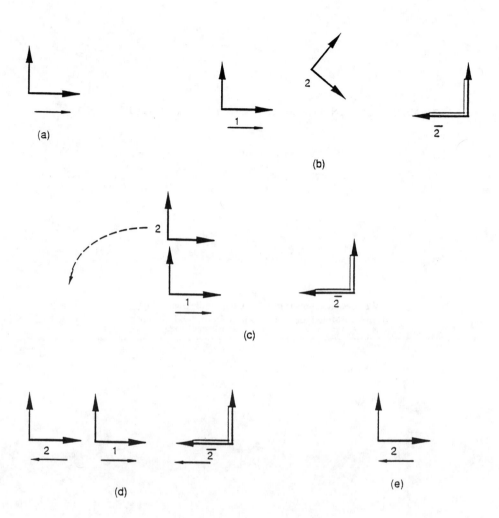

FIG. 5 IS HOMOTOPIC TO THIS FIGURE.

FIG. 8

CP VIOLATION IN THE STANDARD MODEL AND BEYOND

G.C. Branco [*]

Theoretical Physics Division, CERN
CH - 1211 Geneva 23

ABSTRACT

We present an overview of CP violation in the Standard Model and some of its minimal extensions.

---

[*] Permanent address: Departamento de Fisica and CFIF/UTL, Instituto Superior Técnico, Avenida Rovisco Pais, 1096 Lisboa Codex, Portugal.

1 Introduction

Understanding the origin of CP violation is one of the outstanding open questions in particle physics [1]. Although one can incorporate CP violation in the three generations Standard Model (SM) through the Kobayashi-Maskawa (KM) mechanism [2], there is no deep understanding of the origin of CP violation. From the experimental point of view, very little is also known. At present, CP violation has only been observed in the $K^0 - \bar{K}^0$ sector and indeed all the experimental observations are consistent with the original superweak model [3] suggested by Wolfenstein soon after the discovery of the decay $K_L \to \pi^+\pi^-$, by Christensen, Cronin, Fitch and Turlay [4].

On the other hand, the present bounds on the electric dipole moment of the neutron and the electron imply significant restrictions on physics beyond the SM, while the observed baryon asymmetry may provide motivation for extending the SM [5]. Indeed the indication that the amount of CP violation one has in the SM through the KM mechanism is probably not sufficient to generate the observed baryon asymmetry [5] suggests other sources of CP violation beyond the SM.

2 The Standard Model

2.1 The Kobayashi-Maskawa mechanism

In the standard $SU(2) \times U(1)$ electroweak gauge theory with one Higgs doulbet, the only way to introduce CP violation is by having complex Yukawa couplings:

$$\mathcal{L}_Y = g_d(\bar{u}^0 \bar{d}^0)_L \Phi d_R^0 + g_u(\bar{u}^0 \bar{d}^0)_L \tilde{\Phi} u_R^0 + h.c. \tag{1}$$

where u^0, d^0 stand for quark weak-eigenstates, Φ is the weak Higgs doublet, and $\tilde{\Phi} = i\sigma_2 \Phi^*$. We have suppressed the family indices, so for n families, g_d, g_u will be $n \times n$ matrices in flavour space. The charged-current couplings are diagonal at this stage:

$$\mathcal{L}_W = \frac{g}{\sqrt{2}} \left[\bar{u}_L^0 \gamma_\mu d_L^0 W^\mu + h.c. \right] \tag{2}$$

Gauge symmetry breaking generates arbitrary complex quark mass matrices M_u, M_d which can be diagonalized through the bi-unitary transfomations:

$$u_L^0 \to U_L^u u_L; \quad d_L^0 = U_L^d d_L; \quad u_R^0 \to U_R^u u_R; \quad d_r^0 = U_R^d d_R$$

$$U_L^{u\dagger} M_u U_R^u = D_u$$
$$U_L^{d\dagger} M_d U_R^d = D_d \tag{3}$$

where u, d stand for the quark mass eigenstates and D_u, D_d are the diagonal quark mass matrices. When expressed in terms of the mass eigenstates, the charged weak current in the quark sector is no longer diagonal and the interaction is given by:

$$\mathcal{L}_W = \frac{g}{\sqrt{2}} \bar{u}_L \gamma^\mu V^{\text{CKM}} d_L W_\mu + \text{h.c.} \tag{4}$$

where $V^{\text{CKM}} = U_L^{u\dagger} U_L^d$. If one makes a redefinition of the quark field phases:

$$u_i \to \exp(i\alpha_i) u_i; \quad d_j \to \exp(i\beta_j) d_j \tag{5}$$

the CKM matrix transforms as:

$$V_{ij}^{\text{CKM}} \to V_{ij}^{\text{CKM}} \exp i(\beta_j - \alpha_i) \tag{6}$$

By using the freedom to redefine the quark phases, one can remove $(2n-1)$ phases from V^{CKM}, since an overall change of all phases leaves V^{CKM} invariant. Since a n-dimensional unitary matrix has n^2 parameters, $1/2n(n-1)$ of which are used to parametrize the $O(n)$ rotation, the number of physical phases in V^{CKM} is given by:

$$N_\phi = n^2 - \frac{1}{2}n(n-1) - (2n-1) = \frac{1}{2}(n-1)(n-2) \tag{7}$$

From Eq. (7) one concludes that for three or more generations there will be (a) physical phase(s) which lead to CP violation. This is the crucial observation made by Kobayashi and Maskawa [2] in their seminal paper. It is clear that only functions of V^{CKM} which are rephasing invariant can be observable. The simplest invariants are the moduli $|V_{ij}|$ and the quartic products $V_{ij} V_{k\ell} V_{i\ell}^* V_{kj}^*$, which are usually designated "quartets". It can easily be shown that any higher order invariant $(V_{ij} V_{k\ell} V_{mn} \ldots V_{rs} V_{i\ell}^* V_{kn}^* \ldots V_{rj}^*)$ can be expressed in terms of moduli and quartets. For example, in the case of a sextet, one has:

$$V_{ij} V_{k\ell} V_{mn} V_{i\ell}^* V_{kn}^* V_{mj}^* = \frac{(V_{ij} V_{k\ell} V_{i\ell}^* V_{kj}^*)(V_{kj} V_{mn} V_{kn}^* V_{mj}^*)}{|V_{kj}|^2} \tag{8}$$

So far our discussion has been for an arbitrary number of generations. For three generations, the imaginary parts of all quartets have the same absolute value, as a result of unitarity. For example, orthogonality of the first two rows of V^{CKM} implies

$$(V_{11} V_{21}^* + V_{13} V_{23}^*) V_{12}^* V_{22} = -|V_{12}|^2 |V_{22}|^2 \tag{9}$$

and taking the imaginary part of Eq. (9):

$$\text{Im}(V_{11} V_{22} V_{21}^* V_{12}^*) = -\text{Im}(V_{13} V_{22} V_{12}^* V_{23}^*) \tag{10}$$

The fact that in the three generations SM the absolute value of the imaginary part of all quartets is the same, enables one to identify the strength of CP violation with the quantity δ_{KM} defined by:

$$\delta_{\text{KM}} = |\text{Im } V_{cd} V_{ub} V_{cb}^* V_{ud}^*| \tag{11}$$

where $V_{ub} \equiv V_{13}$, etc. The smallness of δ_{KM} just reflects the smallness of some of the CKM matrix elements. Indeed from Eq. (11) and taking into account the experimental values of $|V_{ij}|$, one obtains the bound

$$\delta_{KM} \leq 10^{-4} \tag{12}$$

The strength of CP violation has an interesting geometrical interpretation. If one writes the unitarity relation:

$$V_{ud}V_{ub}^* + V_{cd}V_{cb}^* + V_{td}V_{tb}^* = 0 \tag{13}$$

in the complex plane, it can be ready verified that

$$\delta_{KM} = \frac{1}{2}A \tag{14}$$

where A is the area of the triangle formed by the three complex quantities of Eq. (13). Needless to say, one can form other unitarity triangles, but in the 3gSM, all unitarity triangles have the same area. In the 3gSM, one can express δ_{KM} in terms of four independent moduli [6]

$$\delta_{KM} = \left[U_{11}U_{22}U_{12}U_{21} - R^2 \right]^{1/2} \tag{15}$$

where $U_{ij} \equiv |V_{ij}|^2$ and R is given by:

$$R = \frac{1}{2}\left[1 - U_{11} - U_{22} - U_{12} - U_{21} + U_{11}U_{22} + U_{12}U_{21} \right] \tag{16}$$

This is a remarkable property of the 3gSM, since it implies that if one could measure with sufficient precision four independent moduli (through CP conserving processes) one could get the strength of CP violation in the 3gSM. At present, this is not possible, due to both the relative size of the various $|V_{ij}|$ and the relatively large experimental errors.

As a result of the rephasing freedom of Eq. (6), it is clear that there are many alternative ways to parametrize the CKM matrix. The most used ones are the so-called standard parametrization [7] and the Wolfenstein parametrization [8]. They are well described in the literature and we will not present them here. Instead we would like to refer to alternative ways of parametrizing the CKM matrix which only use directly measurable quantities which have to be rephasing invariant. We will present two examples of rephasing invariant parametrizations, one using three moduli and the argument of a quartet as input parameters and another using four moduli as input parameters.

Three moduli and one invariant phase

This parametrization was proposed by Bjorken an Dunietz [9] who suggest the following basic parameters:

$$|V_{us}|; |V_{ub}|; |V_{cb}|; \arg\left[V_{cb}V_{us}V_{cs}^*V_{ub}^*\right] \tag{17}$$

It is suggested that the reconstruction of the full matrix be done by choosing $V_{ud}, V_{us}, V_{cb}, V_{tb}, V_{cs}$ real and positive.

Four moduli

This parametrization [6] uses four moduli and a possible choice is:

$$|V_{us}|; |V_{ub}|; |V_{cb}|; |V_{td}| \tag{18}$$

The magnitude of δ_{KM} can be derived by using an expression analogous to the ones given in Eqs. (15), (16).

2.2 Invariants and CP violation

In this section we address the following question: given a Lagrangian, how can one verify whether it violates CP invariance? We will show next that the simplest way to study the CP properties of a Lagrangian is by investigating whether it is possible to construct a CP transformation which leaves the Lagrangian invariant. If the existence of such a transformation requires non-trivial restrictions on some of the couplings, then the Lagrangian can violate CP. We will illustrate the application of the method by considering first the SM and then indicating how it can be easily extended to models beyond the SM. The part of the SM Lagrangian which is relevant to the study of the CP properties is:

$$\mathcal{L} = \mathcal{L}_Y + \mathcal{L}_W \tag{19}$$

where $\mathcal{L}_Y, \mathcal{L}_W$ are given by Eqs. (1), (2). The most general CP transformation which leaves the gauge interactions invariant is:

$$\begin{aligned} u_L &\to W_L C u_L^* \quad ; \quad u_R \to W_R^u C u_R^* \\ d_L &\to W_L C d_L^* \quad ; \quad d_R \to W_R^d C d_R^* \end{aligned} \tag{20}$$

where W_L, W_R^u, W_R^d are n-dimensional unitary matrices acting in flavour space. It is clear that the left-handed charged current interactions constrain u_L, d_L to transform in the same way, while the absence of right-handed charged currents in the SM allow u_R, d_R to transform differently under CP. In order for $\mathcal{L}_Y$ to be invariant under Eq. (20) the matrices g_d, g_u are constrained to satify the conditions [10]

$$\begin{aligned} W_L^\dagger g_u W_R^u &= g_u^* \\ W_L^\dagger g_d W_R^d &= g_d^* \end{aligned} \tag{21}$$

From Eqs. (21) one obtains:

$$W_L^\dagger G_u W_L = G_u^* \tag{22}$$

$$W_L^\dagger G_d W_L = G_d^* \tag{23}$$

where $G_u = g_u g_u^\dagger, G_d = g_d g_d^\dagger$. It can be shown [10] that if there is a matrix W_L satisfying simultaneously Eqs. (22), (23), then there always exist matrices W_R^u, W_R^d satisfying Eqs. (21). Therefore, a necessary and sufficient condition for the Lagrangian of the SM to be CP invariant is the existence of a matrix W_L satisfying Eqs. (22), (23). From these equations one can now derive non-trivial CP restrictions on the Yukawa couplings, expressed in terms of weak-basis invariants. From Eqs. (22), (23), one obtains [10]

$$W_L^\dagger \left[G_u^p, G_d^q\right] W_L = - \left[G_u^p, G_d^q\right]^T \qquad p,q \text{ integers} \tag{24}$$

Note that the minus sign in the right-hand side of Eq. (24) is crucial. If one multiplies Eq. (24) by itself an odd number of times and takes traces, one derives:

$$\text{tr}\left[G_u^p, G_d^q\right]^r = 0 \qquad r \text{ odd} \tag{25}$$

Equations (25) are the necessary conditions to have CP invariance, expressed in terms of weak basis invariants and valid for an arbitrary number of generations. It can be readily verified that any hermitian 2×2 matrix automatically satisfies Eqs. (25) and one obtains the well-known result that for two generations and one Higgs doublet the Lagrangian of the SM does not violate CP. For three generations, one can show that there is only one independent condition, corresponding to $p = q = 1$. This results from the fact that for $n = 3$, the traces involving higher powers are proportional to the trace corresponding to $p = q = 1$. One has, for example:

$$\text{tr}\left[G_u^2, G_d\right]^3 = \frac{1}{3}\left[(\text{tr}G_u)^3 - \text{tr}(G_u)^3\right] \text{tr}\left[G_u, G_d\right]^3 \tag{26}$$

It can be easily shown [10] that for three generations the condition

$$\text{tr}\left[G_u, G_d\right]^3 = 0 \tag{27}$$

is not only a necessary but also a sufficient condition to guarantee CP invariance of the Lagrangian. This is readily shown by explicitly constructing the matrix W_L which satisfies Eqs. (22), (23). Note that all our discussion was done prior to gauge symmetry breaking. After $SU(2) \times U(1)$ breaking, quark masses are generated and the conditions of Eq. (25) can be readily written in terms of quark masses, and one obtains, for example:

$$\text{tr}\left[H_u, H_d\right]^3 = 0 \tag{28}$$

where $H_{u,d} = M_{u,d} M_{u,d}^\dagger$. It is instructive to write the above invariant in terms of quark masses and mixings:

$$\text{tr}\left[H_u, H_d\right]^3 = 6(\Delta_{21}\Delta_{31}\Delta_{32})\text{Im}(V_{11}V_{22}V_{12}^*V_{21}^*) \tag{29}$$

where $\Delta_{21} = (m_s^2 - m_d^2)(m_c^2 - m_u^2)$, with analogous expressions for Δ_{31}, Δ_{32}. This result was to be expected, since we have seen in the previous analysis that δ_{KM} can be identified with the strength of CP violation. The quark mass factors were also to be expected, since mass degeneracy in either the up or the down quark sector, immediately implies that for three generations all phases can be removed from the CKM matrix and therefore no CP violation arises.

At this stage, it is worth commenting on how the method we have presented here can be extended to theories beyond the SM. As an example, we will take the left-right (LR) symmetric model [11] whose relevant part of the Lagrangian can be written:

$$\mathcal{L} = \mathcal{L}_Y + \mathcal{L}_{W_L} + \mathcal{L}_{W_R} \tag{30}$$

where $\mathcal{L}_{W_{L,R}}$ stand for the left-handed (LH) and right-handed (RH) charged-current interactions. The most general CP transformation allowed by the gauge interactions is

$$u_L \to W_L C u_L^* \quad ; \quad u_R \to W_R C u_R^*$$
$$d_L \to W_L C u_L^* \quad ; \quad u_R \to W_R C u_R^* \qquad (31)$$

The important point is that due to the presence of both LH and RH charged currents, each of the pairs of chiral fields (u_L, d_L), (u_R, d_R) are required to transform in the same way under CP. In order for the mass terms arising from $\mathcal{L}_Y$, to satisfy the conditions of Eq. (31), they have to satisfy the following equations [12]:

$$W_L^\dagger M_u W_R = M_u^*$$
$$W_L^\dagger M_d W_R = M_d^* \qquad (32)$$

From Eqs. (32), one obtains:

$$W_L^\dagger M_u M_d^\dagger W_L = (M_d M_u^\dagger)^T$$

which leads to the condition:

$$\text{tr}\left[(M_u M_d^\dagger)^p\right] = \text{tr}\left[(M_d M_u^\dagger)^p\right] \qquad (33)$$

The conditions of eq. (33) are necessary conditions for CP invariance in the LR models, written in terms of invariants. One can deduce from Eqs. (32) other necessary conditions for CP invariance in the LR models and for two and three generations it is possible to derive a set of invariant conditions which are necessary and sufficient for CP invariance in LR models [12].

3 Beyond the Standard Model

3.1 Motivation

The success of the SM is truly impressive and therefore one has to carefully justify the motivation for considering extensions of the SM. Actually, the phenomenon of CP violation may provide some motivation to go beyond the SM. Some of these motivations can be:

(i) - Spontaneous T, CP Violation

One may find more appealing, as first suggested by T.D. Lee [13] that the Lagrangian be CP invariant and CP be only broken by the vacuum, in analogy with gauge symmetry breaking.

(ii) - Baryogenesis

It is likely [5] that the amount of CP violation present in the SM is not sufficient to produce adequate baryon asymmetry and the electroweak phase transition.

There are some excellent recent reviews [5] on (ii) and therefore we will concentrate here on (i).

3.2 Spontaneous T, CP violation

We will address the following questions:

(i) What conditions should a set of vacua satisfy in order to violate T, CP?

(ii) What are the minimal extensions of the SM which can lead to spontaneous T, CP violation?

3.2.1 Condition for spontaneous T, CP violation

For definiteness, we will assume the $SU(2) \times U(1)$ model with an arbitrary number of Higgs doublets ϕ_i. However, the condition we will derive can be readily extended for models with other representations of Higgs fields. Since we wish that our analysis be applicable to Lagrangians which may be also invariant under linear transformations which mix the various ϕ_i, we will consider the most general T transformation which leaves the Lagrangian invariant:

$$T\phi_i T^{-1} = U_{ij}\phi_j \qquad (34)$$

where U is a unitary matrix which mixes the fields ϕ_i. From Eq. (34) and taking ionto account the anti-unitarity of the T operator, it follows that if the vacuum is T invariant, i.e.,

$$T|0> = 0 \qquad (35)$$

then the following condition holds [14]:

$$U^*_{ij} < 0|\phi_j|0>^* = < 0|\phi_i|0 > \qquad (36)$$

The simplest way to find whether a given set of vacuum expectation values correspond to a T, CP breaking vacuum, consists of verifying whether there is a unitary matrix U satisfying Eqs. (35), (36). There will be spontaneous T, CP violation if and only if no such unitary matrix exists.

3.2.2 Minimal extensions of the Standard Model

We will analyze here the minimal extensions of the SM based on $SU(2) \times U(1)$ which can lead to spontaneous CP violation. It is convenient to consider separately the cases where no non-standard quarks are introduced and where new quarks are added.

Only Standard Quarks - In the framework of an $SU(2) \times U(1)$ gauge theory, the minimal Higgs structure which can lead to spontaneous CP violation capable of explaining the observed CP violation in the kaon sector, consists of two Higgs doublets [13]. The most general Higgs potential can be written:

$$V(\phi_1, \phi_2) = V_0 + \left[\lambda_1(\phi_1^\dagger\phi_2)(\phi_1^\dagger\phi_2) + \lambda_2(\phi_1^\dagger\phi_2)(\phi_1^\dagger\phi_1) + \lambda_3(\phi_2^\dagger\phi_2)(\phi_1^\dagger\phi_2) + h.c. \right] \qquad (37)$$

where V_0 denotes the part of the Higgs potential without any phase dependence. We will assume that the Lagrangian is T, CP invariant and in this case we may choose, without loss of generality, the coupling constants in V real. It has been shown [13] that for an appropriate finite range of the parameters of the scalar potential, the absolute minimum of the potential is at:

$$\cos\theta = -(4\lambda_1 v_1 v_2)^{-1}\left[\lambda_2 v_1^2 + \lambda_3 v_2^2\right] \tag{38}$$

where

$$<0|\phi_j^0|0> = v_j \exp(i\theta_j) \tag{39}$$

and $\theta = \theta_2 - \theta_1$. In general, the vacuum corresponding to Eq. (39) leads to spontaneous T, CP violation. This can be easily seen by noting that for arbitrary λ_i, the Lagrangian is not invariant under transformations which mix ϕ_1, ϕ_2 and as a result the scalar fields transform under T as:

$$T\phi_i T^{-1} = e^{i\alpha_i}\phi_i \tag{40}$$

The terms with coefficients λ_2, λ_3 lead to the constraint $\alpha_1 = \alpha_2 \equiv \alpha$. It is then clear that for generic values of θ, one cannot find a value of α such that Eq. (36) is satisfied for both ϕ_1, ϕ_2, which proves that T is spontaneously violated. The two-Higgs model was suggested at the time when only two fermion generations were known. In this case, CP violation is solely due to Higgs exchange, the dominant contribution to $k^0 - \bar{k}^0$ mixing arising through tree diagrams mediated by neutral Higgs coupled to strangeness changing scalar currents. For three-fermion generations, it can be verified that the phase θ induces a KM phase in the quark CKM matrix. This is due to the fact that although there is only one phase θ, there are two completely independent matrices of Yukawa couplings, corresponding to each one of the ϕ_i. The simplest way to recover natural flavour conservation (NFC) in the Higgs sector consists of introducing a Z_2 symmetry, under which

$$\phi_2 \to -\phi_2; \quad d_{jR} \to -d_{jR} \tag{41}$$

with all the remaining fields transforming trivially under Z_2. The important effect of the Z_2 symmetry is to forbid couplings of the type $(\phi_1^\dagger \phi_2)(\phi_i^\dagger \phi_i)$. One can then show that all the minimae of the new potential are T, CP conserving. This is shown by noting that for an appropriate finite range of the parameters, the minimum of the scalar potential is at

$$<\phi_1> = v_1; \quad <\phi_2> = v_2 \exp(i\pi/2) \tag{42}$$

Naïvely one could think that this vacuum violates CP. However, this is not the case [15] since a matrix U exists, satisfying Eqs. (35), (36). Namely:

$$T \begin{bmatrix} \phi_1 \\ \phi_2 \end{bmatrix} T^{-1} = U \begin{bmatrix} \phi_1 \\ \phi_2 \end{bmatrix} \tag{43}$$

with $U = \begin{bmatrix} 1 & 0 \\ 0 & -1 \end{bmatrix}$. If one allows for soft symmetry breaking of the Z_2 symmetry through terms like $\phi_1^\dagger \phi_2$, one can then obtain [16] spontaneous CP violation.

The simplest possibility to achieve spontaneous CP violation and keep NFC in the Higgs sector consists of introducing a third Higgs doublet which does not couple to quarks. In this case, it has been shown [1] that for an appropriate range of the parameters of the Higgs potential, one can obtain spontaneous CP violation. Furthermore, it has been pointed out [17] that in this case the CP violating CKM phase does not generate a KM phase in the quark mixing matrix, and as a result, CP violation arises exclusively through Higgs exchange.

Extending the fermion sector

We have seen that in the $SU(2) \times U(1)$ gauge theory with only standard families, two Higgs doublets is the minimal structure required to generate spontaneous CP breaking, capable of accounting for the observed CP violation. We will see next that if one introduces isosinglet quarks one is able to generate spontaneous CP violation [18] with only one Higgs doublet and a Higgs singlet. For definiteness, we will add to the SM only one isosinglet $Q = 1/3$ quark D and a singlet Higgs S. The field content of the model is:

$$(ud)^i_L u^i_R, \quad d^i_R, D_R, D_L, \Phi, S \tag{44}$$

where i is a family index. We will impose CP invariance in the Lagrangian, together with an additional Z_2 symmetry under which all the fields of the SM transform trivially, while the new fields D_L, D_R, S are odd. This symmetry is not required in order to have spontaneous CP violation, but it is crucial in order for the model to provide a possible solution [18] of the strong CP problem [19]. The most general renormalizable $SU(2) \times U(1) \times Z_2$ invariant potential can be written as:

$$\begin{aligned} V &= V_\phi + V_s + V_{\phi,s} \\ V_\phi &= m^2 \phi^\dagger \phi + \lambda(\phi^\dagger \phi)^2 \\ V_s &= S^* S [a_1 + b_1 S^* S] + [S^2 + S^{*2}](a_2 + b_2 S^* S) + b_3(S^4 + S^{*4}) \\ V_{\phi,S} &= \phi^\dagger \phi [c_1(S^2 + S^{*2}) + c_2 S^* S] \end{aligned} \tag{45}$$

It can be readily shown that for a range of the parameters of the Higgs potential, the minimum is at:

$$<\phi^0> = \frac{v}{\sqrt{2}}; <S> = \frac{1}{\sqrt{2}} V e^{i\alpha} \tag{46}$$

which leads to spontaneous CP violation. The $SU(2) \times U(1) \times Z_2$ invariant Yukawa couplings are:

$$\begin{aligned} \mathcal{L}_Y = & -\sqrt{2}(\bar{u}\bar{d})^i_L \left[g_{ij}\phi d^j_R + h_{ij}\tilde{\phi} u^j_R\right] - \mu \bar{D}_L D_R - \\ & -\sqrt{2}(f_i S + f'_i S^*)\bar{D}_L d^i_R + h.c. \end{aligned} \tag{47}$$

where all interaction couplings are real, due to CP invariance. It is convenient to work in the weak basis where the up quark mass matrix is diagonal. The down quark mass matrix can be written as:

$$\mathcal{M}_d = \begin{bmatrix} m_d & 0 \\ M_D & \mu \end{bmatrix} \tag{48}$$

where $(m_d)_{ij} = g_{ij}v$ and $(M_D)_i = [f_i V e^{i\alpha} + f'_i V e^{-i\alpha}]$. The matrix $\mathcal{M}_d$ is diagonalized by the bi-unitary transformation:

$$\mathcal{U}_L^\dagger \mathcal{M}_d \mathcal{U}_R = \begin{bmatrix} \bar{m} & 0 \\ 0 & \bar{M} \end{bmatrix} \qquad (49)$$

where $\bar{m} = \text{diag}(m_d, m_s, m_b)$ and $\bar{M}$ is the mass of the heavy quark. It is convenient to write $\mathcal{U}_L$ in block form:

$$\mathcal{U}_L = \begin{pmatrix} K & R \\ S & T \end{pmatrix} \qquad (50)$$

where K is the usual 3×3 KM matrix connecting the standard quarks. It can be seen that in the limit $\bar{m}_i^2/\bar{M}^2 \ll 1$, one obtains [18]:

$$K\bar{m}^2 K^{-1} \cong m_0^2 \qquad (51)$$

where

$$m_0^2 = m_d m_d^\dagger - \frac{m_d M_D^\dagger M_D m_d^\dagger}{M^2} \qquad (52)$$

with $M^2 = (M_D M_D^\dagger + \mu^2) \cong \bar{M}^2$. It can be readily verified that the CKM matrix will have a non-vanishing KM phase even in the limit of very large V. The rôle of the isosinglet quark D is clear: through its mixing with standard quarks, it enables that CP violation generated by $<S>$ can also appear and unsuppressed in the light quark mixing.

It should be emphasized that this model besides providing a solution [18] to the strong CP problem [19] suggests the exciting possibility of having the observed CP violation at low energies as the result of physics at a higher energy scale.

4 Conclusions

We have reviewed how CP violation arises in electroweak gauge theories. At the Lagrangian level, CP breaking appears when there are at least two sets of couplings which require incompatible CP transformation properties of the fields. We have shown how this can be used to derive weak-basis invariant conditions for having CP invariance in the SM and its extensions. Alternatively, CP may be an exact symmetry of the Lagrangian, spontaneously broken by the vacuum. We have given the conditions which have to be satisfied in order for a given vacuum to violate T, CP, and described the minimal extensions of the SM which can lead to spontaneous CP violation.

We have not covered at all the important topic of how to test experimentally the KM mechanism and how to look for physics beyond the SM, through CP violating processes. These topics have been extensively covered in the literature [1], [20]. Here we would like to remark only that the study of heavy flavour physics and especially B-physics at either $e^+e^- B$ factories or at hadron colliders, provides an excellent opportunity to test some of the least known aspects of the SM, namely those having to do with Yukawa couplings, the breaking of gauge symmetry and the closely related phenomenon of CP violation.

References

[1] For recent reviews, see:
L. Wolfenstein, *Ann.Rev.Nucl.Part.Sci.* **36** (1986);
W. Grimus, *Fortsch.Phys.* **36** (1988) 201;
B. Winstein and L. Wolfenstein, EFI 92-55, to appear in *Rev.Mod.Phys.* ;
J.F. Donoghue, B.R. Holstein and G. Valencia, *Int.J.Mod.Phys.* **A2** (1987) 319;
Y. Nir and H. Quinn, SLAC.PUB.5737 (1992).

[2] M. Kobayashi and T. Maskawa, *Prog.Theor.Phys.* **49** (1973) 652.

[3] L. Wolfenstein, *Phys.Rev.Lett.* **13** (1964) 562.

[4] J.H. Christenson, J.W. Cronin, W.L. Fitch and R. Turlay, *Phys.Rev.Lett.* **13** (1964) 138.

[5] M.B. Gavela, P. Hernandez, J. Orloff and O. Pène, CERN Preprint TH. 7081/93 (1993);
G.R. Farrar and M.E. Shaposhnikov, CERN Preprint TH. 6734/93 (1993);
A.G. Cohen, D.B. Kaplan and A.E. nelson, *Ann.Rev.Nucl.Part.Sci.* **43** (1993).

[6] G.C. Branco and L. Lavoura, *Phys.Lett.* **B208** (1988) 123.

[7] See Particle Data Group, Review of Particle Properties, *Phys.Rev.* **D45** (1992), part 2 and references therein.

[8] L. Wolfenstein, *Phys.Rev.Lett.* **51** (1983) 1945.

[9] J.D. Bjorken and I. Dunietz, *Phys.Rev.* **D36** (1987) 2109.

[10] J. Bernabéu, G.C. Branco and M. Gronau, *Phys.Lett.* **169B** (1986) 243.

[11] J.C. Pati and A Salam, *Phys.Rev.* **D10** (1974) 275;
R.N. Mohapatra and J.C. Pati, *Phys.Rev.* **D11** (1975) 2558;
G. Senjanović and R.N. Mohapatra, *Phys.Rev.* **D12** (1975) 1502.

[12] G.C. Branco and M.N. Rebelo, *Phys.Lett.* **B173** (1986) 313.

[13] T.D. Lee, *Physics Reports* **9C** (1974) 143.

[14] G.C. Branco, J.M. Gérard and W. Grimus, *Phys.Lett.* **B136** (1984) 383.

[15] G.C. Branco, *Phys.Rev.* **D22** (1980) 2901.

[16] G.C. Branco and M.N. Rebelo, *Phys.Lett.* **160B** (1985) 117.

[17] G.C. Branco, *Phys.Rev.Lett.* **44** (1980) 504.

[18] L. Bento, G.C. Branco and P.A. Parada, *Phys.Lett.* **B267** (1991) 95.

[19] A. Nelson, *Phys.Lett.* B136 (1984) 387;
S.M. Barr, *Phys.Rev.Lett.* 53 (1984) 329.
For excellent reviews on the strong CP problem, see:
R.D. Peccei, DESY Report 88-109 (1988), in CP violation (World Scientific, Singapore);
J. Kim, *Physics Reports* 150 (1987) 1;
H.Y. Cheng, *Physics Reports* 158 (1988) 1.

[20] ECFA Workshop on a European B-Meson Factory, ECFA 93/151, Eds. R. Alexan and A. Ali.

INTRODUCTION TO
CHIRAL PERTURBATION THEORY

A. Pich[*][†]

Theory Division, CERN, CH-1211 Geneva 23

ABSTRACT

An introduction to the basic ideas and methods of Chiral Perturbation Theory is presented. Several phenomenological applications of the effective Lagrangian technique to strong, electromagnetic and weak interactions are discussed.

1 EFFECTIVE FIELD THEORIES

Effective Field Theories (EFTs) are the appropriate theoretical tool to describe "low-energy" physics, where "low" is defined with respect to some energy scale Λ. What that means is that they only take explicitly into account the relevant degrees of freedom, i.e. those states with $m \ll \Lambda$, while the heavier excitations with $M \gg \Lambda$ are integrated out from the action. One gets in this way a string of non-renormalizable interactions among the light states, which can be organized as an expansion in powers of energy$/\Lambda$. The information on the heavier degrees of freedom is then contained in the couplings of the resulting low-energy Lagrangian. Although EFTs contain an infinite number of terms, renormalizability is not an issue since, at a given order in the energy expansion, the low-energy theory is specified by a finite number of couplings; this allows for an order-by-order renormalization. Obviously, for this procedure to make sense, it is necessary that the spectrum of the fundamental theory contains a mass gap, separating the light and heavy states.

A simple example of EFT is provided by QED at very low energies, $\omega \ll m_e$, where ω denotes the photon energy. In this limit, one can describe the light-by-light scattering using an effective Lagrangian in terms of the electromagnetic field only. Gauge and Lorentz invariance constrain the possible structures present in the effective Lagrangian:

$$\mathcal{L}_{\text{eff}} = -\frac{1}{4} F^{\mu\nu} F_{\mu\nu} + \frac{a}{m_e^4} \left(F^{\mu\nu} F_{\mu\nu}\right)^2 + \frac{b}{m_e^4} F^{\mu\nu} F_{\nu\sigma} F^{\sigma\rho} F_{\rho\mu} + O(F^6/m_e^8). \quad (1)$$

In the low-energy regime, all the information on the original QED dynamics is embodied in the values of the two low-energy couplings a and b. The values of these constants can be computed, by explicitly integrating out the electron

---

[*]On leave of absence from Departament de Física Teòrica, Universitat de València, and IFIC, Centre Mixte Universitat de València–CSIC, E-46100 Burjassot, València, Spain.

[†]Work supported in part by CICYT (Spain), under grant No. AEN90-0040.

field from the original QED generating functional (or equivalently, by computing the relevant light-by-light box diagrams). One then gets the well-known Euler-Heisenberg result [1]:

$$a = -\frac{\alpha^2}{36}, \qquad b = \frac{7\alpha^2}{90}. \qquad (2)$$

The important point to realize is that, even in the absence of an explicit computation of the couplings a and b, the Lagrangian (1) contains non-trivial information, which is a consequence of the imposed symmetries. The dominant contributions to the amplitudes for different low-energy photon reactions like $\gamma\gamma \to 2\gamma, 4\gamma, \ldots$ can be directly obtained from $\mathcal{L}_{\text{eff}}$. Moreover, the order of magnitude of the constants a, b can also be easily estimated through a naïve counting of powers of the electromagnetic coupling and combinatorial and loop $[1/(16\pi^2)]$ factors.

The previous example is somehow academic, since perturbation theory in powers of α works extremely well in QED. However, the effective Lagrangian (1) would be valid even if the fine structure constant were big; the only difference would then be that we would not be able to perturbatively compute the couplings a and b.

We can mention two generic situations where EFTs become particularly useful:

- The underlying fundamental theory is unknown, but the symmetry properties of the light states can be used to build an effective Lagrangian. The low-energy couplings then parametrize the unknown new physics. A typical example are EFTs at the electroweak scale.

- Even if the underlying fundamental theory is known, sometimes it is not directly applicable in the low-energy region. For instance, due to confinement, the quark and gluon fields of QCD are not asymptotic states. Since we do not know how to solve QCD, we cannot derive the hadronic interactions directly from the original QCD Lagrangian. However, we do know the symmetry properties of the strong interactions; therefore, we can write an EFT in terms of the hadronic asymptotic states, and parametrize the unknown dynamical information in a few couplings.

The theoretical basis of EFTs can be formulated [2] as a "theorem"[1]: for a given set of asymptotic states, perturbation theory with the most general Lagrangian containing all terms allowed by the assumed symmetries will yield the most general S-matrix elements consistent with analyticity, perturbative unitarity and the assumed symmetries.

The purpose of these lectures is to give a pedagogical introduction to Chiral Perturbation Theory (ChPT), the low-energy effective field theory of the Standard Model. The chiral symmetry of the QCD Lagrangian is discussed in Section 2,

---

[1] Although this "theorem" is almost self-evident, it is only "proven" to the extent that no counter-examples are known.

and a toy low-energy model incorporating the right symmetry properties is studied in Section 3. The ChPT formalism is presented in Sections 4 and 5, where the lowest-order and next-to-leading-order terms in the chiral expansion are analysed. Section 6 contains a few selected phenomenological applications. The relation between the effective Lagrangian and the underlying fundamental QCD theory is studied in Section 7, which summarizes recent attempts to calculate the chiral couplings. The effective realization of the non-leptonic $\Delta S = 1$ interactions is described in Section 8, and a brief overview of the application of the chiral techniques to K decays is given in Sections 9, 10 and 11. Section 12 shows how ChPT can be used to work out the low-energy interactions of a possible light Higgs boson. Finally, Section 13 illustrates the use of the chiral techniques to describe the Goldstone dynamics associated with the Standard Model electroweak symmetry breaking. A few summarizing comments are collected in Section 14.

To prepare these lectures I have made extensive use of excellent reviews [3–10] and books [11–14] already existing in the literature. In many cases, I have sacrificed some rigour to simplify the presentation of the subject. A more careful discussion and further details can be found in those references.

2 CHIRAL SYMMETRY

In the absence of quark masses, the QCD Lagrangian [$q = \text{column}(u, d, \ldots)$]

$$\mathcal{L}^0_{QCD} = -\frac{1}{4}\text{Tr}(G_{\mu\nu}G^{\mu\nu}) + i\bar{q}_L\gamma^\mu D_\mu q_L + i\bar{q}_R\gamma^\mu D_\mu q_R \tag{3}$$

is invariant under independent global $G \equiv SU(N_f)_L \otimes SU(N_f)_R$ transformations[2] of the left- and right-handed quarks in flavour space:

$$q_L \xrightarrow{G} g_L\, q_L, \qquad q_R \xrightarrow{G} g_R\, q_R, \qquad g_{L,R} \in SU(N_f)_{L,R}. \tag{4}$$

The Noether currents associated with the chiral group G are [λ_a are Gell-Mann's matrices with $\text{Tr}(\lambda_a\lambda_b) = 2\delta_{ab}$]:

$$J^{a\mu}_X = \bar{q}_X\gamma^\mu\frac{\lambda_a}{2}q_X, \qquad (X = L, R; \quad a = 1, \ldots, 8). \tag{5}$$

The corresponding Noether charges $Q^a_X = \int d^3x J^{a0}_X(x)$ satisfy the familiar commutation relations

$$[Q^a_X, Q^b_Y] = i\delta_{XY}f_{abc}Q^c_X, \tag{6}$$

which were the starting point of the Current Algebra methods of the sixties [17].

---

[2]Actually, the Lagrangian (3) has a larger $U(N_f)_L \otimes U(N_f)_R$ global symmetry. However, the $U(1)_A$ part is broken by quantum effects [$U(1)_A$ anomaly], while the quark-number symmetry $U(1)_V$ is trivially realized in the meson sector. A discussion of the $U(1)_A$ part, within ChPT, is given in refs. [15,16].

This chiral symmetry, which should be approximately good in the light quark sector (u,d,s), is however not seen in the hadronic spectrum. Although hadrons can be nicely classified in $SU(3)_V$ representations, degenerate multiplets with opposite parity do not exist. Moreover, the octet of pseudoscalar mesons happens to be much lighter than all the other hadronic states. To be consistent with this experimental fact, the ground state of the theory (the vacuum) should not be symmetric under the chiral group. The $SU(3)_L \otimes SU(3)_R$ symmetry spontaneously breaks down to $SU(3)_{L+R}$ and, according to Goldstone's theorem [18], an octet of pseudoscalar massless bosons appears in the theory.

More specifically, let us consider a Noether charge Q, and assume the existence of an operator O that satisfies

$$\langle 0|[Q,O]|0\rangle \neq 0; \tag{7}$$

this is clearly only possible if $Q|0\rangle \neq 0$. Goldstone's theorem then tells us that there exists a massless state $|G\rangle$ such that

$$\langle 0|J^0|G\rangle \langle G|O|0\rangle \neq 0. \tag{8}$$

The quantum numbers of the Goldstone boson are dictated by those of J^0 and O. The quantity in the left-hand side of Eq. (7) is called the order parameter of the spontaneous symmetry breakdown.

Since there are eight broken axial generators of the chiral group, $Q_A^a = Q_R^a - Q_L^a$, there should be eight pseudoscalar Goldstone states $|G^a\rangle$, which we can identify with the eight lightest hadronic states (π^+, π^-, π^0, η, K^+, K^-, K^0 and $\bar{K}^0$); their small masses being generated by the quark-mass matrix, which explicitly breaks the global symmetry of the QCD Lagrangian. The corresponding O^a must be pseudoscalar operators. The simplest possibility are $O^a = \bar{q}\gamma_5\lambda_a q$, which satisfy

$$\langle 0|[Q_A^a, \bar{q}\gamma_5\lambda_b q]|0\rangle = -\frac{1}{2}\langle 0|\bar{q}\{\lambda_a,\lambda_b\}q|0\rangle = -\frac{2}{3}\delta_{ab}\langle 0|\bar{q}q|0\rangle. \tag{9}$$

The quark condensate

$$\langle 0|\bar{u}u|0\rangle = \langle 0|\bar{d}d|0\rangle = \langle 0|\bar{s}s|0\rangle \neq 0 \tag{10}$$

is then the natural-order parameter of Spontaneous Chiral Symmetry Breaking (SCSB).

The Goldstone nature of the pseudoscalar mesons implies strong constraints on their interactions, which can be most easily analysed on the basis of an effective Lagrangian.

3 A TOY LAGRANGIAN: THE LINEAR SIGMA MODEL

The linear sigma model [19]

$$\mathcal{L}_\sigma = \frac{1}{2}[\partial_\mu\sigma\partial^\mu\sigma + \partial_\mu\vec{\pi}\partial^\mu\vec{\pi}] - V(\sigma,\vec{\pi}), \tag{11}$$

$$V(\sigma,\vec{\pi}) = \frac{\lambda}{4}\left(\sigma^2 + \vec{\pi}^2 - v^2\right)^2, \qquad (\lambda > 0),$$

provides a very simple example of SCSB. If $v^2 < 0$, the global symmetry $O(4) \sim SU(2) \otimes SU(2)$ is realized in the usual Wigner–Weyl way; one then has degenerate $\sigma, \vec{\pi}$ states with mass $m^2 = -\lambda v^2$. However, for $v^2 > 0$, the potential $V(\sigma, \vec{\pi})$ has a family of minima occurring for all $\sigma, \vec{\pi}$ with $\sigma^2 + \vec{\pi}^2 = v^2$; these minima correspond to degenerate ground states, which transform into each other under chiral rotations. The symmetry is then realized à la Nambu–Goldstone, and three massless states appear, corresponding to the flat directions of $V(\sigma, \vec{\pi})$. Taking

$$\langle 0|\sigma|0\rangle = v, \qquad \langle 0|\vec{\pi}|0\rangle = 0, \tag{12}$$

and making the field redefinition $\hat{\sigma} = \sigma - v$, the Lagrangian takes the form

$$\mathcal{L}_\sigma = \frac{1}{2}\left[\partial_\mu\hat{\sigma}\partial^\mu\hat{\sigma} - 2\lambda v^2 \hat{\sigma}^2 + \partial_\mu\vec{\pi}\partial^\mu\vec{\pi}\right] - \lambda v \hat{\sigma}\left(\hat{\sigma}^2 + \vec{\pi}^2\right) - \frac{\lambda}{4}\left(\hat{\sigma}^2 + \vec{\pi}^2\right)^2, \tag{13}$$

which shows that $\vec{\pi}$ corresponds to the three massless Goldstone modes, while the $\hat{\sigma}$ field acquires a mass $m_{\hat{\sigma}}^2 = 2\lambda v^2$.

To clarify the role of chiral symmetry on the Goldstone dynamics, it is useful to rewrite the sigma-model Lagrangian in a different way. Using the 2×2 matrix notation

$$\Sigma(x) \equiv \sigma(x)I + i\vec{\tau}\vec{\pi}, \tag{14}$$

the Lagrangian (11) takes the compact form

$$\mathcal{L}_\sigma = \frac{1}{4}\langle\partial_\mu\Sigma^\dagger\partial^\mu\Sigma\rangle - \frac{\lambda}{16}\left(\langle\Sigma^\dagger\Sigma\rangle - 2v^2\right)^2, \tag{15}$$

where $\langle A\rangle$ denotes the trace of the matrix A. In this notation the Lagrangian is explicitly invariant under global chiral $G \equiv SU(2)_L \otimes SU(2)_R$ transformations:

$$\Sigma \xrightarrow{G} g_R \Sigma g_L^\dagger, \qquad g_{L,R} \in SU(2)_{L,R}. \tag{16}$$

We can now make the polar decomposition [5]

$$\Sigma(x) = (v + S(x))\, U(\phi(x)), \tag{17}$$
$$U(\phi(x)) = \exp\left(i\vec{\tau}\vec{\phi}(x)/v\right),$$

in terms of a Hermitian scalar field S and pseudoscalar fields $\vec{\phi}$. These fields transform in a non-linear way under the chiral group:

$$S \xrightarrow{G} S, \qquad U(\phi) \xrightarrow{G} g_R U(\phi) g_L^\dagger. \tag{18}$$

The sigma-model Lagrangian then takes the form

$$\mathcal{L}_\sigma = \frac{v^2}{4}\left(1 + \frac{S}{v}\right)^2 \langle\partial_\mu U^\dagger \partial^\mu U\rangle + \frac{1}{2}\left(\partial_\mu S \partial^\mu S - 2\lambda v^2 S^2\right) - \lambda v S^3 - \frac{\lambda}{4}S^4. \tag{19}$$

Equation (19) is very instructive:

- It shows explicitly that the Goldstone bosons have purely derivative couplings, as they should. This was not so obvious in Eq. (13). Of course one should get the same measurable amplitudes from both Lagrangians, which means that the original Lagrangian (13) gives rise to exact (and not very transparent) cancellations among different momentum-independent contributions.

- In the limit $\lambda >> 1$, the scalar field S becomes very heavy and can be integrated out from the Lagrangian. The linear sigma model then reduces to the familiar Lagrangian

$$\mathcal{L}_2 = \frac{v^2}{4} \langle \partial_\mu U^\dagger \partial^\mu U \rangle. \qquad (20)$$

As we will see in the next section, this is a universal model-independent interaction of the Goldstone bosons induced by SCSB. It is often claimed in the literature that the linear sigma model correctly describes the low-energy strong interactions. This is, however, quite a misleading statement. To the extent that one is only looking at the predictions coming from the model-independent lowest-order term (20), the comparison with experiment only tests the assumed pattern of SCSB.

- In order to be sensitive to the particular structure of the linear sigma model, one needs to test the model-dependent part involving the scalar field S. At low momenta ($p << M_S$), the dominant tree-level corrections originate from the exchange of an S particle, which generates the four-derivative term

$$\mathcal{L}_\sigma^4 = \frac{v^2}{8M_S^2} \langle \partial_\mu U^\dagger \partial^\mu U \rangle^2. \qquad (21)$$

It will be shown later that this kind of interaction does not agree with the experimental data. Therefore, the linear sigma model is not a phenomenologically viable EFT of QCD [20].

4 EFFECTIVE CHIRAL LAGRANGIAN AT LOWEST ORDER

We want to get an effective Lagrangian realization of QCD, at low energies, for the light-quark sector (u, d, s). Our basic assumption is the pattern of SCSB:

$$G \equiv SU(3)_L \otimes SU(3)_R \xrightarrow{SCSB} H \equiv SU(3)_V. \qquad (22)$$

The present understanding of the mechanism of SCSB is based in the dynamical generation of a non-zero vacuum expectation value of the scalar quark density, i.e. the vacuum condensate $v \equiv \langle 0|\bar{u}u|0\rangle = \langle 0|\bar{d}d|0\rangle = \langle 0|\bar{s}s|0\rangle \neq 0$. The Goldstone bosons correspond to the zero-energy excitations over this vacuum condensate; their fields can be collected in a 3 × 3 unitary matrix $U(\phi)$,

$$\langle 0|\bar{q}_L^j q_R^i|0\rangle \longrightarrow \frac{v}{2} U^{ij}(\phi), \qquad (23)$$

which parametrizes those excitations. A convenient parametrization is given by

$$U(\phi) \equiv \exp\left(i\sqrt{2}\Phi/f\right), \qquad (24)$$

where

$$\Phi(x) \equiv \frac{\vec{\lambda}}{\sqrt{2}} \vec{\phi} = \begin{pmatrix} \frac{\pi^0}{\sqrt{2}} + \frac{\eta_8}{\sqrt{6}} & \pi^+ & K^+ \\ \pi^- & -\frac{\pi^0}{\sqrt{2}} + \frac{\eta_8}{\sqrt{6}} & K^0 \\ K^- & \bar{K}^0 & -\frac{2\eta_8}{\sqrt{6}} \end{pmatrix}. \qquad (25)$$

The matrix $U(\phi)$ transforms linearly under the chiral group,

$$U(\phi) \xrightarrow{G} g_R U(\phi) g_L^\dagger, \qquad (26)$$

but the induced transformation on the Goldstone fields $\vec{\phi}$ is highly non-linear.

Since there is a mass gap separating the pseudoscalar octet from the rest of the hadronic spectrum, we can build an EFT containing only the Goldstone modes. We should write the more general Lagrangian involving the matrix $U(\phi)$, which is consistent with chiral symmetry. Moreover, we can organize the Lagrangian in terms of increasing powers of momentum or, equivalently, in terms of an increasing number of derivatives (parity conservation requires an even number of derivatives):

$$\mathcal{L}_{\text{eff}}(U) = \sum_n \mathcal{L}_{2n}. \qquad (27)$$

In the low-energy domain we are interested in, the terms with a minimum number of derivatives will dominate.

Due to the unitarity of the U matrix, $UU^\dagger = I$, at least two derivatives are required to generate a non-trivial interaction. To lowest order, the effective chiral Lagrangian is uniquely given by the term

$$\mathcal{L}_2 = \frac{f^2}{4} \langle \partial_\mu U^\dagger \partial^\mu U \rangle. \qquad (28)$$

This is exactly the structure (20), which we derived from the linear sigma model in the last section.

Expanding $U(\phi)$ in a power series in ϕ, one obtains the Goldstone's kinetic terms plus a tower of interactions involving an increasing number of pseudoscalars. The requirement that the kinetic terms are properly normalized fixes the global coefficient $f^2/4$ in Eq. (28). All interactions among the Goldstones can then be predicted in terms of the single coupling f:

$$\mathcal{L}_2 = \frac{1}{2}\langle\partial_\mu\phi\partial^\mu\phi\rangle + \frac{1}{12f^2}\langle(\phi\overleftrightarrow{\partial}_\mu\phi)(\phi\overleftrightarrow{\partial}^\mu\phi)\rangle + O(\phi^6/f^4). \tag{29}$$

To compute the $\pi\pi$ scattering amplitude, for instance, is now a trivial perturbative exercise. One gets the well-known [21] Weinberg result $[t \equiv (p'_+ - p_+)^2]$

$$T(\pi^+\pi^0 \to \pi^+\pi^0) = \frac{t}{f^2}. \tag{30}$$

Similar results can be obtained for $\pi\pi \to 4\pi, 6\pi, 8\pi, \ldots$ It is the non-linearity of the effective Lagrangian that relates amplitudes with different numbers of Goldstone bosons, allowing for absolute predictions in terms of f.

The EFT technique becomes much more powerful if one introduces couplings to external classical fields. Let us consider an extended QCD Lagrangian, with quark couplings to external Hermitian matrix-valued fields v_μ, a_μ, s, p:

$$\mathcal{L}_{QCD} = \mathcal{L}^0_{QCD} + \bar{q}\gamma^\mu(v_\mu + \gamma_5 a_\mu)q - \bar{q}(s - i\gamma_5 p)q. \tag{31}$$

The external fields will allow us to compute the effective realization of general Green functions of quark currents in a very straightforward way. Moreover, they can be used to incorporate the electromagnetic and semileptonic weak interactions, and the explicit breaking of chiral symmetry through the quark masses:

$$\begin{aligned} r_\mu &\equiv v_\mu + a_\mu = eQA_\mu + \ldots \\ \ell_\mu &\equiv v_\mu - a_\mu = eQA_\mu + \frac{e}{\sqrt{2}\sin\theta_W}(W^\dagger_\mu T_+ + \text{h.c.}) + \ldots \\ s &= \mathcal{M} + \ldots \end{aligned} \tag{32}$$

Here, Q and $\mathcal{M}$ denote the quark-charge and quark-mass matrices, respectively,

$$Q = \frac{1}{3}\text{diag}(2,-1,-1), \qquad \mathcal{M} = \text{diag}(m_u, m_d, m_s), \tag{33}$$

and T_+ is a 3×3 matrix containing the relevant Cabibbo–Kobayashi–Maskawa factors

$$T_+ = \begin{pmatrix} 0 & V_{ud} & V_{us} \\ 0 & 0 & 0 \\ 0 & 0 & 0 \end{pmatrix}. \tag{34}$$

Formally, the Lagrangian (31) is invariant under the following set of local $SU(3)_L \otimes SU(3)_R$ transformations:

$$\begin{aligned} q_L &\longrightarrow g_L\, q_L, \\ q_R &\longrightarrow g_R\, q_R, \\ \ell_\mu &\longrightarrow g_L \ell_\mu g^\dagger_L + ig_L\partial_\mu g^\dagger_L, \\ r_\mu &\longrightarrow g_R r_\mu g^\dagger_R + ig_R\partial_\mu g^\dagger_R, \\ s+ip &\longrightarrow g_R(s+ip)g^\dagger_L. \end{aligned} \tag{35}$$

We can use this formal symmetry to build a generalized effective Lagrangian for the Goldstone bosons, in the presence of external sources. Note that to respect the local invariance, the gauge fields v_μ, a_μ can only appear through the covariant derivative

$$D_\mu U = \partial_\mu U - i r_\mu U + i U \ell_\mu, \qquad D_\mu U^\dagger = \partial_\mu U^\dagger + i U^\dagger r_\mu - i \ell_\mu U^\dagger, \qquad (36)$$

and through the field strength tensors

$$F_L^{\mu\nu} = \partial^\mu \ell^\nu - \partial^\nu \ell^\mu - i[\ell^\mu, \ell^\nu], \qquad F_R^{\mu\nu} = \partial^\mu r^\nu - \partial^\nu r^\mu - i[r^\mu, r^\nu]. \qquad (37)$$

At lowest order in momenta, the more general effective Lagrangian consistent with Lorentz invariance and with (local) chiral symmetry is of the form [15]

$$\mathcal{L}_2 = \frac{f^2}{4} \langle D_\mu U^\dagger D^\mu U + U^\dagger \chi + \chi^\dagger U \rangle, \qquad (38)$$

where

$$\chi = 2 B_0 (s + ip), \qquad (39)$$

and B_0 is a constant, which, like f, is not fixed by symmetry requirements alone.

Once special directions in flavour space, like the ones in Eq. (32), are selected for the external fields, chiral symmetry is of course explicitly broken. The important point is that (38) then breaks the symmetry in exactly the same way as the fundamental short-distance Lagrangian (31) does.

The power of the external field technique becomes obvious when computing the chiral Noether currents. Formally, the physical Green functions are obtained as functional derivatives of the generating functional $Z[v, a, s, p]$, defined via the path-integral formula

$$\exp\{iZ\} = \int \mathcal{D}q \, \mathcal{D}\bar{q} \, \mathcal{D}G_\mu \exp\left\{i \int d^4x \, \mathcal{L}_{QCD}\right\} = \int \mathcal{D}U(\phi) \exp\left\{i \int d^4x \, \mathcal{L}_{\text{eff}}\right\}. \qquad (40)$$

At lowest order in momenta, the generating functional reduces to the classical action $S_2 = \int d^4x \, \mathcal{L}_2$; therefore, the currents can be trivially computed by taking the appropriate derivatives with respect to the external fields:

$$J_L^\mu \doteq \frac{\delta S_2}{\delta \ell_\mu} = \frac{i}{2} f^2 D_\mu U^\dagger U = \frac{f}{\sqrt{2}} D_\mu \phi - \frac{i}{2} \left(\phi \overset{\leftrightarrow}{D^\mu} \phi\right) + O(\phi^3/f),$$

$$J_R^\mu \doteq \frac{\delta S_2}{\delta r_\mu} = \frac{i}{2} f^2 D_\mu U U^\dagger = -\frac{f}{\sqrt{2}} D_\mu \phi - \frac{i}{2} \left(\phi \overset{\leftrightarrow}{D^\mu} \phi\right) + O(\phi^3/f). \qquad (41)$$

The physical meaning of the chiral coupling f is now obvious; at $O(p^2)$, f equals the pion decay constant, $f = f_\pi = 93.2$ MeV, defined as

$$\langle 0 | (J_A^\mu)^{12} | \pi^+ \rangle \equiv i\sqrt{2} f_\pi p^\mu. \qquad (42)$$

Similarly, by taking a derivative with respect to the external scalar source s, we learn that the constant B_0 is related to the quark condensate

$$\langle 0|\bar{q}^j q^i|0\rangle = -f^2 B_0 \delta^{ij}. \tag{43}$$

Taking $s = \mathcal{M}$ and $p = 0$, the χ term in Eq. (38) gives rise to a quadratic pseudoscalar-mass term plus additional interactions proportional to the quark masses. Expanding in powers of ϕ (and dropping an irrelevant constant), one has

$$\frac{f^2}{4} 2B_0 \langle \mathcal{M}(U + U^\dagger)\rangle = B_0 \left\{ -\langle \mathcal{M}\phi^2\rangle + \frac{1}{6f^2}\langle \mathcal{M}\phi^4\rangle + O(\phi^6/f^4)\right\}. \tag{44}$$

The explicit evaluation of the trace in the quadratic mass term provides the relation between the physical meson masses and the quark masses:

$$\begin{aligned}
M_{\pi^\pm}^2 &= 2\hat{m} B_0, \\
M_{\pi^0}^2 &= 2\hat{m} B_0 - \varepsilon + O(\varepsilon^2), \\
M_{K^\pm}^2 &= (m_u + m_s) B_0, \\
M_{K^0}^2 &= (m_d + m_s) B_0, \\
M_{\eta_8}^2 &= \frac{2}{3}(\hat{m} + 2m_s) B_0 + \varepsilon + O(\varepsilon^2),
\end{aligned} \tag{45}$$

where[3]

$$\hat{m} = \frac{1}{2}(m_u + m_d), \qquad \varepsilon = \frac{B_0}{4}\frac{(m_u - m_d)^2}{(m_s - \hat{m})}. \tag{46}$$

Chiral symmetry relates the magnitude of the meson and quark masses to the size of the quark condensate. Using the result (43), one gets from the first equation in (45) the well-known Gell-Mann–Oakes–Renner relation [22]

$$f_\pi^2 M_\pi^2 = -\hat{m}\langle 0|\bar{u}u + \bar{d}d|0\rangle. \tag{47}$$

Taking out the common B_0 factor, Eqs. (45) imply the old Current Algebra mass ratios [22,23],

$$\frac{M_{\pi^\pm}^2}{2\hat{m}} = \frac{M_{K^+}^2}{(m_u + m_s)} = \frac{M_{K^0}}{(m_d + m_s)} \approx \frac{3M_{\eta_8}^2}{(2\hat{m} + 4m_s)}, \tag{48}$$

and (up to $O(m_u - m_d)$ corrections) the Gell-Mann–Okubo mass relation [24],

$$3M_{\eta_8}^2 = 4M_K^2 - M_\pi^2. \tag{49}$$

---

[3] The $O(\varepsilon)$ corrections to $M_{\pi^0}^2$ and $M_{\eta_8}^2$ originate from a small mixing term between the π^0 and η_8 fields,

$$-B_0\langle \mathcal{M}\phi^2\rangle \longrightarrow -(B_0/\sqrt{3})(m_u - m_d)\,\pi^0 \eta_8.$$

The diagonalization of the quadratic π^0, η_8 mass matrix, gives the mass eigenstates, $\pi^0 = \cos\delta\,\phi^3 + \sin\delta\,\phi^8$ and $\eta_8 = -\sin\delta\,\phi^3 + \cos\delta\,\phi^8$, where $\tan(2\delta) = \sqrt{3}(m_d - m_u)/(2(m_s - \hat{m}))$.

Note that the chiral Lagrangian automatically implies the successful quadratic Gell-Mann–Okubo mass relation, and not a linear one. Since $B_0 m_q \propto M_\phi^2$, the external field χ is counted as $O(p^2)$ in the chiral expansion.

Although chiral symmetry alone cannot fix the absolute values of the quark masses, it gives information about quark-mass ratios. Neglecting the tiny $O(\varepsilon)$ effects, one gets the relations

$$\frac{m_d - m_u}{m_d + m_u} = \frac{(M_{K^0}^2 - M_{K^+}^2) - (M_{\pi^0}^2 - M_{\pi^+}^2)}{M_{\pi^0}^2} = 0.29, \qquad (50)$$

$$\frac{m_s - \hat{m}}{2\hat{m}} = \frac{M_{K^0}^2 - M_{\pi^0}^2}{M_{\pi^0}^2} = 12.6. \qquad (51)$$

In Eq. (50) we have subtracted the pion square-mass difference, to take into account the electromagnetic contribution to the pseudoscalar-meson self-energies; in the chiral limit ($m_u = m_d = m_s = 0$), this contribution is proportional to the square of the meson charge and it is the same for K^+ and π^+ [25]. The mass formulae (50) and (51) imply the quark ratios advocated by Weinberg [23]:

$$m_u : m_d : m_s = 0.55 : 1 : 20.3. \qquad (52)$$

Quark-mass corrections are therefore dominated by m_s, which is large compared with m_u, m_d. Notice that the difference $m_d - m_u$ is not small compared with the individual up- and down-quark masses; in spite of that, isospin turns out to be an extremely good symmetry, because isospin-breaking effects are governed by the small ratio $(m_d - m_u)/m_s$.

The ϕ^4 interactions in Eq. (44) introduce mass corrections to the $\pi\pi$ scattering amplitude (30),

$$T(\pi^+ \pi^0 \to \pi^+ \pi^0) = \frac{t - M_\pi^2}{f_\pi^2}, \qquad (53)$$

in perfect agreement with the Current Algebra result [21]. Since $f \approx f_\pi$ is fixed from pion decay, this result is now an absolute prediction of chiral symmetry!

The lowest-order chiral Lagrangian (38) encodes in a very compact way all the Current Algebra results obtained in the sixties [17]. The nice feature of the chiral approach is its elegant simplicity. Moreover, as we will see in the next section, the EFT method allows us to estimate higher-order corrections in a systematic way.

5 CHPT AT $O(p^4)$

At next-to-leading order in momenta, $O(p^4)$, the computation of the generating functional $Z[v, a, s, p]$ involves three different ingredients:

1. The most general effective chiral Lagrangian of $O(p^4)$, $\mathcal{L}_4$, to be considered at tree level.

2. One-loop graphs associated with the lowest-order Lagrangian $\mathcal{L}_2$.

5.1 $O(p^4)$ Lagrangian

At $O(p^4)$, the most general[4] Lagrangian, invariant under parity, charge conjugation and the local chiral transformations (35), is given by [15]

$$\begin{aligned}\mathcal{L}_4 =\ & L_1 \langle D_\mu U^\dagger D^\mu U\rangle^2 + L_2 \langle D_\mu U^\dagger D_\nu U\rangle \langle D^\mu U^\dagger D^\nu U\rangle \\ & + L_3 \langle D_\mu U^\dagger D^\mu U D_\nu U^\dagger D^\nu U\rangle + L_4 \langle D_\mu U^\dagger D^\mu U\rangle \langle U^\dagger \chi + \chi^\dagger U\rangle \\ & + L_5 \langle D_\mu U^\dagger D^\mu U \left(U^\dagger \chi + \chi^\dagger U\right)\rangle + L_6 \langle U^\dagger \chi + \chi^\dagger U\rangle^2 \\ & + L_7 \langle U^\dagger \chi - \chi^\dagger U\rangle^2 + L_8 \langle \chi^\dagger U \chi^\dagger U + U^\dagger \chi U^\dagger \chi\rangle \\ & - i L_9 \langle F_R^{\mu\nu} D_\mu U D_\nu U^\dagger + F_L^{\mu\nu} D_\mu U^\dagger D_\nu U\rangle + L_{10} \langle U^\dagger F_R^{\mu\nu} U F_{L\mu\nu}\rangle \\ & + H_1 \langle F_{R\mu\nu} F_R^{\mu\nu} + F_{L\mu\nu} F_L^{\mu\nu}\rangle + H_2 \langle \chi^\dagger \chi\rangle. \end{aligned} \qquad (54)$$

The terms proportional to H_1 and H_2 do not contain the pseudoscalar fields and are therefore not directly measurable. Thus, at $O(p^4)$ we need ten additional coupling constants L_i to determine the low-energy behaviour of the Green functions. These constants parametrize our ignorance about the details of the underlying QCD dynamics. In principle, all the chiral couplings are calculable functions of Λ_{QCD} and the heavy-quark masses. At the present time, however, our main source of information about these couplings is low-energy phenomenology.

5.2 Chiral loops

ChPT is a quantum field theory, perfectly defined through Eq. (40). As such, we must take into account quantum loops with Goldstone-boson propagators in the internal lines. The chiral loops generate non-polynomial contributions, with logarithms and threshold factors, as required by unitarity.

The loop integrals are homogeneous functions of the external momenta and the pseudoscalar masses occurring in the propagators. A simple dimensional counting shows that, for a general connected diagram with N_d vertices of $O(p^d)$ ($d = 2, 4, \ldots$), L loops and I internal lines, the overall chiral dimension is given by [2]

$$D = 2L + 2 + \sum_d N_d(d-2). \qquad (55)$$

Each loop adds two powers of momenta; this power suppression of loop diagrams is at the basis of low-energy expansions, such as ChPT. The leading $D = 2$ contributions are obtained with $L = 0$ and $d = 2$, i.e. only tree-level graphs with

---

[4]Since we will only need $\mathcal{L}_4$ at tree level, the general expression of this Lagrangian has been simplified, using the $O(p^2)$ equations of motion obeyed by U. Moreover, a 3×3 matrix relation has been used to reduce the number of independent terms. For the two-flavour case, not all of these terms are independent [20, 15].

$\mathcal{L}_2$ insertions. At $O(p^4)$, we have tree-level contributions from $\mathcal{L}_4$ ($L = 0$, $d = 4$, $N_4 = 1$) and one-loop graphs with the lowest-order Lagrangian $\mathcal{L}_2$ ($L = 1$, $d = 2$).

ChPT is an expansion in powers of momenta over some typical hadronic scale, usually called the scale of chiral symmetry breaking Λ_χ. Since each chiral loop generates a geometrical factor $(4\pi)^{-2}$, plus a factor of $1/f^2$ to compensate the additional dimensions, one could expect [29] Λ_χ to be about $4\pi f_\pi \sim 1.2\,\text{GeV}$.

The Goldstone loops are divergent and need to be renormalized. Although EFTs are non-renormalizable (i.e. an infinite number of counter-terms is required), order by order in the momentum expansion they define a perfectly renormalizable theory. If we use a regularization which preserves the symmetries of the Lagrangian, such as dimensional regularization, the counter-terms needed to renormalize the theory will be necessarily symmetric. Since by construction the full effective Lagrangian contains all terms permitted by the symmetry, the divergences can then be absorbed in a renormalization of the coupling constants occurring in the Lagrangian. At one loop (in $\mathcal{L}_2$), the ChPT divergences are $O(p^4)$ and are therefore renormalized by the low-energy couplings in Eq. (54):

$$L_i = L_i^r(\mu) + \Gamma_i \lambda, \qquad H_i = H_i^r(\mu) + \tilde{\Gamma}_i \lambda, \qquad (56)$$

where

$$\lambda = \frac{\mu^{d-4}}{16\pi^2} \left\{ \frac{1}{d-4} - \frac{1}{2} \left[\log(4\pi) + \Gamma'(1) + 1 \right] \right\}. \qquad (57)$$

The explicit calculation of the one-loop generating functional Z_4 [15] gives:

$$\begin{aligned}
&\Gamma_1 = \frac{3}{32}, \quad \Gamma_2 = \frac{3}{16}, \quad \Gamma_3 = 0, \quad \Gamma_4 = \frac{1}{8}, \\
&\Gamma_5 = \frac{3}{8}, \quad \Gamma_6 = \frac{11}{144}, \quad \Gamma_7 = 0, \quad \Gamma_8 = \frac{5}{48}, \\
&\Gamma_9 = \frac{1}{4}, \quad \Gamma_{10} = -\frac{1}{4}, \quad \tilde{\Gamma}_1 = -\frac{1}{8}, \quad \tilde{\Gamma}_2 = \frac{5}{24}.
\end{aligned} \qquad (58)$$

The renormalized couplings $L_i^r(\mu)$ depend on the arbitrary scale of dimensional regularization μ. This scale dependence is of course cancelled by that of the loop amplitude, in any physical, measurable quantity.

A typical $O(p^4)$ amplitude will then consist of a non-polynomial part, coming from the loop computation, plus a polynomial in momenta and pseudoscalar masses, which depends on the unknown constants L_i. The non-polynomial part (the so-called chiral logarithms) is completely predicted as a function of the lowest-order coupling f and the Goldstone masses. This chiral structure can be easily understood in terms of dispersion relations. Given the lowest-order Lagrangian $\mathcal{L}_2$, the non-trivial analytic behaviour associated with some physical intermediate state is calculable without the introduction of new arbitrary chiral coefficients. Analiticity then allows us to reconstruct the full amplitude, through a dispersive integral, up to a subtraction polynomial. ChPT generates (perturbatively) the correct dispersion integrals and organizes the subtraction polynomials in a derivative expansion.

5.3 THE CHIRAL ANOMALY

Although the QCD Lagrangian (31) is formally invariant under local chiral transformations, this is no longer true for the associated generating functional. The anomalies of the fermionic determinant break chiral symmetry at the quantum level [27, 28]. The anomalous change of the generating functional under an infinitesimal chiral transformation

$$g_{L,R} = 1 + i\alpha \mp i\beta + \ldots \tag{59}$$

is given by [28]:

$$\delta Z[v,a,s,p] = -\frac{N_C}{16\pi^2} \int d^4x \, \langle \beta(x)\, \Omega(x) \rangle, \tag{60}$$

$$\Omega(x) = \varepsilon^{\mu\nu\sigma\rho} \left[v_{\mu\nu} v_{\sigma\rho} + \frac{4}{3} \nabla_\mu a_\nu \nabla_\sigma a_\rho + \frac{2}{3} i \{v_{\mu\nu}, a_\sigma a_\rho\} \right.$$
$$\left. + \frac{8}{3} i\, a_\sigma v_{\mu\nu} a_\rho + \frac{4}{3} a_\mu a_\nu a_\sigma a_\rho \right],$$

$$v_{\mu\nu} = \partial_\mu v_\nu - \partial_\nu v_\mu - i[v_\mu, v_\nu], \qquad \nabla_\mu a_\nu = \partial_\mu a_\nu - i[v_\mu, a_\nu].$$

($N_C = 3$ is the number of colours, and $\varepsilon_{0123} = 1$.) This anomalous variation of Z is an $O(p^4)$ effect, in the chiral counting.

So far, we have been imposing chiral symmetry to construct the effective ChPT Lagrangian. Since chiral symmetry is explicitly violated by the anomaly at the fundamental QCD level, we need to add a functional Z_A with the property that its change under a chiral gauge transformation reproduces (60). Such a functional was constructed by Wess and Zumino [30], and reformulated in a nice geometrical way by Witten [31]. It has the explicit form:

$$S[U,\ell,r]_{WZW} = -\frac{iN_C}{240\pi^2} \int d\sigma^{ijklm} \left\langle \Sigma_i^L \Sigma_j^L \Sigma_k^L \Sigma_l^L \Sigma_m^L \right\rangle$$
$$- \frac{iN_C}{48\pi^2} \int d^4x\, \varepsilon_{\mu\nu\alpha\beta} \left(W(U,\ell,r)^{\mu\nu\alpha\beta} - W(1,\ell,r)^{\mu\nu\alpha\beta} \right), \tag{61}$$

$$W(U,\ell,r)_{\mu\nu\alpha\beta} = \Big\langle U\ell_\mu \ell_\nu \ell_\alpha U^\dagger r_\beta + \frac{1}{4} U\ell_\mu U^\dagger r_\nu U\ell_\alpha U^\dagger r_\beta + iU\partial_\mu \ell_\nu \ell_\alpha U^\dagger r_\beta$$
$$+ i\partial_\mu r_\nu U\ell_\alpha U^\dagger r_\beta - i\Sigma_\mu^L \ell_\nu U^\dagger r_\alpha U\ell_\beta + \Sigma_\mu^L U^\dagger \partial_\nu r_\alpha U\ell_\beta$$
$$- \Sigma_\mu^L \Sigma_\nu^L U^\dagger r_\alpha U\ell_\beta + \Sigma_\mu^L \ell_\nu \partial_\alpha \ell_\beta + \Sigma_\mu^L \partial_\nu \ell_\alpha \ell_\beta$$
$$- i\Sigma_\mu^L \ell_\nu \ell_\alpha \ell_\beta + \frac{1}{2} \Sigma_\mu^L \ell_\nu \Sigma_\alpha^L \ell_\beta - i\Sigma_\mu^L \Sigma_\nu^L \Sigma_\alpha^L \ell_\beta \Big\rangle$$
$$- (L \leftrightarrow R), \tag{62}$$

where

$$\Sigma_\mu^L = U^\dagger \partial_\mu U, \qquad \Sigma_\mu^R = U\partial_\mu U^\dagger, \tag{63}$$

and $(L \leftrightarrow R)$ stands for the interchanges $U \leftrightarrow U^\dagger$, $\ell_\mu \leftrightarrow r_\mu$ and $\Sigma_\mu^L \leftrightarrow \Sigma_\mu^R$. The integration in the first term of Eq. (61) is over a five-dimensional manifold whose boundary is four-dimensional Minkowski space. The integrand is a surface term; therefore both the first and the second terms of S_{WZW} are $O(p^4)$, according to the chiral counting rules.

Since anomalies have a short-distance origin, their effect is completely calculable. The translation from the fundamental quark–gluon level to the effective chiral level is unaffected by hadronization problems. In spite of its considerable complexity, the anomalous action (61) has no free parameters.

The anomaly functional gives rise to interactions that break the intrinsic parity. It is responsible for the $\pi^0 \to 2\gamma$, $\eta \to 2\gamma$ decays, and the $\gamma 3\pi$, $\gamma \pi^+ \pi^- \eta$ interactions. The five-dimensional surface term generates interactions among five or more Goldstone bosons.

6 LOW-ENERGY PHENOMENOLOGY AT $O(p^4)$

At lowest order in momenta, the predictive power of the chiral Lagrangian was really impressive: with only two low-energy couplings, it was possible to describe all Green functions associated with the pseudoscalar-meson interactions. The symmetry constraints become less powerful at higher orders. Ten additional constants appear in the $\mathcal{L}_4$ Lagrangian, and many more would be needed at $O(p^6)$. Higher-order terms in the chiral expansion are much more sensitive to the non-trivial aspects of the underlying QCD dynamics.

With $p \lesssim M_K (M_\pi)$, we expect $O(p^4)$ corrections to the lowest-order amplitudes at the level of $p^2/\Lambda_\chi^2 \lesssim 20\% (2\%)$. We need to include those corrections if we aim to increase the accuracy of the ChPT predictions beyond this level. Although the number of free constants in $\mathcal{L}_4$ looks quite big, only a few of them contribute to a given observable. In the absence of external fields, for instance, the Lagrangian reduces to the first three terms; elastic $\pi\pi$ and πK scatterings are then sensitive to $L_{1,2,3}$. The two-derivative couplings $L_{4,5}$ generate mass corrections to the meson decay constants (and mass-dependent wave-function renormalizations). Pseudoscalar masses are affected by the non-derivative terms $L_{6,7,8}$; L_9 is mainly responsible for the charged-meson electromagnetic radius and L_{10}, finally, only contributes to amplitudes with at least two external vector or axial-vector fields, like the radiative semileptonic decay $\pi \to e\nu\gamma$.

Table 1, taken from ref. [32], summarizes the present status of the phenomenological determination [15,33] of the constants L_i. The quoted numbers correspond to the renormalized couplings, at a scale $\mu = M_\rho$. The values of these couplings at any other renormalization scale can be trivially obtained, through the logarithmic running implied by Eq. (56):

$$L_i^r(\mu_2) = L_i^r(\mu_1) + \frac{\Gamma_i}{(4\pi)^2} \log\left(\frac{\mu_1}{\mu_2}\right). \qquad (64)$$

Comparing the Lagrangians $\mathcal{L}_2$ and $\mathcal{L}_4$, one can make an estimate of the expected size of the couplings L_i in terms of the scale of SCSB. Taking $\Lambda_\chi \sim$

Table 1: Phenomenological values of the renormalized couplings $L_i^r(M_\rho)$. The last column shows the source used to extract this information.

i	$L_i^r(M_\rho) \times 10^3$	Source
1	0.7 ± 0.5	K_{e4}, $\pi\pi \to \pi\pi$
2	1.2 ± 0.4	K_{e4}, $\pi\pi \to \pi\pi$
3	-3.6 ± 1.3	K_{e4}, $\pi\pi \to \pi\pi$
4	-0.3 ± 0.5	Zweig rule
5	1.4 ± 0.5	$F_K : F_\pi$
6	-0.2 ± 0.3	Zweig rule
7	-0.4 ± 0.2	Gell-Mann–Okubo, L_5, L_8
8	0.9 ± 0.3	$M_{K^0} - M_{K^+}$, L_5, $(m_s - \hat{m}) : (m_d - m_u)$
9	6.9 ± 0.7	$\langle r^2 \rangle_{em}^\pi$
10	-5.5 ± 0.7	$\pi \to e\nu\gamma$

$4\pi f_\pi \sim 1.2\,\text{GeV}$, one would get

$$L_i \sim \frac{f_\pi^2/4}{\Lambda_\chi^2} \sim \frac{1}{4(4\pi)^2} \sim 2 \times 10^{-3}, \qquad (65)$$

in reasonable agreement with the phenomenological values quoted in Table 1. This indicates a good convergence of the momentum expansion below the resonance region, i.e. $p < M_\rho$.

The chiral Lagrangian allows us to make a good book-keeping of phenomenological information with a few couplings. Once these couplings have been fixed, we can predict many other quantities. Moreover, the information contained in Table 1 is very useful to easily test different QCD-inspired models. Given any particular model aiming to correctly describe QCD at low energies, we no longer need to make an extensive phenomenological analysis of all the predictions of the model, in order to test its degree of reliability; we only need to calculate the predicted low-energy couplings, and compare them with the values in Table 1. For instance, if one integrates out the heavy scalar of the linear sigma model described in Section 3, the resulting Goldstone Lagrangian only contains the L_1 term [see Eq. (21)] at $O(p^4)$; obviously this is not a satisfactory approximation to the physical world[5].

An exhaustive description of the chiral phenomenology at $O(p^4)$ is beyond the scope of these lectures. Instead, I will just present a few examples to illustrate

---

[5] A more detailed study of the renormalizable linear sigma model can be found in ref. [20]. The conclusion is that this model is clearly ruled out by the data.

both the power and limitations of the ChPT techniques.

6.1 Decay constants

In the isospin limit ($m_u = m_d = \hat{m}$), the $O(p^4)$ calculation of the meson-decay constants gives [15]:

$$\begin{aligned}
f_\pi &= f\left\{1 - 2\mu_\pi - \mu_K + \frac{4M_\pi^2}{f^2}L_5^r(\mu) + \frac{8M_K^2 + 4M_\pi^2}{f^2}L_4^r(\mu)\right\}, \\
f_K &= f\left\{1 - \frac{3}{4}\mu_\pi - \frac{3}{2}\mu_K - \frac{3}{4}\mu_{\eta_8} + \frac{4M_K^2}{f^2}L_5^r(\mu) + \frac{8M_K^2 + 4M_\pi^2}{f^2}L_4^r(\mu)\right\}, \quad (66)\\
f_{\eta_8} &= f\left\{1 - 3\mu_K + \frac{4M_{\eta_8}^2}{f^2}L_5^r(\mu) + \frac{8M_K^2 + 4M_\pi^2}{f^2}L_4^r(\mu)\right\},
\end{aligned}$$

where

$$\mu_P \equiv \frac{M_P^2}{32\pi^2 f^2} \log\left(\frac{M_P^2}{\mu^2}\right). \qquad (67)$$

The result depends on two $O(p^4)$ couplings, L_4 and L_5. The L_4 term generates a universal shift of all meson-decay constants, $\delta f^2 = 16 L_4 B_0 \langle \mathcal{M} \rangle$, which can be eliminated taking ratios. From the experimental value [34]

$$\frac{f_K}{f_\pi} = 1.22 \pm 0.01, \qquad (68)$$

one can then fix $L_5(\mu)$; this gives the result quoted in Table 1. Moreover, one gets the absolute prediction [15]

$$\frac{f_{\eta_8}}{f_\pi} = 1.3 \pm 0.05. \qquad (69)$$

Taking into account isospin violations, one can also predict [15] a tiny difference between $f_{K^\pm}$ and f_{K^0}, proportional to $m_d - m_u$.

6.2 Electromagnetic form factors

At $O(p^2)$ the electromagnetic coupling of the Goldstone bosons is just the minimal one, obtained through the covariant derivative. The next-order corrections generate a momentum-dependent form factor

$$F^{\phi^\pm}(q^2) = 1 + \frac{1}{6}\langle r^2 \rangle^{\phi^\pm} q^2 + \cdots ; \qquad F^{\phi^0}(q^2) = \frac{1}{6}\langle r^2 \rangle^{\phi^0} q^2 + \cdots \qquad (70)$$

The meson electromagnetic radius $\langle r^2 \rangle^\phi$ gets local contributions from the L_9 term, plus logarithmic loop corrections [15]:

$$\begin{aligned}
\langle r^2 \rangle^{\pi^\pm} &= \frac{12 L_9^r(\mu)}{f^2} - \frac{1}{32\pi^2 f^2}\left\{2\log\left(\frac{M_\pi^2}{\mu^2}\right) + \log\left(\frac{M_K^2}{\mu^2}\right) + 3\right\}, \\
\langle r^2 \rangle^{K^0} &= -\frac{1}{16\pi^2 f^2}\log\left(\frac{M_K}{M_\pi}\right), \qquad (71)\\
\langle r^2 \rangle^{K^\pm} &= \langle r^2 \rangle^{\pi^\pm} + \langle r^2 \rangle^{K^0}.
\end{aligned}$$

Since neutral bosons do not couple to the photon at tree level, $\langle r^2 \rangle^{K^0}$ only gets a loop contribution, which is moreover finite (there cannot be any divergence because there exists no counter-term to renormalize it!). The predicted value, $\langle r^2 \rangle^{K^0} = -0.04 \pm 0.03 \,\text{fm}^2$, is in perfect agreement with the experimental determination [35] $\langle r^2 \rangle^{K^0} = -0.054 \pm 0.026 \,\text{fm}^2$.

The measured electromagnetic pion radius [36], $\langle r^2 \rangle^{\pi^\pm} = 0.439 \pm 0.008 \,\text{fm}^2$, is used as input to estimate the coupling L_9. This observable provides a good example of the importance of higher-order local terms in the chiral expansion. If one tries to ignore the L_9 contribution, using instead some "physical" cut-off p_{max} to regularize the loops, one needs [37] $p_{\text{max}} \sim 60 \,\text{GeV}$, in order to reproduce the experimental value; this is clearly nonsense. The pion charge radius is dominated by the $L_9^r(\mu)$ contribution, for any reasonable value of μ.

The measured K^+ charge radius [38], $\langle r^2 \rangle^{K^\pm} = 0.28 \pm 0.07 \,\text{fm}^2$, has a larger experimental uncertainty. Within present errors, it is in agreement with the parameter-free relation in Eq. (71).

6.3 K_{l3} DECAYS

The semileptonic decays $K^+ \to \pi^0 l^+ \nu_l$ and $K^0 \to \pi^- l^+ \nu_l$ are governed by the corresponding hadronic matrix element of the vector current [$t \equiv (P_K - P_\pi)^2$],

$$\langle \pi | \bar{s} \gamma^\mu u | K \rangle = C_{K\pi} \left[(P_K + P_\pi)^\mu f_+^{K\pi}(t) + (P_K - P_\pi)^\mu f_-^{K\pi}(t) \right], \quad (72)$$

where $C_{K^+\pi^0} = 1/\sqrt{2}$, $C_{K^0\pi^-} = 1$. At lowest order, the two form factors reduce to trivial constants: $f_+^{K\pi}(t) = 1$ and $f_-^{K\pi}(t) = 0$. There is however a sizeable correction to $f_+^{K^+\pi^0}(t)$, due to $\pi^0\eta$ mixing, which is proportional to $(m_d - m_u)$,

$$f_+^{K^+\pi^0}(0) = 1 + \frac{3}{4} \frac{m_d - m_u}{m_s - \hat{m}} = 1.017. \quad (73)$$

This number should be compared with the experimental ratio

$$\frac{f_+^{K^+\pi^0}(0)}{f_+^{K^0\pi^-}(0)} = 1.028 \pm 0.010. \quad (74)$$

The $O(p^4)$ corrections to $f_+^{K\pi}(0)$ can be expressed in a parameter-free manner in terms of the physical meson masses [15]. Including those contributions, one gets the more precise values

$$f_+^{K^0\pi^-}(0) = 0.977, \qquad \frac{f_+^{K^+\pi^0}(0)}{f_+^{K^0\pi^-}(0)} = 1.022, \quad (75)$$

which are in perfect agreement with the experimental result (74). The accurate ChPT calculation of these quantities allows us to extract [34] the most precise determination of the Kobayashi–Maskawa matrix element V_{us}:

$$|V_{us}| = 0.2196 \pm 0.0023. \quad (76)$$

At $O(p^4)$, the form factors get momentum-dependent contributions. Since L_9 is the only unknown chiral coupling occurring in $f_+^{K\pi}(t)$ at this order, the slope λ_+ of this form factor can be fully predicted. Alternatively, we can use the measured slope [39],

$$\lambda_+ \equiv \frac{1}{6} \langle r^2 \rangle^{K\pi} M_\pi^2 = 0.0300 \pm 0.0016, \tag{77}$$

as an input to get an independent determination of L_9. The value (77) corresponds [15] to $L_9^r(M_\rho) = (6.6 \pm 0.4) \times 10^{-3}$, in excellent agreement with the determination from the pion-charge radius, quoted in Table 1.

Instead of $f_-^{K\pi}(t)$, it is usual to parametrize the experimental results in terms of the so-called scalar form factor

$$f_0^{K\pi}(t) = f_+^{K\pi}(t) + \frac{t}{M_K^2 - M_\pi^2} f_-^{K\pi}(t). \tag{78}$$

The slope of this form factor is determined by the constant L_5, which in turn is fixed by f_K/f_π. One gets the result [15]:

$$\lambda_0 \equiv \frac{1}{6} \langle r^2 \rangle_S^{K\pi} M_\pi^2 = 0.017 \pm 0.004. \tag{79}$$

The experimental situation concerning the value of this slope is far from clear; while an older high-statistics measurement [40], $\lambda_0 = 0.019 \pm 0.004$, confirmed the theoretical expectations, more recent experiments find higher values, which disagree with this result. Reference [41], for instance, report $\lambda_0 = 0.046 \pm 0.006$, which differs from (79) by more than 4 standard deviations. The Particle Data Group [39] quote a world average $\lambda_0 = 0.025 \pm 0.006$.

6.4 Meson masses

The relations (45) get modified at $O(p^4)$. The additional contributions depend on the low-energy constants L_4, L_5, L_6, L_7 and L_8. It is possible, however, to obtain one relation between the quark and meson masses, which does not contain any of the $O(p^4)$ couplings. The dimensionless ratios

$$Q_1 \equiv \frac{M_K^2}{M_\pi^2}, \qquad Q_2 \equiv \frac{(M_{K^0}^2 - M_{K^+}^2) - (M_{\pi^0}^2 - M_{\pi^+}^2)}{M_K^2 - M_\pi^2}, \tag{80}$$

get the same $O(p^4)$ correction [15]:

$$Q_1 = \frac{m_s + \hat{m}}{2\hat{m}} \{1 + \Delta_M\}, \qquad Q_2 = \frac{m_d - m_u}{m_s - \hat{m}} \{1 + \Delta_M\}, \tag{81}$$

where

$$\Delta_M = -\mu_\pi + \mu_{\eta_8} + \frac{8}{f^2} (M_K^2 - M_\pi^2) \left[2L_8^r(\mu) - L_5^r(\mu) \right]. \tag{82}$$

Therefore, at this order, the ratio Q_1/Q_2 is just given by the corresponding ratio of quark masses,

$$Q^2 \equiv \frac{Q_1}{Q_2} = \frac{m_s^2 - \hat{m}^2}{m_d^2 - m_u^2}. \tag{83}$$

To a good approximation, Eq. (83) can be written as an ellipse,

$$\left(\frac{m_u}{m_d}\right)^2 + \frac{1}{Q^2}\left(\frac{m_s}{m_d}\right)^2 = 1, \tag{84}$$

which constrains the quark-mass ratios. The observed values of the meson masses give $Q = 24$.

Obviously, the quark-mass ratios (52), obtained at $O(p^2)$, satisfy this elliptic constraint. At $O(p^4)$, however, it is not possible to make a separate determination of m_u/m_d and m_s/m_d without having additional information on some of the L_i couplings.

A useful quantity is the deviation of the Gell-Mann–Okubo relation,

$$\Delta_{GMO} \equiv \frac{4M_K^2 - 3M_{\eta_8}^2 - M_\pi^2}{M_{\eta_8}^2 - M_\pi^2}. \tag{85}$$

Neglecting the mass difference $m_d - m_u$, one gets [15]

$$\Delta_{GMO} = \frac{-2\left(4M_K^2\mu_K - 3M_{\eta_8}^2\mu_{\eta_8} - M_\pi^2\mu_\pi\right)}{M_{\eta_8}^2 - M_\pi^2}$$
$$- \frac{6}{f^2}(M_{\eta_8}^2 - M_\pi^2)\left[12L_7^r(\mu) + 6L_8^r(\mu) - L_5^r(\mu)\right]. \tag{86}$$

Experimentally, correcting the masses for electromagnetic effects, $\Delta_{GMO} = 0.21$. Since L_5 is already known, this allows the combination $2L_7 + L_8$ to be fixed.

In order to determine the individual quark-mass ratios from Eqs. (81), we would need to fix the constant L_8. However, there is no way to find an observable that isolates this coupling. The reason is an accidental symmetry of the Lagrangian $\mathcal{L}_2 + \mathcal{L}_4$. The chiral Lagrangian remains invariant under the following simultaneous change [42] of the quark-mass matrix and some of the chiral couplings:

$$\mathcal{M}' = \alpha\mathcal{M} + \beta(\mathcal{M}^\dagger)^{-1}\det\mathcal{M}, \quad B_0' = B_0/\alpha, \tag{87}$$
$$L_6' = L_6 - \zeta, \quad L_7' = L_7 - \zeta, \quad L_8' = L_8 + 2\zeta,$$

where α and β are arbitrary constants, and $\zeta = \beta f^2/(32\alpha B_0)$. The only information on the quark-mass matrix $\mathcal{M}$ that we used to construct the effective Lagrangian was that it transforms as $\mathcal{M} \to g_R \mathcal{M} g_L^\dagger$. The matrix $\mathcal{M}'$ transforms in the same manner; therefore, symmetry alone does not allow us to distinguish between $\mathcal{M}$ and $\mathcal{M}'$. Since only the product $B_0\mathcal{M}$ appears in the Lagrangian,

α merely changes the value of the constant B_0. The term proportional to β is a correction of $O(\mathcal{M}^2)$; when inserted in $\mathcal{L}_2$, it generates a contribution to $\mathcal{L}_4$, which is reabsorbed by the redefinition of the $O(p^4)$ couplings. All chiral predictions will be invariant under the transformation (87); therefore it is not possible to separately determine the values of the quark masses and the constants B_0, L_6, L_7 and L_8. We can only fix those combinations of chiral couplings and masses that remain invariant under (87).

Notice that (87) is certainly not a symmetry of the underlying QCD Lagrangian. The accidental symmetry arises in the effective theory because we are not making use of the explicit form of the QCD Lagrangian; only its symmetry properties under chiral rotations have been taken into account. Therefore, we can resolve the ambiguity by obtaining one additional information from outside the pseudoscalar-meson chiral Lagrangian framework. For instance, by analysing the isospin breaking in the baryon mass spectrum and the ρ-ω mixing [43], it is possible to fix the ratio $(m_s - \hat{m})/(m_d - m_u) = 43.7 \pm 2.7$. Inserting this number in Eq. (83), one gets [15]

$$\frac{m_s}{\hat{m}} = 25.7 \pm 2.6. \tag{88}$$

Moreover, one can now determine L_8 from Eqs. (81), and therefore fix L_7 with Eq. (86); one then gets the values quoted in Table 1. Other ways of resolving the ambiguity, by using different additional inputs [44,45], lead to similar estimates of the quark-mass ratios and the low-energy couplings.

7 INFORMATION ENCODED IN THE CHIRAL COUPLINGS

The effective theory takes explicitly into account the poles and cuts generated by the Goldstone bosons. Given the non-trivial analytic structure associated with those physical intermediate states, the full amplitudes are reconstructed up to a subtraction polynomial. Obviously, the subtraction constants L_i contain all the information on the heavy degrees of freedom, which do not appear in the low-energy Lagrangian.

It seems rather natural to expect that the lowest-mass resonances, such as ρ mesons, should have an important impact on the physics of the pseudoscalar bosons. In particular, the low-energy singularities due to the exchange of those resonances should generate sizeable contributions to the chiral couplings. This can be easily understood, making a Taylor expansion of the ρ propagator:

$$\frac{1}{p^2 - M_\rho^2} = \frac{-1}{M_\rho^2}\left\{1 + \frac{p^2}{M_\rho^2} + \ldots\right\}, \qquad (p^2 < M_\rho^2). \tag{89}$$

Below the ρ-mass scale, the singularity associated with the pole of the resonance propagator is replaced by the corresponding momentum expansion. The exchange of virtual ρ mesons should result in derivative Goldstone couplings proportional to powers of $1/M_\rho^2$.

It is well known, for instance, that the electromagnetic form factor of the charged pion is well reproduced by the vector-meson dominance (VMD) formula

$$F^{\pi^\pm}(t) \approx \frac{M_\rho^2}{M_\rho^2 - t}, \qquad (90)$$

i.e. $\langle r^2 \rangle^{\pi^\pm} \approx 6/M_\rho^2 = 0.4\,\text{fm}^2$, to be compared with the measured value $\langle r^2 \rangle^{\pi^\pm} = 0.439 \pm 0.008\,\text{fm}^2$.

Writing a chiral-invariant $\rho\pi\pi$ interaction, with coupling G_V, one can compute the effect of a ρ-exchange diagram at low energies; the leading contribution [20,46] is a π^4 local interaction, with a coupling constant proportional to G_V^2/M_ρ^2. Since G_V can be directly measured from the $\rho \to 2\pi$ decay width, $|G_V| = 69$ MeV, the size of this contribution is fully predicted. Similarly, one can write a chiral invariant $\rho^0 \gamma$ interaction, with coupling F_V; this coupling can be extracted from the $\rho^0 \to e^+ e^-$ decay width, $|F_V| = 154$ MeV. The exchange of a ρ meson between the G_V and F_V vertices, generates a contribution to the electromagnetic form factor of the charged pion [46]:

$$\langle r^2 \rangle^{\pi^\pm} = \frac{6 F_V G_V}{f^2 M_\rho^2}. \qquad (91)$$

From the success of the naïve VMD formula (90), one could expect $F_V G_V/f^2 \approx 1$, which is indeed approximately satisfied (one obtains 1.2 with the measured F_V and G_V values).

A systematic analysis of the role of resonances in the ChPT Lagrangian has been performed[6] in ref. [46]. One writes first a general chiral-invariant Lagrangian $\mathcal{L}(U, V, A, S, P)$, describing the couplings between meson resonances of the type V, A, S, P and the Goldstone bosons, at lowest-order in derivatives. The coupling constants of this Lagrangian are phenomenologically extracted from physics at the resonance-mass scale. One has then an effective chiral theory defined in the intermediate-energy region. Formally, the generating functional (40) is given in this theory by the path-integral formula

$$\exp\{iZ\} = \int \mathcal{D}U(\phi)\,\mathcal{D}V\,\mathcal{D}A\,\mathcal{D}S\,\mathcal{D}P\,\exp\left\{i \int d^4x\,\mathcal{L}(U, V, A, S, P)\right\}. \qquad (92)$$

The integration of the resonance fields results in a low-energy theory with only Goldstones, i.e. the usual ChPT Lagrangian. At lowest-order this integration can be explicitly performed, expanding around the classical solution for the resonance fields. The resulting L_i couplings [46] are summarized in Table 2, which compares the different resonance-exchange contributions with the phenomenologically determined values of $L_i^r(M_\rho)$. For vector and axial-vector mesons only the $SU(3)$ octets contribute, whereas both octets and singlets are relevant in the case of scalar and pseudoscalar resonances.

---

[6] Related work can be found in ref. [47].

Table 2: V, A, S, S_1 and η_1 contributions to the coupling constants L_i^r in units of 10^{-3}. The last column shows the results obtained with the relations in Eq. (95).

i	$L_i^r(M_\rho)$	V	A	S	S_1	η_1	Total	Total[c]
1	0.7 ± 0.5	0.6	0	-0.2	$0.2^{b)}$	0	0.6	0.9
2	1.2 ± 0.4	1.2	0	0	0	0	1.2	1.8
3	-3.6 ± 1.3	-3.6	0	0.6	0	0	-3.0	-4.9
4	-0.3 ± 0.5	0	0	-0.5	$0.5^{b)}$	0	0.0	0.0
5	1.4 ± 0.5	0	0	$1.4^{a)}$	0	0	1.4	1.4
6	-0.2 ± 0.3	0	0	-0.3	$0.3^{b)}$	0	0.0	0.0
7	-0.4 ± 0.2	0	0	0	0	-0.3	-0.3	-0.3
8	0.9 ± 0.3	0	0	$0.9^{a)}$	0	0	0.9	0.9
9	6.9 ± 0.7	$6.9^{a)}$	0	0	0	0	6.9	7.3
10	-5.5 ± 0.7	-10.0	4.0	0	0	0	-6.0	-5.5

a) Input.   b) Large-N_C estimate.   c) With (95)

At lowest order, the most general interaction of the V octet to the Goldstone bosons contains two terms, corresponding to the couplings G_V and F_V described before. Due to the different parity, only one term with coupling F_A is present for the axial octet A. While V exchange generates contributions to L_1, L_2, L_3, L_9 and L_{10}, A exchange only contributes to L_{10} [46]:

$$L_1^V = \frac{G_V^2}{8M_V^2}, \quad L_2^V = 2L_1^V, \quad L_3^V = -6L_1^V, \qquad (93)$$

$$L_9^V = \frac{F_V G_V}{2M_V^2}, \quad L_{10}^{V+A} = -\frac{F_V^2}{4M_V^2} + \frac{F_A^2}{4M_A^2}.$$

To obtain the numbers in Table 2, the value of $L_9^r(M_\rho)$ has been fitted to determine $|G_V| = 53 \,\mathrm{MeV}$; nevertheless, the qualitative conclusion would be the same with the $\rho \to 2\pi$ determination mentioned before. The axial parameters have been fixed using the old Weinberg sum rules [48]: $F_A^2 = F_V^2 - f_\pi^2 = (123 \,\mathrm{MeV})^2$ and $M_A^2 = M_V^2 F_V^2/F_A^2 = (968 \,\mathrm{MeV})^2$. The results shown in the table clearly establish a chiral version of vector (and axial-vector) meson dominance: whenever they can contribute at all, V and A exchange seem to completely dominate the relevant coupling constants.

There are different phenomenologically successful models in the literature for V and A resonances (tensor-field description [20, 46], massive Yang–Mills [49], hidden gauge formulations [50], etc.). It can be shown [51] that all models are equivalent (i.e. give the same contributions to the L_i), provided they incorporate the appropriate QCD constraints at high energies. Moreover, with additional QCD-inspired assumptions of high-energy behaviour, such as an unsubtracted dispersion relation for the pion electromagnetic form factor, all V and A couplings can be expressed in terms of f and M_V only [51]:

$$F_V = \sqrt{2} f_\pi, \qquad G_V = f_\pi/\sqrt{2}, \qquad F_A = f_\pi, \qquad M_A = \sqrt{2} M_V. \qquad (94)$$

In that case, one has

$$L_1^V = L_2^V/2 = -L_3^V/6 = L_9^V/8 = -L_{10}^{V+A}/6 = f_\pi^2/(16 M_V^2). \qquad (95)$$

The last column in Table 2 shows the predicted numerical values of the L_i couplings, using the relations (95).

The analysis of scalar exchange is very similar [46]. Since the experimental information is quite scarce in the scalar sector, one needs to assume that the couplings L_5 and L_8 are due exclusively to scalar-octet exchange, to determine the scalar-octet couplings. The scalar-octet contributions to the other L_i ($i = 1, 3, 4, 6$) are then fixed. Moreover, one can then predict $\Gamma(a_0 \to \eta\pi)$, in good agreement with experiment. The scalar-singlet-exchange contributions can be expressed in terms of the octet parameters using large-N_C arguments. For $N_C = \infty$, octet- and singlet-scalar exchange cancel in L_1, L_4 and L_6. Although the results in Table 2 cannot be considered as a proof for scalar dominance, they provide at least a convincing demonstration of its consistency.

Neglecting the higher-mass 0^- resonances, the only remaining meson-exchange is the one associated with the η_1, which generates a sizeable contribution to L_7 [15,46]. The magnitude of this contribution can be calculated from the quark-mass expansion of M_η^2. The result for L_7 is in close agreement with its phenomenological value.

The combined resonance contributions appear to saturate the L_i^r almost entirely [46]. Within the uncertainties of the approach, there is no need for invoking any additional contributions. Although the comparison has been made for $\mu = M_\rho$, a similar conclusion would apply for any value of μ in the low-lying resonance region between 0.5 and 1 GeV.

All chiral couplings are in principle calculable from QCD. They are functions of Λ_{QCD} and the heavy-quark masses m_c, m_b, m_t. Unfortunately, we are not able at present to make such a first-principle computation. Although the integral over the quark fields in Eq. (40) can be done explicitly, we do not know how to perform analytically the remaining integration over the gluon fields. A perturbative evaluation of the gluonic contribution would obviously fail in reproducing the correct dynamics of SCSB. A possible way out is to parametrize phenomenologically the SCSB and make a weak gluon-field expansion around the resulting physical vacuum.

Table 3: Leading-order ($\alpha_s = 0$) predictions for the L_i's, within the QCD-inspired model (96). The phenomenological values are shown in the second row for comparison. All numbers are given in units of 10^{-3}.

	L_1	L_2	L_3	L_9	L_{10}
$L_i^{th}(\alpha_s = 0)$	0.79	1.58	-3.17	6.33	-3.17
$L_i^r(M_\rho)$	0.7 ± 0.5	1.2 ± 0.4	-3.6 ± 1.3	6.9 ± 0.7	-5.5 ± 0.7

The simplest parametrization [52] is obtained by adding to the QCD Lagrangian the term

$$\Delta \mathcal{L}_{QCD} = -M_Q \left(\bar{q}_R U q_L + \bar{q}_L U^\dagger q_R \right), \tag{96}$$

which serves to introduce the U field, and a mass parameter M_Q, which regulates the infra-red behaviour of the low-energy effective action. In the presence of this term the operator $\bar{q}q$ acquires a vacuum expectation value; therefore, (96) is an effective way to generate the order parameter due to SCSB. Making a chiral rotation of the quark fields, $Q_L = \xi q_L$, $Q_R = \xi^\dagger q_R$, with ξ chosen such that $U = \xi^2$, the interaction (96) reduces to a mass-term for the "dressed" quarks Q; the parameter M_Q can then be interpreted as a "constituent-quark mass".

The derivation of the low-energy effective chiral Lagrangian within this framework has been extensively discussed in ref. [52]. In the chiral and large-N_C limits, and including the leading gluonic contributions, one gets:

$$8L_1 = 4L_2 = L_9 = \frac{N_C}{48\pi^2} \left[1 + O\left(1/M_Q^6\right) \right], \tag{97}$$

$$L_3 = L_{10} = -\frac{N_C}{96\pi^2} \left[1 + \frac{\pi^2}{5N_C} \frac{\langle \frac{\alpha_s}{\pi} GG \rangle}{M_Q^4} + O\left(1/M_Q^6\right) \right].$$

Due to dimensional reasons, the leading contributions to the $O(p^4)$ couplings only depend on N_C and geometrical factors. It is remarkable that L_1, L_2 and L_9 do not get any gluonic correction at this order; this result is independent of the way SCSB has been parametrized (M_Q can be taken to be infinite). Table 3 compares the predictions obtained with only the leading term in Eqs. (97) (i.e. neglecting the gluonic correction) with the phenomenological determination of the L_i couplings. The numerical agreement is quite impressive; both the order of magnitude and the sign are correctly reproduced (notice that this is just a free-quark result!). Moreover, the gluonic corrections shift the values of L_3 and L_{10} in the right direction, making them more negative.

The results (97) obey almost all the short-distance relations (95). Comparing the predictions for $L_{1,2,9}$ in the VMD approach of Eq. (95) with the QCD-inspired ones in Eq. (97), one gets a quite good estimate of the ρ mass:

$$M_V = 2\sqrt{2}\pi f = 830 \,\text{MeV}. \tag{98}$$

Is it quite easy to prove that the interaction (96) is equivalent to the mean-field approximation of the Nambu–Jona–Lasinio (NJL) model, where SCSB is triggered by four-quark operators. It has been conjectured recently [53] that integrating out the quark and gluon fields of QCD, down to some intermediate scale Λ_χ, gives rise to an extended NJL Lagrangian. By introducing collective fields (to be identified later with the Goldstone fields and S, V, A resonances) the model can be transformed into a Lagrangian bilinear in the quark fields, which can therefore be integrated out. One then gets an effective Lagrangian, describing the couplings of the pseudoscalar bosons to vector, axial-vector and scalar resonances. Extending the analysis beyond the mean-field approximation, ref. [53] obtains predictions for 20 measurable quantities, including the L_i's, in terms of only 4 parameters. The quality of the fits is quite impressive. Since the model contains all resonances that are known to saturate the L_i couplings, it is not surprising that one gets an improvement of the mean-field-approximation results, specially for the constants L_5 and L_8, which are sensitive to scalar exchange. What is more important, this analysis clarifies a potential problem of double-counting: in certain limits the model approches either the pure quark-loop predictions of Eqs. (97) or the VMD results (95), but in general it interpolates between these two cases.

8 $\Delta S = 1$ NON-LEPTONIC WEAK INTERACTIONS

The Standard Model predicts strangeness-changing transitions with $\Delta S = 1$ via W-exchange between two weak charged currents. At low energies ($E \ll M_W$), the heavy fields W, Z, t, b, c can be integrated out. Using standard operator-product-expansion techniques, the $\Delta S = 1$ weak interactions are described by an effective Hamiltonian [54]

$$\mathcal{H}_{\text{eff}}^{\Delta S=1} = \frac{G_F}{\sqrt{2}} V_{ud} V_{us}^* \sum_i C_i(\mu) Q_i + \text{h.c.}, \tag{99}$$

which is a sum of local four-quark operators, constructed with the light (u, d, s) quark fields only,

$$Q_1 \equiv 4 \left(\bar{s}_L \gamma^\mu d_L\right)\left(\bar{u}_L \gamma_\mu u_L\right), \qquad Q_2 \equiv 4\left(\bar{s}_L \gamma^\mu u_L\right)\left(\bar{u}_L \gamma_\mu d_L\right),$$

$$Q_3 \equiv 4\left(\bar{s}_L \gamma^\mu d_L\right) \sum_{q=u,d,s} (\bar{q}_L \gamma_\mu q_L), \qquad Q_4 \equiv 4 \sum_{q=u,d,s} (\bar{s}_L \gamma^\mu q_L)(\bar{q}_L \gamma_\mu d_L), \tag{100}$$

$$Q_5 \equiv 4\left(\bar{s}_L \gamma^\mu d_L\right) \sum_{q=u,d,s} (\bar{q}_R \gamma_\mu q_R), \qquad Q_6 \equiv -8 \sum_{q=u,d,s} (\bar{s}_L q_R)(\bar{q}_R d_L),$$

modulated by Wilson coefficients $C_i(\mu)$, which are functions of the heavy W, t, b, c masses and an overall renormalization scale μ. Only five of these operators are independent, since $Q_4 = -Q_1+Q_2+Q_3$. From the point of view of chiral $SU(3)_L \otimes SU(3)_R$ and isospin quantum numbers, $Q_- \equiv Q_2 - Q_1$ and Q_i ($i = 3,4,5,6$) transform as $(8_L, 1_R)$ and induce $|\Delta I| = 1/2$ transitions, while $Q_1+2/3Q_2-1/3Q_3$ transforms like $(27_L, 1_R)$ and induces both $|\Delta I| = 1/2$ and $|\Delta I| = 3/2$ transitions.

In the absence of strong interactions, $C_2(\mu) = 1$ and all other Wilson coefficients vanish. The Standard Electroweak Model then gives rise to $|\Delta I| = 1/2$ and $|\Delta I| = 3/2$ amplitudes of nearly equal size, while experimentally the ratio between the two amplitudes is a factor of 20. To solve this large discrepancy, QCD effects should be enormous. The leading α_s corrections indeed give, for μ-values around 1 GeV, an enhancement by a factor of 2 to 3 of the Q_- Wilson coefficient with respect to the $Q_+ \equiv Q_2 + Q_1$ one. Moreover, the gluonic exchanges generate the additional $|\Delta I| = 1/2$ operators Q_i ($i = 3,4,5,6$). Nevertheless, this by itself is not enough to explain the experimentally observed rates, without simultaneously appealing to a further enhancement in the hadronic matrix elements of at least some of the isospin 1/2 four-quark operators. The computation of hadronic matrix elements at the K-mass scale is however a very difficult non-perturbative problem.

The effect of $\Delta S = 1$ non-leptonic weak interactions can be incorporated in the low-energy chiral theory, as a perturbation to the strong effective Lagrangian $\mathcal{L}_{\text{eff}}(U)$. At lowest order in the number of derivatives, the most general effective bosonic Lagrangian, with the same $SU(3)_L \otimes SU(3)_R$ transformation properties as the four-quark Hamiltonian in Eqs. (99) and (100), contains two terms[7]:

$$\mathcal{L}_2^{\Delta S=1} = -\frac{G_F}{\sqrt{2}} V_{ud} V_{us}^* \left\{ g_8 \langle \lambda L_\mu L^\mu \rangle + g_{27} \left(L_{\mu 23} L_{11}^\mu + \frac{2}{3} L_{\mu 21} L_{13}^\mu \right) + \text{h.c.} \right\}, \quad (101)$$

where

$$\lambda = (\lambda_6 - i\lambda_7)/2, \qquad L_\mu = if^2 U^\dagger D_\mu U. \quad (102)$$

The chiral couplings g_8 and g_{27} measure the strength of the two parts in the effective Hamiltonian (99) transforming as $(8_L, 1_R)$ and $(27_L, 1_R)$, respectively, under chiral rotations. Their values can be extracted from $K \to 2\pi$ decays [55]:

$$|g_8| \approx 5.1, \qquad g_{27}/g_8 \approx 1/18. \quad (103)$$

The huge difference between these two couplings shows the enhancement of the octet $|\Delta I| = 1/2$ transitions.

Using the effective Lagrangian (101), the calculation of hadronic weak transitions becomes a straightforward perturbative problem. The highly non-trivial QCD dynamics has been parametrized in terms of the two chiral couplings. Of course, the interesting problem that remains to be solved is to compute g_8 and

---

[7] One can build an additional octet term with the external χ field, $\langle \lambda (U^\dagger \chi + \chi^\dagger U) \rangle$; however, this term does not contribute to on-shell amplitudes.

g_{27} from the underlying QCD theory, and therefore to gain a dynamical understanding of the so-called $|\Delta I| = 1/2$ rule. Although this is a very difficult task, considerable progress has been achieved recently. Applying the QCD-inspired model of Eq. (96) to the weak sector, a quite successful estimate of these two couplings has been obtained [56]. A very detailed description of this calculation, and a comparison with other approaches, can be found in ref. [56].

Once the couplings g_8 and g_{27} have been phenomenologically fixed to the values in Eq. (103), other decays like $K \to 3\pi$ or $K \to 2\pi\gamma$ can be easily predicted at $O(p^2)$. As in the strong sector, one reproduces in this way the successful soft-pion relations of Current Algebra. However, the data are already accurate enough for the next-order corrections to be sizeable. Moreover, many transitions do not occur at $O(p^2)$. For instance, due to a mismatch between the minimum number of powers of momenta required by gauge invariance and the powers of momenta that the lowest-order effective Lagrangian can provide, the amplitude for any non-leptonic radiative K-decay with at most one pion in the final state $(K \to \gamma\gamma, K \to \gamma l^+l^-, K \to \pi\gamma\gamma, K \to \pi l^+l^-, ...)$ vanishes to lowest order in ChPT [57–59]. These decays are then sensitive to the non-trivial quantum field theory aspects of ChPT.

Unfortunately, at $O(p^4)$ there is a very large number of possible terms, satisfying the appropriate $(8_L, 1_R)$ and $(27_L, 1_R)$ transformation properties [60]. Using the $O(p^2)$ equations of motion obeyed by U to reduce the number of terms, 35 independent structures (plus 2 contact terms involving external fields only) remain in the octet sector alone [60,61]. Restricting the attention to those terms that contribute to non-leptonic amplitudes where the only external gauge fields are photons, still leaves 22 relevant octet terms [62]. Clearly, the predictive power of a completely general chiral analysis, using only symmetry constraints, is rather limited. Nevertheless, as we are going to see in the next sections, it is still possible to make predictions.

Due to the complicated interplay of electroweak and strong interactions, the low-energy constants of the weak non-leptonic chiral Lagrangian encode a much richer information than in the pure strong sector. These chiral couplings contain both long- and short-distance contributions, and some of them (like g_8) have in addition a CP-violating imaginary part. Genuine short-distance physics, such as the electroweak penguin operators, have their corresponding effective realization in the chiral Lagrangian. Moreover, there are four $O(p^4)$ terms containing an $\varepsilon_{\mu\nu\alpha\beta}$ tensor, which get a direct (probably dominant) contribution from the chiral anomaly [63,64].

In recent years, there have been several attempts to estimate these low-energy couplings using different approximations, such as factorization [56,65], weak-deformation model [66], effective-action approach [56,67], or resonance exchange [62,68]. Although more work in this direction is certainly needed, a qualitative picture of the size of the different couplings is already emerging.

9 $K \to 2\pi, 3\pi$ DECAYS

Imposing isospin and Bose symmetries, and keeping terms up to $O(p^4)$, a general parametrization of the $K \to 3\pi$ amplitudes involves ten measurable parameters [69], α_i, β_i, ζ_i, ξ_i, γ_3 and ξ'_3, where $i = 1,3$ refers to the $\Delta I = 1/2, 3/2$ pieces. At $O(p^2)$, the quadratic slope parameters ζ_i, ξ_i and ξ'_3 vanish; therefore the lowest-order Lagrangian (101) predicts five $K \to 3\pi$ parameters in terms of the two couplings g_8 and g_{27}, extracted from $K \to 2\pi$. These predictions give the right qualitative pattern, but there are sizeable differences with the measured parameters. Moreover, non-zero values for some of the slope parameters have been clearly established experimentally.

The agreement is substantially improved at $O(p^4)$ [70]. In spite of the large number of unknown couplings in the general effective $\Delta S = 1$ Lagrangian, only 7 combinations of these weak chiral constants are relevant for describing the $K \to 2\pi$ and $K \to 3\pi$ amplitudes [71]. Therefore, one has 7 parameters for 12 observables, which results in 5 relations. The extent to which these relations are satisfied provides a non-trivial test of chiral symmetry at the four-derivative level. The results of such a test [71] are shown in Table 4, where the 5 conditions have been formulated as predictions for the 5 slope parameters. The comparison is very successful for the two $\Delta I = 1/2$ parameters. The data are not good enough to say anything conclusive about the other three $\Delta I = 3/2$ predictions; moreover, the possible discrepancy in the value of ξ_3 is not very significant, because this parameter is expected to be rather sensitive to electromagnetic effects, which have been omitted in the analysis.

Table 4: Predicted and measured values of the quadratic slope parameters in the $K \to 3\pi$ amplitudes [71]. All values are given in units of 10^{-8}.

Parameter	Experimental value	Prediction
ζ_1	-0.47 ± 0.15	-0.47 ± 0.18
ξ_1	-1.51 ± 0.30	-1.58 ± 0.19
ζ_3	-0.21 ± 0.08	-0.011 ± 0.006
ξ_3	-0.12 ± 0.17	0.092 ± 0.030
ξ'_3	-0.21 ± 0.51	-0.033 ± 0.077

The $O(p^4)$ analysis of these decays has also clarified the role of long-distance effects ($\pi\pi$ rescattering) in the dynamical enhancement of $\Delta I = 1/2$ amplitudes. The $O(p^4)$ corrections give indeed a sizeable constructive contribution, which results [70] in a fitted value for $|g_8|$ that is about 30% smaller than the lowest-order determination (103). While this certainly goes in the right direction, it also shows

10 RADIATIVE K DECAYS

Owing to the constraints of electromagnetic gauge invariance, radiative K decays with at most one pion in the final state do not occur at $O(p^2)$ [57–59]. Moreover, only a few terms of the $O(p^4)$ Lagrangian are relevant for these kinds of processess [57–59]:

$$\mathcal{L}_4^{\Delta S=1,\text{em}} \doteq -\frac{G_F}{\sqrt{2}} V_{ud} V_{us}^* g_8 \left\{ -\frac{ie}{f^2} F^{\mu\nu} \left\{ w_1 \langle Q\lambda L_\mu L_\nu \rangle + w_2 \langle Q L_\mu \lambda L_\nu \rangle \right\} \right.$$
$$\left. + e^2 f^2 w_4 F^{\mu\nu} F_{\mu\nu} \langle \lambda Q U^\dagger Q U \rangle + \text{h.c.} \right\}. \tag{104}$$

The small number of unknown chiral couplings allows us to derive useful relations among different processes and to obtain definite predictions. Moreover, the absence of a tree-level $O(p^2)$ contribution makes the final results very sensitive to the loop structure of the amplitudes.

10.1 $K_S \to \gamma\gamma$

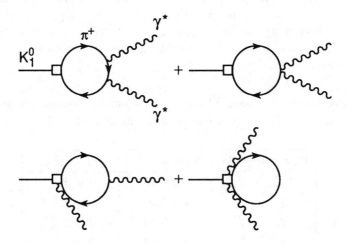

Figure 1: Feynman diagrams for $K_1^0 \to \gamma^*\gamma^*$.

The symmetry constraints do not allow any direct tree-level $K_1^0 \gamma\gamma$ coupling at $O(p^4)$ ($K_{1,2}^0$ refer to the CP-even and CP-odd eigenstates, respectively). This decay proceeds then through one loop of charged pions as shown in Fig. 1 (there are similar diagrams with charged kaons in the loop, but their sum gives a zero contribution to the decay amplitude). Moreover, since there are no possible counter-terms to renormalize divergences, the one-loop amplitude is necessarily finite. Although each of the four diagrams in Fig. 1 is quadratically divergent, these divergences cancel in the sum. The resulting prediction [72] is in very good agreement with the experimental measurement [73]

$$Br(K_S \to \gamma\gamma) = \begin{cases} 2.0 \times 10^{-6} & \text{(theory)} \\ (2.4 \pm 1.2) \times 10^{-6} & \text{(experiment)} \end{cases}. \quad (105)$$

10.2 $K_{L,S} \to \mu^+\mu^-$

There are well-known short-distance contributions (electroweak penguins and box diagrams) to the decay $K_L \to \mu^+\mu^-$. However, this transition is dominated by long-distance physics. The main contribution proceeds through a two-photon intermediate state: $K_2^0 \to \gamma^*\gamma^* \to \mu^+\mu^-$. Contrary to $K_1^0 \to \gamma\gamma$, the prediction for the $K_2^0 \to \gamma\gamma$ decay is very uncertain, because the first non-zero contribution occurs[8] at $O(p^6)$. That makes very difficult any attempt to predict the $K_L \to \mu^+\mu^-$ amplitude.

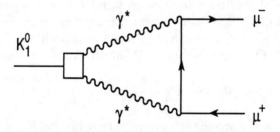

Figure 2: Feynman diagram for the $K_1^0 \to \mu^+\mu^-$ decay. The $K_1^0\gamma^*\gamma^*$ vertex is generated through the one-loop diagrams shown in Fig. 1

The situation is completely different for the K_S decay. A straightforward chiral analysis [74] shows that, at lowest order in momenta, the only allowed tree-level $K^0\mu^+\mu^-$ coupling corresponds to the CP-odd state K_2^0. Therefore, the $K_1^0 \to \mu^+\mu^-$ transition can only be generated by a finite non-local loop contribution. The two-loop calculation has been performed recently [74], with the result:

$$\frac{\Gamma(K_S \to \mu^+\mu^-)}{\Gamma(K_S \to \gamma\gamma)} = 1.9 \times 10^{-6}, \qquad \frac{\Gamma(K_S \to e^+e^-)}{\Gamma(K_S \to \gamma\gamma)} = 7.9 \times 10^{-9}, \quad (106)$$

---

[8]At $O(p^4)$, this decay proceeds through a tree-level $K_2^0 \to \pi^0, \eta$ transition, followed by $\pi^0, \eta \to \gamma\gamma$ vertices. Because of the Gell-Mann–Okubo relation, the sum of the π^0 and η contributions cancels exactly to lowest order. The decay amplitude is then very sensitive to $SU(3)$ breaking.

well below the present experimental upper limits [39]. Although, in view of the smallness of the predicted ratios, this calculation seems quite academic, it has important implications for CP-violation studies.

The longitudinal muon polarization $\mathcal{P}_L$ in the decay $K_L \to \mu^+\mu^-$ is an interesting measure of CP violation. As for every CP-violating observable in the neutral kaon system, there are in general two different kinds of contributions to $\mathcal{P}_L$: indirect CP violation through the small K_1^0 admixture of the K_L (ε effect), and direct CP violation in the $K_2^0 \to \mu^+\mu^-$ decay amplitude.

In the Standard Model, the direct-CP-violating amplitude is induced by Higgs exchange with an effective one-loop flavour-changing $\bar{s}dH$ coupling [75]. The present lower bound [76] on the Higgs mass $m_H > 60$ GeV (95% C.L.), implies [75,77] a conservative upper limit $|\mathcal{P}_{L,\text{Direct}}| < 10^{-4}$. A much larger value $\mathcal{P}_L \sim O(10^{-2})$ appears quite naturally in various extensions of the Standard Model [78]. It is worth emphasizing that $\mathcal{P}_L$ is especially sensitive to the presence of light scalars with CP-violating Yukawa couplings. Thus, $\mathcal{P}_L$ seems to be a good signature to look for new physics beyond the Standard Model; for this to be the case, however, it is very important to have a good quantitative understanding of the Standard Model prediction to allow us to infer, from a measurement of $\mathcal{P}_L$, the existence of a new CP-violation mechanism.

The chiral calculation of the $K_1^0 \to \mu^+\mu^-$ amplitude allows us to make a reliable estimate[9] of the contribution to $\mathcal{P}_L$ due to K^0-$\bar{K}^0$ mixing [74]:

$$1.9 < |\mathcal{P}_{L,\varepsilon}| \times 10^3 \left(\frac{2 \times 10^{-6}}{Br(K_S \to \gamma\gamma)} \right)^{1/2} < 2.5. \qquad (107)$$

Taking into account the present experimental errors in $Br(K_S \to \gamma\gamma)$ and the inherent theoretical uncertainties due to uncalculated higher-order corrections, one can conclude that experimental indications for $|\mathcal{P}_L| > 5 \times 10^{-3}$ would constitute clear evidence for additional mechanisms of CP violation beyond the Standard Model.

10.3 $K_L \to \pi^0 \gamma\gamma$

Assuming CP conservation, the most general form of the amplitude for $K_2^0 \to \pi^0 \gamma\gamma$ depends on two independent invariant amplitudes A and B [59],

$$\mathcal{A}[K_L(p_K) \to \pi^0(p_0)\gamma(q_1)\gamma(q_2)] =$$
$$\epsilon_\mu(q_1)\epsilon_\nu(q_2) \left\{ \frac{A(y,z)}{M_K^2} \left(q_2^\mu q_1^\nu - q_1 \cdot q_2\, g^{\mu\nu} \right) + \frac{2B(y,z)}{M_K^4} \left(p_K \cdot q_1\, q_2^\mu p_K^\nu \right. \right.$$
$$\left. \left. + p_K \cdot q_2\, q_1^\nu p_K^\mu - p_K^\mu p_K^\nu\, q_1 \cdot q_2 - p_K \cdot q_1\, p_K \cdot q_2\, g^{\mu\nu} \right) \right\}, \qquad (108)$$

---

[9]Taking only the absorptive parts of the $K_{1,2} \to \mu^+\mu^-$ amplitudes into account, a value $|\mathcal{P}_{L,\varepsilon}| \approx 7 \times 10^{-4}$ was estimated previously [79]. However, this is only one out of four contributions to $\mathcal{P}_L$ [74], which could all interfere constructively with unknown magnitudes.

where $y \equiv |p_K \cdot (q_1 - q_2)|/M_K^2$ and $z = (q_1 + q_2)^2/M_K^2$.

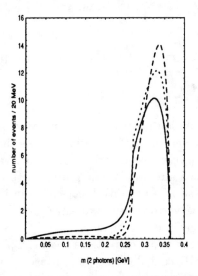

Figure 3: 2γ-invariant-mass distribution for $K_L \to \pi^0 \gamma\gamma$: $O(p^4)$ (dotted curve), $O(p^6)$ with $a_V = 0$ (dashed curve), $O(p^6)$ with $a_V = -0.9$ (full curve). The spectrum is normalized to the 50 unambiguous events of NA31 (without acceptance corrections).

Figure 4: Measured [81] 2γ-invariant-mass distribution for $K_L \to \pi^0 \gamma\gamma$ (solid line). The dashed line shows the estimated background. The experimental acceptance is given by the crosses. The dotted line simulates the $O(p^4)$ ChPT prediction.

Only the amplitude $A(y,z)$ is non-vanishing to lowest non-trivial order, $O(p^4)$, in ChPT. Again, the symmetry constraints do not allow any tree-level contribution from $O(p^4)$ terms in the Lagrangian. The $A(y,z)$ amplitude is therefore determined by a finite-loop calculation [58]. The relevant Feynman diagrams are analogous to the ones in Fig. 1, but with an additional π^0 line emerging from the weak vertex; charged kaon loops also give a small contribution in this case. Due to the large absorptive $\pi^+\pi^-$ contribution, the spectrum in the invariant mass of the two photons is predicted [58, 80] to have a very characteristic behaviour (dotted line in Fig. 3), peaked at high values of $m_{\gamma\gamma}$. The agreement with the measured two-photon distribution [81], shown in Fig. 4, is remarkably good. However, the $O(p^4)$ prediction [58, 80] for the rate, $Br(K_L \to \pi^0\gamma\gamma) = 0.67 \times 10^{-6}$, is smaller than the experimental value [81, 82]:

$$Br(K_L \to \pi^0\gamma\gamma) = \begin{cases} (1.7 \pm 0.3) \times 10^{-6} & \text{NA31 [81]}, \\ (2.2 \pm 1.0) \times 10^{-6} & \text{E731 [82]}. \end{cases} \qquad (109)$$

Since the effect of the amplitude $B(y,z)$ first appears at $O(p^6)$, one could worry about the size of the next-order corrections. In fact, a naïve VMD estimate through the decay chain $K_L \to \pi^0, \eta, \eta' \to V\gamma \to \pi^0\gamma\gamma$ [83] results in a sizeable contribution to $B(y,z)$ [66],

$$A(y,z)|_{\text{VMD}} = a_V \frac{G_8 M_K^2 \alpha}{\pi} \left(3 - z + \frac{M_\pi^2}{M_K^2} \right), \qquad (110)$$

$$B(y,z)|_{\text{VMD}} = -2a_V \frac{G_8 M_K^2 \alpha}{\pi},$$

with $a_V \approx 0.32$. However, this type of calculation predicts a photon spectrum peaked at low values of $m_{\gamma\gamma}$, in strong disagreement with experiment. As first emphasized in ref. [66], there are also so-called direct weak contributions associated with V exchange, which cannot be written as a strong VMD amplitude with an external weak transition. Model-dependent estimates of this direct contribution [66] suggest a strong cancellation with the naïve vector-meson-exchange effect, i.e. $|a_V| < 0.32$; but the final result is unfortunately quite uncertain.

A detailed calculation of the most important $O(p^6)$ corrections has been performed recently [84]. In addition to the VMD contribution, the unitarity corrections associated with the two-pion intermediate state (i.e. $K_L \to \pi^0 \pi^+ \pi^- \to \pi^0 \gamma\gamma$) have been included[10]. Figure 3 shows the resulting photon spectrum for $a_V = 0$ (dashed curve) and $a_V = -0.9$ (full curve). The predicted branching ratio is:

$$BR(K_L \to \pi^0 \gamma\gamma) = \begin{cases} 0.67 \times 10^{-6}, & O(p^4), \\ 0.83 \times 10^{-6}, & O(p^6), a_V = 0, \\ 1.60 \times 10^{-6}, & O(p^6), a_V = -0.9. \end{cases} \qquad (111)$$

The unitarity corrections by themselves raise the rate only moderately. Moreover, they produce an even more pronounced peaking of the spectrum at large $m_{\gamma\gamma}$, which tends to ruin the success of the $O(p^4)$ prediction. The addition of the V exchange contribution restores again the agreement. Both the experimental rate and the spectrum can be simultaneously reproduced with $a_V = -0.9$.

10.4 $K \to \pi l^+ l^-$

In contrast to the previous processes, the $O(p^4)$ calculation of $K^+ \to \pi^+ l^+ l^-$ and $K_S \to \pi^0 l^+ l^-$ involves a divergent loop, which is renormalized by the $O(p^4)$ Lagrangian. The decay amplitudes can then be written [57] as the sum of a calculable loop contribution plus an unknown combination of chiral couplings,

$$w_+ = -\frac{1}{3}(4\pi)^2[w_1^r + 2w_2^r - 12L_9^r] - \frac{1}{3}\log\left(M_K M_\pi/\mu^2\right), \qquad (112)$$

$$w_S = -\frac{1}{3}(4\pi)^2[w_1^r - w_2^r] - \frac{1}{3}\log\left(M_K^2/\mu^2\right),$$

---

[10]The charged-pion loop has also been computed in ref. [85].

where w_+, w_S refer to the decay of the K^+ and K_S respectively. These constants are expected to be of order 1 by naïve power-counting arguments. The logarithms have been included to compensate the renormalization-scale dependence of the chiral couplings, so that w_+, w_S are observable quantities. If the final amplitudes are required to transform as octets, then $w_2 = 4L_9$, implying $w_S = w_+ + \log(M_\pi/M_K)/3$ [57]. It should be emphasized that this relation goes beyond the usual requirement of chiral invariance.

The measured $K^+ \to \pi^+ e^+ e^-$ decay rate determines two possible solutions for w_+ [57]. The same parameter w_+ regulates [57] the shape of the invariant-mass distribution of the final lepton pair. A fit to the recent BNL E777 data [86] gives

$$w_+ = 0.89^{+0.24}_{-0.14}, \tag{113}$$

solving the previous two-fold ambiguity in favour of the positive solution, as expected from model-dependent theoretical estimates [66]. Once w_+ has been fixed, one can make predictions [57] for the rates and Dalitz-plot distributions of the related modes $K^+ \to \pi^+ \mu^+ \mu^-$, $K_S \to \pi^0 e^+ e^-$ and $K_S \to \pi^0 \mu^+ \mu^-$.

The rare decay $K_L \to \pi^0 e^+ e^-$ is an interesting process in looking for new CP-violating signatures. If CP were an exact symmetry, only the CP-even state K_1^0 could decay via one-photon emission, while the decay of the CP-odd state K_2^0 would proceed through a two-photon intermediate state and, therefore, its decay amplitude would be suppressed by an additional power of α. When CP-violation is taken into account, however, an $O(\alpha)$ $K_L \to \pi^0 e^+ e^-$ decay amplitude is induced, both through the small K_1^0 component of the K_L (ε effect) and through direct CP-violation in the $K_2^0 \to \pi^0 e^+ e^-$ transition. The electromagnetic suppression of the CP-conserving amplitude then makes it plausible that this decay is dominated by the CP-violating contributions.

The short-distance analysis of the product of weak and electromagnetic currents allows a reliable estimate of the direct CP-violating $K_2^0 \to \pi^0 e^+ e^-$ amplitude. The corresponding branching ratio induced by this amplitude has been estimated [87] to be around

$$Br(K_L \to \pi^0 e^+ e^-)\Big|_{\text{Direct}} \simeq 5 \times 10^{-12}, \tag{114}$$

the exact number depending on the values of m_t and the quark-mixing angles.

The indirect CP-violating amplitude induced by the K_1^0 component of the K_L is given by the $K_S \to \pi^0 e^+ e^-$ amplitude times the CP-mixing parameter ε. Using the octet relation between w_+ and w_S, the determination of the parameter ω_+ in Eq. (113) implies

$$Br(K_L \to \pi^0 e^+ e^-)\Big|_{\text{Indirect}} \leq 1.6 \times 10^{-12}. \tag{115}$$

Comparing this value with the one in Eq. (114), we see that the interesting direct CP-violating contribution is expected to be bigger than the indirect one. This is very different from the situation in $K \to \pi\pi$, where the contribution due to mixing completely dominates.

The present experimental upper bound [88] (90% C.L.)

$$Br(K_L \to \pi^0 e^+ e^-)\Big|_{\text{Exp}} < 5.5 \times 10^{-9}, \tag{116}$$

is still far away from the expected Standard Model signal, but the prospects for getting the needed sensitivity of around 10^{-12} in the next few years are rather encouraging. In order to be able to interpret a future experimental measurement of this decay as a CP-violating signature, it is first necessary, however, to pin down the actual size of the two-photon exchange CP-conserving amplitude.

Using the computed $K_L \to \pi^0 \gamma\gamma$ amplitude, one can estimate the two-photon exchange contribution to $K_L \to \pi^0 e^+ e^-$, by taking the absorptive part due to the two-photon discontinuity as an educated guess of the actual size of the complete amplitude. At $O(p^4)$, the $K_L \to \pi^0 e^+ e^-$ decay amplitude is strongly suppressed (it is proportional to m_e), owing to the helicity structure of the $A(y,z)$ term [59]:

$$Br(K_L \to \pi^0 \gamma^* \gamma^* \to \pi^0 e^+ e^-)\Big|_{O(p^4)} \sim 5 \times 10^{-15}. \tag{117}$$

This helicity suppression is, however, no longer true at the next order in the chiral expansion. The $O(p^6)$ estimate of the amplitude $B(y,z)$ [84] gives rise to

$$Br(K_L \to \pi^0 \gamma^* \gamma^* \to \pi^0 e^+ e^-)\Big|_{O(p^6)} \sim \begin{cases} 0.3 \times 10^{-12}, & a_V = 0, \\ 1.8 \times 10^{-12}, & a_V = -0.9. \end{cases} \tag{118}$$

Although the rate increases of course with $|a_V|$, there is some destructive interference between the unitarity corrections of $O(p^6)$ and the V-exchange contribution (for $a_V = -0.9$). In order to get a more accurate estimate, it would be necessary to make a careful fit to the $K_L \to \pi^0 \gamma\gamma$ data, taking the experimental acceptance into account, to extract the actual value of a_V.

11 THE CHIRAL ANOMALY IN NON-LEPTONIC K DECAYS

The chiral anomaly also appears in the non-leptonic weak interactions. A systematic study of all non-leptonic K decays where the anomaly contributes at leading order, $O(p^4)$, has been performed recently [63]. Only radiative K decays are sensitive to the anomaly in the non-leptonic sector.

The manifestations of the anomaly can be grouped in two different classes of anomalous amplitudes: reducible and direct contributions. The reducible amplitudes arise from the contraction of meson lines between a weak $\Delta S = 1$ vertex and the Wess–Zumino–Witten functional (61). In the octet limit, all reducible anomalous amplitudes of $O(p^4)$ can be predicted in terms of the coupling g_8. The direct anomalous contributions are generated through the contraction of the W boson field between a strong Green function on one side and the Wess–Zumino–Witten functional on the other. Their computation is not straightforward, because of the presence of strongly interacting fields on both sides of the W. Nevertheless, due to the non-renormalization theorem of the chiral anomaly [89], the bosonized form of the direct anomalous amplitudes can be fully predicted [64]. In spite of its anomalous origin, this contribution is chiral-invariant. The anomaly turns out to contribute to all possible octet terms of $\mathcal{L}_4^{\Delta S=1}$ proportional to the $\varepsilon_{\mu\nu\alpha\beta}$ tensor. Unfortunately, the coefficients of these terms get also non-factorizable contributions of non-anomalous origin, which cannot be computed in a model-independent way. Therefore, the final predictions can only be parametrized in terms of four dimensionless chiral couplings, which are expected to be positive and of order one.

The most frequent "anomalous" decays $K_L \to \pi^+\pi^-\gamma$ and $K^+ \to \pi^+\pi^0\gamma$ share the remarkable feature that the normally dominant bremsstrahlung amplitude is strongly suppressed, making the experimental verification of the anomalous amplitude substantially easier. This suppression has different origins: $K^+ \to \pi^+\pi^0$ proceeds through the small 27-plet part of the non-leptonic weak interactions, whereas $K_L \to \pi^+\pi^-$ is CP-violating. The remaining non-leptonic K decays with direct anomalous contributions are either suppressed by phase space $[K^+ \to \pi^+\pi^0\pi^0\gamma(\gamma), K^+ \to \pi^+\pi^+\pi^-\gamma(\gamma), K_L \to \pi^+\pi^-\pi^0\gamma, K_S \to \pi^+\pi^-\pi^0\gamma(\gamma)]$ or by the presence of an extra photon in the final state $[K^+ \to \pi^+\pi^0\gamma\gamma, K_L \to \pi^+\pi^-\gamma\gamma]$. A detailed phenomenological analysis of these decays can be found in ref. [63].

12 INTERACTIONS OF A LIGHT HIGGS

The hadronic couplings of a light Higgs particle are fixed by low-energy theorems [90–93], which relate the $\phi \to \phi' h^0$ transition with a zero-momentum Higgs to the corresponding $\phi \to \phi'$ coupling. Although, within the Standard Model, the possibility of a light Higgs boson is already excluded [76], an extended scalar sector with additional degrees of freedom could easily avoid the present experimental limits, leaving the question of a light Higgs open to any speculation.

The quark–Higgs interaction can be written down in the general form

$$\mathcal{L}_{h^0\bar{q}q} = -\frac{h^0}{u}\left\{k_d\,\bar{d}M_d d + k_u\,\bar{u}M_u u\right\}, \qquad (119)$$

where $u = (\sqrt{2}G_F)^{-1/2} \approx 246\,\text{GeV}$, M_u and M_d are the diagonal mass matrices for up- and down-type quarks respectively, and the couplings k_u and k_d depend on the model considered. In the Standard Model, $k_u = k_d = 1$, while in the usual two-Higgs-doublet models (without tree-level flavour-changing neutral currents) $k_d = k_u = \cos\alpha/\sin\beta$ (model I) or $k_d = -\sin\alpha/\cos\beta$, $k_u = \cos\alpha/\sin\beta$ (model II), where α and β are functions of the parameters of the scalar potential.

The couplings of h^0 to the octet of pseudoscalar mesons can be easily worked out, using ChPT techniques. The Yukawa interactions of the light-quark flavours can be trivially incorporated through the external scalar field s, together with the light-quark-mass matrix $\mathcal{M}$:

$$s = \mathcal{M}\left\{1 + \frac{h^0}{u}(k_d A + k_u B)\right\}, \qquad (120)$$

where $A \equiv \text{diag}(0,1,1)$ and $B \equiv \text{diag}(1,0,0)$. It remains to compute the contribution from the heavy flavours c, b, t. Their Yukawa interactions induce a Higgs–gluon coupling through heavy-quark loops,

$$\mathcal{L}_{h^0 GG} = \frac{\alpha_s}{12\pi}(n_d k_d + n_u k_u)\frac{h^0}{u} G^a_{\mu\nu} G^{\mu\nu}_a. \qquad (121)$$

Here, $n_d = 1$ and $n_u = 2$ are the number of heavy quarks of type down and up respectively. The operator $G^a_{\mu\nu} G^{\mu\nu}_a$ can be related to the trace of the energy-momentum tensor; in the three light-flavour theory, one has

$$\Theta^\mu_\mu = -\frac{b\alpha_s}{8\pi} G^a_{\mu\nu} G^{\mu\nu}_a + \bar{q}\mathcal{M}q, \qquad (122)$$

where $b = 9$ is the first coefficient of the QCD β-function. To obtain the low-energy representation of $\mathcal{L}_{h^0 GG}$ it therefore suffices to replace Θ^μ_μ and $\bar{q}\mathcal{M}q$ by their corresponding expressions in the effective chiral Lagrangian theory. One gets [90–92],

$$\mathcal{L}^{\text{eff}}_{h^0 GG} = \xi \frac{h^0}{u}\frac{f^2}{2}\left\{\langle D_\mu U^\dagger D^\mu U\rangle + 3B_0\langle U^\dagger \mathcal{M} + \mathcal{M} U\rangle\right\}. \qquad (123)$$

The information on the heavy quarks, which survives in the low-energy limit, is contained in the coefficient $\xi \equiv 2(n_d k_d + n_u k_u)/(3b) = 2(k_d + 2k_u)/27$.

Using the chiral formalism, the present experimental constraints on a very light neutral scalar have been investigated in refs. [92] and [93], in the context of two-Higgs-doublet models. A Higgs in the mass range $2m_\mu < m_{h^0} < 2M_\pi$ can be excluded (within model II), analysing the decay $\eta \to \pi^0 h^0$ [92]. A more general analysis [93], using the light-Higgs production channels $Z \to Z^* h^0$, $\eta' \to \eta h^0$, $\eta \to \pi^0 h^0$ and $\pi \to e\nu h^0$, allows us to exclude a large area in the parameter space (α, β, m_{h^0}) of both models (I and II) for $m_{h^0} < 2m_\mu$.

13 EFFECTIVE THEORY AT THE ELECTROWEAK SCALE

In spite of the spectacular success of the Standard Model, we still do not really understand the dynamics underlying the electroweak symmetry breaking $SU(2)_L \otimes U(1)_Y \to U(1)_{\text{em}}$. The Higgs mechanism provides a renormalizable way to generate the W and Z masses and, therefore, their longitudinal degrees of freedom. However, an experimental verification of this mechanism is still lacking.

The scalar sector of the Standard Model Lagrangian can be written in the form

$$\mathcal{L}(\Phi) = \frac{1}{2}\langle D^\mu \Sigma^\dagger D_\mu \Sigma\rangle - \frac{\lambda}{16}\left(\langle \Sigma^\dagger \Sigma\rangle - u^2\right)^2, \qquad (124)$$

where

$$\Sigma \equiv \begin{pmatrix} \Phi^0 & \Phi^+ \\ \Phi^- & \Phi^{0*} \end{pmatrix} \qquad (125)$$

and $D_\mu \Sigma$ is the usual gauge-covariant derivative

$$D_\mu \Sigma \equiv \partial_\mu \Sigma + ig \frac{\vec{\tau}}{2} \vec{W}_\mu \Sigma - ig' \Sigma \frac{\tau_3}{2} B_\mu. \qquad (126)$$

In the limit where the coupling g' is neglected, $\mathcal{L}(\Phi)$ is invariant under global $G \equiv SU(2)_L \otimes SU(2)_C$ transformations,

$$\Sigma \xrightarrow{G} g_L \Sigma g_C^\dagger, \qquad g_{L,C} \in SU(2)_{L,C} \qquad (127)$$

($SU(2)_C$ is the so-called custodial-symmetry group). The symmetry properties of $\mathcal{L}(\Phi)$ are very similar to the ones of the linear-sigma-model Lagrangian (15). Performing an analogous polar decomposition [see Eqs. (17)],

$$\Sigma(x) = \frac{1}{\sqrt{2}}(u + H(x)) U(\phi(x)), \qquad (128)$$

$$U(\phi(x)) = \exp\left(i\vec{\tau}\vec{\phi}(x)/u\right),$$

in terms of the Higgs field H and the Goldstones $\vec{\phi}$, and taking the limit $\lambda \gg 1$ (heavy Higgs), we can rewrite $\mathcal{L}(\Phi)$ in the standard chiral form:

$$\mathcal{L}(\Phi) = \frac{u^2}{4} \langle D_\mu U^\dagger D^\mu U \rangle + O(H). \qquad (129)$$

In the unitary gauge $U = 1$, this $O(p^2)$ Lagrangian reduces to the usual bilinear gauge-mass term.

As we know already, (129) is the universal model-independent interaction of the Goldstone bosons induced by the assumed pattern of SCSB, $SU(2)_L \otimes SU(2)_C \longrightarrow SU(2)_{L+C}$. The scattering of electroweak Goldstone bosons (or equivalently longitudinal gauge bosons) is then described by the same formulae as the scattering of pions, changing f by u [94]. To the extent that the present data are still not sensitive to the virtual Higgs effects, we have only tested up to now the symmetry properties of the scalar sector encoded in Eq. (129).

In order to really prove the particular scalar dynamics of the Standard Model, we need to test the model-dependent part involving the Higgs field H. If the Higgs turns out to be too heavy to be directly produced (or if it does not exist at all!), one could still investigate the higher-order effects [95-103] by applying the standard chiral-expansion techniques in a completely straightforward way. The Standard Model gives definite predictions for the corresponding chiral couplings of the $O(p^4)$ Lagrangian, which could be tested in future high-precision experiments. It remains to be seen if the experimental determination of the higher-order electroweak chiral couplings will confirm the renormalizable Standard Model Lagrangian, or will constitute an evidence of new physics

14 SUMMARY

ChPT is a powerful tool to study the low-energy interactions of the pseudoscalar-meson octet. This effective Lagrangian framework incorporates all the constraints implied by the chiral symmetry of the underlying Lagrangian at the quark–gluon level, allowing for a clear distinction between genuine aspects of the Standard Model and additional assumptions of variable credibility, usually related to the problem of long-distance dynamics. The low-energy amplitudes of the Standard Model are calculable in ChPT, except for some coupling constants which are not restricted by chiral symmetry. These constants reflect our lack of understanding of the QCD confinement mechanism and must be determined experimentally for the time being. Further progress in QCD can only improve our knowledge of these chiral constants, but it cannot modify the low-energy structure of the amplitudes.

ChPT provides a convenient language to improve our understanding of the long-distance dynamics. Once the chiral couplings are experimentally known, one can test different dynamical models, by comparing the predictions that they give for those couplings with their phenomenologically determined values. The final goal would be, of course, to derive the low-energy chiral constants from the Standard Model Lagrangian itself. Although this is a very difficult problem, the recent attempts done in this direction look quite promising.

It is important to emphasize that:

1. ChPT is not a model. The effective Lagrangian generates the more general S-matrix elements consistent with analyticity, perturbative unitarity and the assumed symmetries. Therefore, ChPT is the effective theory of the Standard Model at low energies.

2. The experimental verification of the ChPT predictions does not provide a test of the detailed dynamics of the Standard Model; only the implications of the underlying symmetries are being proved. Any other model with identical chiral-symmetry properties would give rise to the same low-energy structure.

3. The dynamical information on the underlying fundamental Lagrangian is encoded in the chiral couplings. Different short-distance models with identical symmetry properties will result in the same effective Lagrangian, but with different values for the low-energy couplings. In order to actually test the non-trivial low-energy dynamics of the Standard Model, one needs first to know the Standard Model predictions for the chiral couplings.

In these lectures I have presented the basic formalism of ChPT and some selected phenomenological applications. There are many more applications of the chiral framework. Any system which contains Goldstone bosons can be studied in a similar way. A discussion of further topics in ChPT can be found in refs. [3–14].

REFERENCES

[1] E. Euler, Ann. Phys. (Leipzig) 26 (1936) 398;
E. Euler and W. Heisenberg, Z. Phys. 98 (1936) 714.

[2] S. Weinberg, Physica 96A (1979) 327.

[3] J. Bijnens, Int. J. Mod. Phys. A8 (1993) 3045.

[4] J.F. Donoghue, "Chiral symmetry as an experimental science", in "Medium energy antiprotons and the quark–gluon structure of hadrons", eds. R. Landua, J.M. Richard and R. Klapisch (Plenum Press, New York, 1991) p. 39.

[5] G. Ecker, "Chiral Perturbation Theory", in "Quantitative Particle Physics", eds. M. Levy et al. (Plenum Publ. Co.,New York, 1993);
"The Standard Model at low energies", lectures given at the 6th Summer School on Intermediate Energy Physics, Prague, 1993, Wien preprint UWThPh-1993-31 (to appear in Czech J. Phys.);
"Chiral realization of the non-leptonic weak interactions", in Proc. 24th Int. Symposium on the theory of elementary particles, ed. G. Weigt (Zeuthen, 1991).

[6] J. Gasser, "The QCD vacuum and chiral symmetry", in "Hadrons and Hadronic Matter", eds. D. Vautherin et al. (Plenum Press, New York, 1990) p. 87;
"Chiral dynamics", in Proc. Workshop on Physics and Detectors for DAΦNE, ed. G. Pancheri (Frascati, 1991) p. 291.

[7] H. Leutwyler, "Chiral Effective Lagrangians", in "Recent Aspects of Quantum Fields", eds. H. Mitter and M. Gausterer, Lecture Notes in Physics, vol. 396 (Springer Verlag, Berlin, 1991).

[8] U.G. Meißner, "Recent Developments in Chiral Perturbation Theory", Bern preprint BUTP-93/01.

[9] A. Pich, Nucl. Phys. B (Proc. Suppl.) 23A (1991) 399;
"η decays and chiral Lagrangians", in Proc. Workshop on Rare Decays of Light Mesons", ed. B. Mayer (Gif-sur-Yvette, 1990), p. 43.

[10] E. de Rafael, Nucl. Phys. B (Proc. Suppl.) 10A (1989) 37.

[11] A.J. Buras, J.-M. Gerard and W. Huber (eds.), Proc. of the Ringberg Workshop on Hadronic Matrix Elements and Weak Decays, Ringberg Castle, Germany, 1988, Nucl. Phys. B (Proc. Suppl.) 7A (1989).

[12] U.-G. Meißner (ed.), Proc. of the Workshop on Effective Field Theories of the Standard Model, Dobogókö, Hungary, 1991 (World Sientific, Singapore, 1992).

[13] H. Georgi, "Weak Interactions and Modern Particle Theory" (Benjamin / Cummings, Menlo Park, 1984).

[14] J.F. Donoghue, E. Golowich and B.R. Holstein, "Dynamics of the Standard Model" (Cambridge Univ. Press, Cambridge, 1992).

[15] J. Gasser and H. Leutwyler, Nucl. Phys. B250 (1985) 465; 517; 539.

[16] A. Pich and E. de Rafael, Nucl. Phys. B367 (1991) 313.

[17] S. Adler and R.F. Dashen, "Current Algebras", (Benjamin, New York, 1968); V. de Alfaro, S. Fubini, G. Furlan and C. Rossetti, "Currents in Hadron Physics" (North-Holland, Amsterdam, 1973).

[18] J. Goldstone, Nuovo Cim. 19 (1961) 154.

[19] J. Schwinger, Ann. Phys. 2 (1957) 407;
M. Gell-Mann and M. Lévy, Nuovo Cim. 16 (1960) 705.

[20] J. Gasser and H. Leutwyler, Ann. Phys. 158 (1984) 142.

[21] S. Weinberg, Phys. Rev. Lett. 17 (1966) 616.

[22] M. Gell-Mann, R.J. Oakes and B. Renner, Phys. Rev. 175 (1968) 2195.

[23] S. Weinberg, in "A Festschrift for I.I. Rabi", ed. L. Motz (Acad. of Sciences, New York, 1977) p. 185.

[24] M. Gell-Mann, Phys. Rev. 106 (1957) 1296;
S. Okubo, Prog. Theor. Phys. 27 (1962) 949.

[25] R. Dashen, Phys. Rev. 183 (1969) 1245.

[26] J. Wess and B. Zumino, Phys. Lett. 37B (1971) 95;
E. Witten, Nucl. Phys. B223 (1983) 422.

[27] S.L. Adler, Phys. Rev. 177 (1969) 2426;
J.S. Bell and R. Jackiw, Nuovo Cim. 60A (1969) 47.

[28] W.A. Bardeen, Phys. Rev. 184 (1969) 1848.

[29] A. Manohar and H. Georgi, Nucl. Phys. B234 (1984) 189.

[30] J. Wess and B. Zumino, Phys. Lett. 37B (1971) 95.

[31] E. Witten, Nucl. Phys. B223 (1983) 422.

[32] J. Bijnens, G. Ecker and J. Gasser, "Introduction to Chiral Symmetry", in "The DAΦNE Physics Handbook", eds. L. Maiani, G. Pancheri and N. Paver (Frascati 1992) Vol. I, p. 107.

[33] J. Bijnens and F. Cornet, Nucl. Phys. B296 (1988) 557;
J. Bijnens, Nucl. Phys. B337 (1990) 635;
C. Riggenbach, J. Gasser, J.F. Donoghue and B.R. Holstein, Phys. Rev. D43 (1991) 127.

[34] H. Leutwyler and M. Roos, Z. Phys. C25 (1984) 91.

[35] W.R. Molzon et al., Phys. Rev. Lett. 41 (1978) 1213.

[36] S. R. Amendolia et al., Nucl. Phys. B277 (1986) 168.

[37] H. Leutwyler, Nucl. Phys. B (Proc. Suppl.) 7A (1989) 42.

[38] E.B. Dally et al., Phys. Rev. Lett. 45 (1980) 232; 48 (1982) 375.

[39] Particle Data Group, "Review of Particle Properties", Phys. Rev. D45 (1992) Part II.

[40] G. Donaldson et al., Phys. Rev. D9 (1974) 2960.

[41] Y. Cho et al., Phys. Rev. D22 (1980) 2688.

[42] D.B. Kaplan and A.V. Manohar, Phys. Rev. Lett. 56 (1986) 2004.

[43] J. Gasser and H. Leutwyler, Phys. Rep. 87C (1982) 77.

[44] H. Leutwyler, Nucl. Phys. B337 (1990) 108.

[45] J.F. Donoghue and D. Wyler, Phys. Rev. D45 (1992) 892.

[46] G. Ecker, J. Gasser, A. Pich and E. de Rafael, Nucl. Phys. B321 (1989) 311.

[47] J.F. Donoghue, C. Ramirez and G. Valencia, Phys. Rev. D39 (1989) 1947.

[48] S. Weinberg, Phys. Rev. Lett. 18 (1967) 507.

[49] U.-G. Meißner, Phys. Rep. 161 (1988) 213.

[50] M. Bando, T. Kugo and K. Yamawaki, Phys. Rep. 164 (1988) 115.

[51] G. Ecker, J. Gasser, H. Leutwyler, A. Pich and E. de Rafael, Phys. Lett. B223 (1989) 425.

[52] D. Espriu, E. de Rafael and J. Taron, Nucl. Phys. B345 (1990) 22.

[53] J. Bijnens, C. Bruno and E. de Rafael, Nucl. Phys. B390 (1993) 501.

[54] F.J. Gilman and M.B. Wise, Phys. Rev. D20 (1979) 2392.

[55] A. Pich, B. Guberina and E. de Rafael, Nucl. Phys. B277 (1986) 197.

[56] A. Pich and E. de Rafael, Nucl. Phys. B358 (1991) 311.

[57] G. Ecker, A. Pich and E. de Rafael, Nucl. Phys. B291 (1987) 692.

[58] G. Ecker, A. Pich and E. de Rafael, Phys. Lett. B189 (1987) 363.

[59] G. Ecker, A. Pich and E. de Rafael, Nucl. Phys. B303 (1988) 665.

[60] J. Kambor, J. Missimer and D. Wyler, Nucl. Phys. B346 (1990) 17.

[61] G. Ecker, "Geometrical aspects of the non-leptonic weak interactions of mesons", in Proc. of the IX Int. Conference on the Problems of Quantum Field Theory, Dubna, 1990, ed. M.K. Volkov (JINR, Dubna, 1990);
G. Esposito-Farèse, Z. Phys. C50 (1991) 255.

[62] G. Ecker, J. Kambor and D. Wyler, Nucl. Phys. B394 (1993) 101.

[63] G. Ecker, H. Neufeld and A. Pich, "Non-Leptonic Kaon Decays and the Chiral Anomaly", CERN-TH.6920/93; Phys. Lett. B278 (1992) 337.

[64] J. Bijnens, G. Ecker and A. Pich, Phys. Lett. B286 (1992) 341.

[65] S. Fajfer and J.-M. Gérard, Z. Phys. C42 (1989) 425;
H.-Y. Cheng, Phys. Rev. D42 (1990) 72.

[66] G. Ecker, A. Pich and E. de Rafael, Phys. Lett. B237 (1990) 481.

[67] C. Bruno and J. Prades, Z. Phys. C57 (1993) 585.

[68] G. Isidori and A. Pugliese, Nucl. Phys. B385 (1992) 437.

[69] T.J. Devlin and J.O. Dickey, Rev. Mod. Phys. 51 (1979) 237.

[70] J. Kambor, J. Missimer and D. Wyler, Phys. Lett. B261 (1991) 496.

[71] J. Kambor, J.F. Donoghue, B.R. Holstein, J. Missimer and D. Wyler, Phys. Rev. Lett. 68 (1992) 1818.

[72] G. D'Ambrosio and D. Espriu, Phys. Lett. B175 (1986) 237;
J.L. Goity, Z. Phys. C34 (1987) 341.

[73] H. Burkhardt et al. (CERN NA31), Phys. Lett. B199 (1987) 139.

[74] G. Ecker and A. Pich, Nucl. Phys. B366 (1991) 189.

[75] F.J. Botella and C.S. Lim, Phys. Rev. Lett. 56 (1986) 1651.

[76] T. Mori, "Searches for the Standard Model Higgs Boson at LEP", in Proc. XXVI Int. Conf. on High Energy Physics, Dallas, 1992, ed. J.R. Sanford, AIP Conf. Proceedings No. 272 (AIP, New York, 1993) Vol. II, p. 1321.

[77] C.Q. Geng and J.N. Ng, Phys. Rev. D39 (1989) 3330.

[78] R.N. Mohapatra, Prog. Part. Nucl. Phys. 31 (1993) 39;
C.Q. Geng and J.N. Ng, Phys. Rev. D42 (1990) 1509.

[79] P. Herczeg, Phys. Rev. D27 (1983) 1512.

[80] L. Cappiello and G. D'Ambrosio, Nuovo Cim. 99A (1988) 153.

[81] G.D. Barr et al. (CERN NA31), Phys. Lett. B284 (1992) 440; B242 (1990) 523.

[82] V. Papadimitriou et al. (FNAL E731), Phys. Rev. D44 (1991) 573.

[83] L.M. Sehgal, Phys. Rev. D38 (1988) 808; D41 (1990) 161;
T. Morozumi and H. Iwasaki, Progr. Theor. Phys. 82 (1989) 371;
J. Flynn and L. Randall, Phys. Lett. B216 (1989) 221.

[84] A. Cohen, G. Ecker and A. Pich, Phys. Lett. B304 (1993) 347.

[85] L. Cappiello, G. D'Ambrosio and M. Miragliuolo, Phys. Lett. B298 (1993) 423.

[86] C. Alliegro et al. (BNL E777), Phys. Rev. Lett. 68 (1992) 278.

[87] A.J. Buras and M.K. Harlander, "A Top Quark Story: Quark Mixing, CP Violation and Rare Decays in the Standard Model", in "Heavy Flavours", eds. A.J. Buras and M. Lindner, Advanced Series on Directions in High Energy Physics (World Scientific, Singapore 1992);
G. Buchalla, A.J. Buras and M.K. Harlander, Nucl. Phys. B349 (1991) 1;
C. Dib, I. Dunietz and F.J. Gilman, Phys. Lett. B218 (1989) 487; Phys. Rev. D39 (1989) 2639;
J. Flynn and L. Randall, Nucl. Phys. B326 (1989) 31.

[88] K.E. Ohl et al. (BNL E845), Phys. Rev. Lett. 64 (1990) 2755;
A. Barker et al. (FNAL E731), Phys. Rev. D41 (1990) 3546.

[89] S.L. Adler and W.A. Bardeen, Phys. Rev. 182 (1969) 1517.

[90] J.F. Gunion, H.E. Haber, G.L. Kane and S. Dawson, "The Higgs hunter's guide", Frontiers in Physics Lecture Note Series (Addison-Wesley, New York, 1990).

[91] R.S. Chivukula, A. Cohen, H. Georgi, B. Grinstein and A.V. Manohar, Ann. Phys. (N.Y.) 192 (1989) 93;
R.S. Chivukula, A. Cohen, H. Georgi and A.V. Manohar, Phys. Lett. B222 (1989) 258;
S. Dawson, Phys. Lett. B222 (1989) 143;
H. Leutwyler and M.A. Shifman, Phys. Lett. B221 (1989) 384; Nucl. Phys. B343 (1990) 369.

[92] J. Prades and A. Pich, Phys. Lett. B245 (1990) 117.

[93] A. Pich, J. Prades and P. Yepes, Nucl. Phys. B388 (1992) 31.

[94] M.S. Chanowitz and M.K. Gaillard, Nucl. Phys. B261 (1985) 379;
J.M. Cornwall, D.N. Levin and G. Tiktopoulos, Phys. Rev. D10 (1974) 1145.

[95] A. Longhitano, Nucl. Phys. B188 (1981) 118; Phys. Rev. D22 (1980) 1166;
T. Appelquist and C. Bernard, Phys. Rev. D22 (1980) 200.

[96] A. Dobado, D. Espriu and M.J. Herrero, Phys. Lett. B255 (1991) 405;
D. Espriu and M.J. Herrero, Nucl. Phys. B373 (1992) 117;
A. Dobado and M.J. Herrero, Phys. Lett. B228 (1989) 495; B233 (1989) 505.

[97] J.F. Donoghue and C. Ramirez, Phys. Lett. B234 (1990) 361.

[98] B. Holdon and J. Terning, Phys. Lett. B247 (1990) 88;
B. Holdon, Phys. Lett. B259 (1991) 329; B258 (1991) 156.

[99] M. Golden and L. Randall, Nucl. Phys. B361 (1991) 3.

[100] H. Georgi, Nucl. Phys. B363 (1991) 301.

[101] A. Falk, M. Luke and E. Simmons, Nucl. Phys. B365 (1991) 523.

[102] S. Dawson and G. Valencia, Nucl. Phys. B348 (1991) 23; B352 (1991) 27.

[103] A. De Rújula, M.B. Gavela, P. Hernández and E. Massó, Nucl. Phys. B384 (1992) 3.

Knot Theory and Quantum Gravity in Loop Space: A Primer

Jorge Pullin

Center for Gravitational Physics and Geometry
The Pennsylvania State University, University Park, PA 16802

Abstract

These notes summarize the lectures delivered in the V Mexican School of Particle Physics, at the University of Guanajuato. We give a survey of the application of Ashtekar's variables to the quantization of General Relativity in four dimensions with special emphasis on the application of techniques of analytic knot theory to the loop representation. We discuss the role that the Jones Polynomial plays as a generator of nondegenerate quantum states of the gravitational field.

1 Quantum Gravity: why and how?

I wish to thank the organizers for inviting me to speak here. This may well be a sign of our times, that a person generally perceived as a "General Relativist" would be invited to speak at a Particle Physics School. It just reflects the higher degree of interplay these two fields have enjoyed over the last years. In these lectures we will see more reasons for this enhanced interplay. We will see several notions from Gauge Theories, as Wilson Loops for instance, playing a central role in gravitation. An even greater interplay takes place with Topological Field Theories. We will see the important role that the Chern-Simons form, the Jones Polynomial and other notions of knot theory seem to play in General Relativity.

The quantization of General Relativity is a problem that has defied resolution for the last sixty years. In spite of the long time that has been invested in trying to solve it, we believe that several people do not necessarily fully appreciate the reasons of our failure and the magnitude of the problem. It is a general perception –especially among particle physicists– that "General Relativity is nonrenormalizable" and that is the basic problem with the theory. This statement is misleading in three ways:

a) The fact that a theory is nonrenormalizable does not necessarily mean that the theory has an intrinsic problem or is "bad" in any way. It merely says that perturbation theory does not apply to the problem in question. As we will see in c) there are actually good reasons to believe that ordinary perturbation theory *should* fail for General Relativity.

b) Deciding if a theory is or is not renormalizable can be quite tricky. The prime example is 2+1 dimensional gravity, which most people thought to share the renormalizability pathologies of 3+1 gravity until Witten [1] pointed out that it could be exactly solved. A posteriori it was of course found that the theory is in fact renormalizable [2].

c) There actually are very good reasons why we should expect General Relativity to have "problems" (we would rather call them "subtleties" or "challenges") in quantization. Prime among them is the issue of diffeomorphism invariance, which in turn implies other problems as the lack of observables for the theory. This last problem clearly reflects the nature of the issue: even if we were somehow able to make General Relativity renormalizable, we would not know *what* to compute with such a perturbatively well behaved theory. Even if we were able to compute something we would not know how to interpret it.

Due to these and other arguments, we believe that perturbative quantization of General Relativity may well be a red herring. So too may also be the idea of abandoning General Relativity in favour of other theories that present some particular better behaviour (usually only apparent) when perturbatively quantized. We repeat: even if we had a perfectly renormalizable theory of quantum gravity, it is little what we could actually *do* with it until we address the fundamental questions of what kind of physics can we do in a diffeomorphism invariant context.

If one is interested in these kind of questions, in particular the issue of diffeomorphism invariance, nonperturbative quantization seems the way to go. There are several options if one wants to attempt a nonperturbative quantization of gravity, ranging from quite radical to very conservative ones. Of all these, probably the most conservative is Canonical Quantization. After all, this was the first method of quantization ever invented and is the one most physicists feel comfortable with. Canonical Quantization therefore seems an attractive approach to study the issues that arise in the quantization of General Relativity. If the resulting theory makes sense, one expects that other quantization techniques would in the end give the same results.

We will discuss in these lectures the Canonical Quantization of General Relativity in four dimensions. As we have argued, the use of other theories or number of dimensions seems at the moment superfluous. We do not even understand –for good reasons– arguably the simplest theory (General Relativity). Therefore it seems to make little sense for our purposes to embark on the study of more complicated theories. It may well be that at some point it becomes apparent that General Relativity does not furnish a suitable base for a theory of Quantum Gravity. Until that point is reached we think it is useful as the simplest theory of gravity that has all the desired features one would expect in such a theory.

We will therefore proceed with a very conservative Canonical Quantization scheme but we will pursue a slight variant from the traditional approaches. We will use a new set of canonical variables for the treatment of Hamiltonian General Relativity, the Ashtekar Variables [3]. These variables have the advantage of casting General Relativity in a fashion that closely resembles Yang-Mills theories. This will allow us to introduce several useful techniques from Yang-Mills theories into General Relativity.

The plan of these lectures is as follows: in section 2 we discuss the traditional canonical formulation of General Relativity. In section 3 we introduce the Ashtekar New Variables and discuss the classical theory. In section 4 we discuss the quantum theory in the connection representation, and point out the role that Wilson Loops and the Chern-Simons form play in the theory. In section 5 we present the Loop Representation and develop technology for dealing with the constraints and the wavefunctions written in terms of loops. In section 6 we discuss various aspects of Knot Theory. In section 7 we make use of the results of sections 5 and 6 to construct a family of nondegenerate physical states of Quantum Gravity in terms of knot invariants. We end in section 8 with some final remarks and a general discussion of the present status of the program.

2 Brief Summary of General Relativity and Canonical Quantization

2.1 Classical General Relativity

General Relativity is a theory of gravity in which the gravitational interaction is accounted for by a deformation of spacetime. The fundamental variable for the theory is the spacetime metric g_{ab}. The action for the theory is given by,

$$S = \int d^4x \sqrt{-g} R(g_{ab}) + \int d^4x \sqrt{-g} \mathcal{L}(\text{matter}) \tag{1}$$

where g is the determinant of g_{ab}, $R(g_{ab})$ is the curvature scalar and we have included also a term to take into account possible couplings to matter, although in these lectures we will largely concentrate on the theory in vacuum. The equations of motion for this action, obtained by varying the action with respect to g_{ab} are,

$$R_{ab} - \tfrac{1}{2} g_{ab} R = \frac{\delta S_{\text{matter}}}{\delta g^{ab}} \tag{2}$$

and are called Einstein Equations. The theory is invariant under diffeomorphisms on the four manifold (coordinate transformations). This means it has a symmetry. The Einstein Equations are in principle ten equations (all the tensors are symmetric and therefore only have ten independent components). However, due to the presence of the diffeomorphism symmetry, several of the equations are redundant. This issue is best seen in the canonical formalism. We will briefly set it up in the next two subsections. The reader wanting a more detailed treatment can find it in references [4, 5].

2.2 Canonical formulation, the 3+1 split

To cast General Relativity in a canonical form, we need to split space-time into space and time. Without a notion of time, we do not have a notion of evolution,

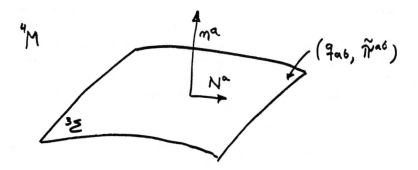

Figure 1: 3+1 foliation of spacetime and variables of the canonical formalism

and therefore no notion of "Hamiltonian". This may seem strange at first. One of the great accomplishments of Relativity was to put space and time into an equal footing, and now we seem to be destroying it. We will show that this is not the case. Although the Canonical formalism manifestly breaks the covariance of the theory by singling out a particular time direction, in the end the formalism tells us that it really did not matter which time direction we took. That is, the covariance is recovered implicitly in the theory, and the time picked is only a "fiducial" one for constructional purposes. We will see more details of this fairly soon.

So we foliate spacetime ^4M into a spatial manifold $^3\Sigma$ and a time direction t^a, as shown in figure 1. We now decompose the time direction into components normal and perpendicular to the three surface,

$$t^a = Nn^a + N^a \tag{3}$$

where n^a is the normal to the three surfaces and N^a is tangent to the three surfaces, and is called the shift vector. The scalar N is called the lapse function. Given the metric g_{ab} and the timelike vector n^a one can define a positive-definite (Euclidean) metric on the three surface $q^{ab} = g^{ab} + n^a n^b$. Another important quantity on the three surface is its extrinsic curvature, defined by $K_{ab} = q_a^m q_b^n \nabla_m n_n$, where ∇ is the covariant derivative compatible with g_{ab}. To clarify the role of the extrinsic curvature it is enough to compute the "time derivative" of the three metric, given by its Lie derivative with respect to t^a (exercise),

$$\dot{q}_{ab} = \mathcal{L}_{\vec{t}}\, q_{ab} = 2NK_{ab} + \mathcal{L}_{\vec{N}}\, q_{ab} \tag{4}$$

So we see that the role of the extrinsic curvature is roughly that of "time derivative" of the three-metric, giving an idea of how the three dimensional surface is deformed with respect to the ambient four dimensional spacetime. One can rewrite the action of General Relativity in terms of these quantities in the following form,

$$S = \int dt\, L(q,\dot{q}) \tag{5}$$

$$L(q,\dot{q}) = \int d^3x\, N\sqrt{q}(^3R + K_{ab}K^{ab} - K^2) \tag{6}$$

where t is a parameter along the integral curves of the vector t^a. ^3R is the scalar curvature of the three dimensional surface. To achieve this form of the action one needs to neglect surface terms. Along these lectures we will always assume that the spatial three-surface is compact, as in the case of some cosmologies. One could treat the asymptotically-flat case (which includes for instance, stars and black-holes) by imposing appropriate boundary conditions at infinity. This can be done in a reasonable straightforward manner, although we will not discuss it here for the sake of brevity.

We now have the action of General Relativity in a reasonable form to formulate a canonical analysis. We have expressed it in terms of variables that are functions of "space" (functions of the three-surface) and that "evolve in time". This is the usual setup for doing canonical formulations.

We pick as a canonical variable the three metric q^{ab} and compute its conjugate momentum,

$$\tilde{\pi}_{ab} = \frac{\delta L}{\delta \dot{q}^{ab}} = \sqrt{q}(K_{ab} - K q_{ab}) \tag{7}$$

(throughout these lectures we will denote tensor densities of weight +1 with a tilde and those of weight -1 with an undertilde). We see that the "conjugate momentum" to q_{ab} is roughly related to the extrinsic curvature ("time derivative").

We are now in position to perform the Legendre transform and obtain the Hamiltonian of the theory,

$$H(\pi, q) = \int d^3x\, (\tilde{\pi}_{ab}\dot{q}^{ab} - L) \tag{8}$$

and replacing $\dot{q}$ in terms of $\tilde{\pi}$, we get (exercise),

$$H(\pi, q) = \int d^3x\, (N(-q^{1/2}R + q^{-1/2}(\tilde{\pi}^{ab}\tilde{\pi}_{ab} - \tfrac{1}{2}\tilde{\tilde{\pi}}^2)) - 2N^b D_a \tilde{\pi}^a_b) \tag{9}$$

where D_a is the covariant derivative on the three surface compatible with q_{ab}.

2.3 The constraints

Having done this, let us step back a minute and analyze the formalism we built. We started from a four dimensional metric g^{ab} and we now have in its place the three dimensional q^{ab} and the "lapse" and "shift" functions N and N^a. We defined a conjugate momentum for q_{ab}. However, notice that nowhere in the formalism does a time derivative of the lapse or shift appear. That means their conjugate momenta are zero. That is, our theory has constraints. In fact, if we rewrite the action using the expression for the Hamiltonian we give above, we get,

$$S = \int dt \int d^3x \left((\tilde{\pi}_{ab}\dot{q}^{ab} + \underset{\sim}{N}(-qR + (\tilde{\pi}^{ab}\tilde{\pi}_{ab} - \tfrac{1}{2}\tilde{\tilde{\pi}}^2)) - 2N^b D_a \tilde{\pi}^a_b \right) \quad (10)$$

and if we vary it with respect to $\underset{\sim}{N}$ and N^b in order to get their respective equations of motion, we get four expressions, functions of $\tilde{\pi}$ and q which should vanish identically, and are usually called $\tilde{C}^a$ and $\tilde{\tilde{C}}$,

$$\tilde{C}_a(\pi, q) = 2D_b \tilde{\pi}^b_a \quad (11)$$
$$\tilde{\tilde{C}}(\pi, q) = -\tilde{q}R + (\tilde{\pi}^{ab}\tilde{\pi}_{ab} - \tfrac{1}{2}\tilde{\pi}^2) \quad (12)$$

For calculational simplicity, these equations are usually "smoothed out" with arbitrary test fields on the three manifold, $C(\vec{N}) = \int d^3x N^a \tilde{C}_a$, $C(\underset{\sim}{N}) = \int d^3x \underset{\sim}{N}\tilde{\tilde{C}}$. (Notice that the notation is unambiguous. One can write the constraints now as $C(\underset{\sim}{N}) = 0$ and $C(\vec{N}) = 0$, due to the arbitrariness of the test fields).

Notice that these equations are "instantaneous" laws, i.e. they must be satisfied *on each hypersurface*. They tell us that if we want to prescribe data for a gravitational field, not every pair of $\tilde{\pi}$ and q will do, eqs. (11, 12) should be satisfied. (Notice that there are six degrees of freedom in q^{ab}, subjected to four constraints, this leaves us with two degrees of freedom for the gravitational field, as expected).

These equations have the same character as the Gauss Law has for electromagnetism, which tells us that any vector field would not necessarily work as an electric field, it must have vanishing divergence in vacuum. As is well known, the Gauss Law appears as a consequence of the $U(1)$ invariance of the Maxwell equations. An analogous situation appears here. To understand this, consider the Poisson bracket of any quantity with the constraint $C(\vec{N})$. It is easy to check that (exercise),

$$\{f(\tilde{\pi}, q), C(\vec{N})\} = \mathcal{L}_{\vec{N}} f(\tilde{\pi}, q). \quad (13)$$

Therefore we see that the constraint $C(\vec{N})$ "Lie drags" the function $f(\pi, q)$ along the vector $\vec{N}$. Technically, it is the infinitesimal generator of diffeomorphisms of the three manifold in phases space. As the Gauss law (in the canonical formulation of Maxwell's theory) is the infinitesimal generator of $U(1)$ gauge transformations, the constraint here is the infinitesimal generator of spatial diffeomorphisms. This clearly shows why we have this constraint in the theory: it is the canonical representation of the fact that the theory is invariant under spatial diffeomorphisms. An analogous situation stands for the constraint $C(\underset{\sim}{N})$, although we will not discuss it in detail for reasons of space: it is the generator of "time" diffeomorphisms, and it is usually called the Hamiltonian constraint.

We can now work out the equations of motion of the theory either varying the action with respect to q^{ab} and $\tilde{\pi}_{ab}$ or taking the Poisson bracket of these quantities with the Hamiltonian constraint.

2.4 Quantization

Having set up the theory in canonical form we can now proceed and attempt a Canonical Quantization. The process can be roughly summarized in the following sequence of steps. The reader may notice that at each step there are many possible choices. Differente choices will yield different quantizations.

1. Pick up a complete set of canonical quantities that form an algebra under Poisson brackets. In many systems one simply takes p and q, but one can pick other "noncanonical" quantities. We will actually do so in the next chapters.

2. Represent these quantities as operators acting on a space of (wave)functionals of "half" of the elements of the algebra. These operators will in turn give a "quantum" representation of the algebra defined in the previous point. At this point the space of functions is not a Hilbert space. One usual choice is to take functionals of q, $\Psi(q)$, and represent $\hat{q}$ as a multiplicative operator and $\hat{p}$ as a functional derivative, $\hat{p}\Psi(q) = \frac{\delta}{\delta q}\Psi(q)$.

3. One wants the wavefunctions to be invariant under the symmetries of the theory. As we saw the symmetries are represented in this language as constraints. The requirement that the wavefunctions be annihilated by the constraints (promoted to operatorial equations) implements this requirement. Not all functionals chosen in the previous steps will be annihilated by the quantum constraint equations. We call those who are "physical states" (notice that we still do not have a Hilbert space).

4. One introduces an inner product on the space of physical states in order to compute expectation values and make physical predictions. Only at this point does one have an actual Hilbert Space. How to find this inner product is not prescribed by standard canonical quantization (we will discuss this in the next section). Under this inner product the physical states should be normalizable. The expectation values, by the way, only make sense for quantities that are invariant under the symmetries of the theory (quantities that classically have vanishing Poisson brackets with all the constraints). We call them physical observables. For the gravitational case *none* is known for compact spacetimes (we will return to this issue later). The observables of the theory should be self adjoint operators with respect to the inner product in order to yield real expectation values.

Let us start by step one. As it is said, one can pick various choices of algebras. Let us concentrate on the simplest one, just picking the canonical variables we introduced q^{ab} and $\tilde{\pi}_{ab}$. Their Poisson bracket is, being canonically conjugate variables, $\{q^{ab}(x), \tilde{\pi}_{cd}(y)\} = \delta(x-y)\delta_c^a\delta_d^b$. Step 2 can be fulfilled taking wavefunctions $\Psi(q^{ab})$ and representing q^{ab} as a multiplicative operator and $\tilde{\pi}_{ab}$ as a functional derivative.

It is in step 3 that we run into trouble. We have to promote the constraints we discussed in the last subsection to quantum operators. This in itself is a troublesome issue, since being General Relativity a field theory, issues of regularization and factor ordering appear. One can, —at least formally— find factor orderings in which the diffeomorphism constraint becomes the infinitesimal generator of diffeomorphisms on the wavefunctions. Therefore the requirement that a wavefunction be annihilated by it just translates itself in the fact that the wavefunction has to be invariant under diffeomorphisms. This is not difficult to accomplish (formally!). One simply requires that the wavefunctions not actually be functionals of the three metric q^{ab}, but of the "three geometry" (by this meaning the properties of the three geometry invariant under diffeomorphisms). That is, what we are saying is just a restatement of the fact that the functional should be invariant under diffeomorphisms. One can come up with several examples of functionals that meet this requirement. The real trouble appears when we want to make the wavefunctions annihilated by the Hamiltonian constraint. This constraint does not have a simple geometrical interpretation in terms of three dimensional quantities (remember that the idea that it represent "diffeomorphisms in time" does not help here, since we are always talking about equations that hold *on the three surface* without any explicit reference to time). Therefore we are just forced to proceed crudely: promote the constraint to a wave equation, use some factor ordering (hopefully with some physical motivation), pick some regularization and try to solve the resulting equation. It turns out that this task was never accomplished in general (it was in simplified minisuperspace examples). Among the difficulties that conspire in this direction is the fact that the constraint is a nonpolynomial function of the basic variables (remember it involves the scalar curvature, a nonpolynomial function of the three-metric).

Therefore the program of canonical quantization stalls here. Having been unable to find the physical states of the theory we are in a bad position to introduce an inner product (since we do not know on what space of functionals to act) and actually make physical predictions. This issue is compounded by the fact that we do not know any observables for the system, which puts us in a more clueless situation with respect to the inner product. This status of affairs was reached already in the work of DeWitt in the 60's [6] and little improvement was made until recently. We will see in the next chapter that the use of a new set of variables improves the situation with respect to the Hamiltonian constraint, giving hopes of maybe allowing us to attack the problem of the inner product.

3 The Ashtekar New Variables

The following three subsections follow closely the treatment of [5]. The reader is referred to it for more detailed explanations.

3.1 Tetradic General Relativity

To introduce the New Variables, we first need to introduce the notion of tetrads. In a nutshell, a tetrad is a vector basis in terms of which the metric of spacetime looks locally flat. Mathematically,

$$g_{ab} = e_a^I e_b^J \eta_{IJ} \tag{14}$$

where $\eta_{IJ} = \text{diag}(-1,1,1,1)$ is the Minkowski metric, and equation (14) simply expresses that g_{ab}, when written in terms of the basis e_a^I, is locally flat. If spacetime were truly flat, one could perform such a transformation globally, integrating the basis vectors into a coordinate transformation $e_a^I = \frac{\partial x^I}{\partial x'^a}$. In a curved spacetime these equations cannot be integrated and the transformation to a flat space only works locally, the flat space in question being the "tangent space". From equation (14) it is immediate to see that given a tetrad, one can reconstruct the metric of spacetime. One can also see that although g_{ab} has only ten independent components, the e_a^I have sixteen. This is due to the fact that eq. (14) is invariant under Lorentz transformations on the indices $I, J \ldots$. That is, these indices behave as if living in flat space. In summary, tetrads have all the information needed to reconstruct the metric of spacetime but there are extra degrees of freedom in them, and this will have a reflection in the canonical formalism.

3.2 The Palatini action

We now write the Einstein action in terms of tetrads. We introduce a covariant derivative via $D_a K_I = \partial_a K_I + \omega_{aI}{}^J K_J$. $\omega_{aI}{}^J$ is a Lorentz connection (the derivative annihilates the Minkowski metric). We define a curvature by $\Omega_{ab}{}^{IJ} = \partial_{[a}\omega_{b]}{}^{IJ} + [\omega_a, \omega_b]^{IJ}$, where $[\,,\,]$ is the commutator in the Lorentz Lie algebra. The Ricci scalar of this curvature can be expressed as $e_I^a e_J^b \Omega_{ab}^{IJ}$ (indices I, J are raised and lowered with the Minkowski metric). The Einstein action can be written,

$$S(e,\omega) = \int d^4x \, e \, e_I^a e_J^b \Omega_{ab}^{IJ} \tag{15}$$

where e is the determinant of the tetrad (equal to $\sqrt{-g}$).

We will now derive the Einstein Equations by varying this action with respect to e and ω as independent quantities. To take the metric and connection as independent variables in the action principle was first considered by Palatini [7].

As a shortcut to performing the calculation (this derivation is taken from [5]), we introduce a (torsion-free) connection compatible with the tetrad via $\nabla_a e_I^b = 0$. The difference between the two connections we have introduced is a field $C_{aI}{}^J$ defined by $C_{aI}{}^J V_J = (D_a - \nabla_a) V_I$. We can compute the difference between the curvatures (R_{ab}^{IJ} is the curvature of ∇_a), $\Omega_{ab}{}^{IJ} - R_{ab}{}^{IJ} = \nabla_{[a} C_{b]}{}^{IJ} + C_{[a}{}^{IM} C_{b]M}{}^J$. The reason for performing this intermediate calculation is that it is easier to compute the variation by reexpressing the action in terms of ∇ and $C_a{}^{IJ}$ and then noting that the variation with respect to $\omega_a{}^{IJ}$ is the same as the variation with respect to C_a^{IJ}. The action, therefore is,

$$S = \int d^4x \; e \; e_I^a e_J^b (R_{ab}{}^{IJ} + \nabla_{[a} C_{b]}{}^{IJ} + C_{[a}{}^{IM} C_{b]M}{}^{J}). \tag{16}$$

The variation with respect to $C_a{}^{IJ}$ is easy to compute: the first term simply does not contain $C_a{}^{IJ}$ so it does not contribute. The second term is a total divergence (notice that ∇ is defined so that it annihilates the tetrad), the last term yields $e_M^{[a} e_N^{b]} \delta_{[I}^M \delta_{J]}^K C_{bK}{}^N$. It is easy to check that the prefactor in this expression is nondegenerate and therefore the vanishing of this expression is equivalent to the vanishing of $C_{bK}{}^N$. So this equation basically tells us that ∇ coincides with D when acting on objects with only internal indices. Thus the connection D is completely determined by the tetrad and Ω coincides with R (some authors refer to this fact as the vanishing of the torsion of the connection). We now compute the second equation, straightforwardly varying with respect to the tetrad. We get, (after substituting $\Omega_{ab}{}^{IJ}$ by $R_{ab}{}^{IJ}$ as given by the previous equation of motion),

$$e_I^c R_{cb}{}^{IJ} - \tfrac{1}{2} R_{cd}{}^{MN} e_M^c e_N^d e_b^J \tag{17}$$

which, after multiplication by e_{Ja} just tells us that the Einstein tensor $R_{ab} - \tfrac{1}{2} R g_{ab}$ of the metric defined by the tetrads vanishes. We have therefore proved that the Palatini variation of the action in tetradic form yields the usual Einstein Equations.

There is a slight difference between the first order (Palatini) tetradic form of the theory and the usual one. One easily sees that a solution to the Einstein Equations we presented above is simply $e_b^J = 0$. This solution would correspond to a vanishing metric and is therefore forbidden in the traditional formulation since quantities as the Ricci or Riemann tensor are not defined for a vanishing metric. However, the first order action and equation of motion are well defined for vanishing triads. We therefore see that strictly speaking the first order tetradic formulation is a "generalization" of General Relativity that contains the traditional theory in the case of nondegenerate triads. We will see this subtlety playing a role in the future chapters. It should be noticed that the possibility of allowing vanishing metrics in General Relativity is quite attractive since one could envisage the formalism "going through", say, the formation of singularities. It also allows for topology change [8].

3.3 The self-dual action

Up to now the treatment has been totally traditional. We will now take a conceptual step that allows the introduction of the Ashtekar variables. We will reconstruct the tetradic formalism of the previous subsection but we will introduce a change. Instead of considering the connection $\omega_a{}^{IJ}$ we will consider its self dual part with respect to the internal indices and we will call it $A_a{}^{IJ}$, that is $iA_a{}^{IJ} = \tfrac{1}{2} \epsilon_{MN}{}^{IJ} A_a{}^{MN}$. Now, to really be able to do this, the connection must be complex (or one should work in an Euclidean signature). Therefore for the moment being we will consider *complex General Relativity* and we will then specify appropriately how to recover the traditional real theory. The connection now takes values in the (complex) self-dual subalgebra of the Lie algebra of the Lorentz group. We will propose as action,

$$S(e,A) = \int d^4x \, e \, e^a_J e^b_K F_{ab}{}^{JK} \quad (18)$$

where $F_{ab}{}^{JK}$ is the curvature of the self dual connection and it can be checked that it corresponds to the self-dual part of the curvature of the usual connection.

We can now repeat the calculations of the previous subsection for the selfdual case. When one varies the self-dual action with respect to the connection $A_a{}^{IJ}$ again one obtains that this connection should annihilate the triad (if one repeated step by step the previous subsection argument, one now finds that the self-dual part of $C_a{}^{IJ}$ vanishes). The variation with respect to the tetrad goes along very similar lines only that $\Omega_{ab}{}^{IJ}$ gets everywhere replaced by $F_{ab}{}^{IJ}$. The final equation one arrives to (exercise) again tells us that the Ricci tensor vanishes. Remarkably, the self-dual action leads to the (complex) Einstein Equations. This essentially be understood in the fact that the two actions differ by boundary terms. We postpone the issue of how to recover the real theory to subsection 3.6.

3.4 The New Variables

If one took the Palatini action of subsection 3.2 and made a canonical 3+1 decomposition, the formalism basically returns to the traditional one [9]. A quite different thing happens if one decomposes the self-dual action. Let us therefore proceed to do the 3+1 split. As before, we introduce a vector $t^a = Nn^a + N^a$ (which, since we are actually dealing with complex Relativity, may be complex). Taking the action,

$$S(e,A) = \int d^4x \, e \, e^a_I e^b_J F_{ab}{}^{IJ} \quad (19)$$

and defining the vector fields orthogonal to n^a, $E^a_I = q^a_b e^b_I$ (where $q^a_b = \delta^a_b + n^a n_b$ is the projector on the three-surface) we have,

$$S(e,A) = \int d^4x \, (e \, E^a_I E^b_J F_{ab}{}^{IJ} - 2 \, e \, E^a_I e^d_J n_d n^b F_{ab}{}^{IJ}). \quad (20)$$

We now define $\tilde{E}^a_I = \sqrt{q} E^a_I$, which is a density on the three manifold. The determinant of the triad can be written as $e = N\sqrt{q}$. We also introduce the vector in the "internal space" induced by n^a, defined by $n_I = e^d_I n_d$. With these definitions, and exploiting the self-duality of $F_{ab}{}^{IJ}$ to write $F_{ab}{}^{IJ} = -i\frac{1}{2}\epsilon^{IJ}{}_{MN} F_{ab}{}^{MN}$, we get,

$$S(e,A) = \int d^4x \, (-\tfrac{i}{2} N \tilde{E}^a_I \tilde{E}^b_J \epsilon^{IJ}{}_{MN} F_{ab}{}^{MN} - 2N n^b \tilde{E}^a_I n_J F_{ab}{}^{IJ}). \quad (21)$$

We now pick a gauge in which $E_0^a = 0$ and $n_I = (1,0,0,0)$[1]. We also exploit the relation between Levi-Civita densities in three and four space, $\epsilon^{IJKL} n_L = \epsilon^{IJK}$, which in our gauge reads $\epsilon^{IJK0} = \epsilon^{IJK}$. Therefore, we have,

$$S(e,A) = \int d^4x \, (-\tfrac{i}{2} \underset{\sim}{N} \tilde{E}_I^a \tilde{E}_J^b \epsilon^{IJ}{}_M F_{ab}{}^{M0} - 2N n^b \tilde{E}_I^a F_{ab}{}^{I0}). \tag{22}$$

Notice that the indices $I, J \ldots$ in the above expression now only range from 1 to 3. We denote them with lowercase letters to highlight this fact and rename, $A_a^{I0} = A_a^i$ and similarly for $F_{ab}^{I0} = F_{ab}^i$. The effect of the presence of the vector in internal space n_I has been to split the two copies of $SU(2)$ present in the Lorentz group in such a way that our new indices i, j, k are $SU(2)$ indices. We now replace in the second term $N n^b$ by $t^b - N^b$ and use the identity (exercise) $t^a F_{ab}^i = \mathcal{L}_t A_b^i - \mathcal{D}_b (t^a A_a)^i$, where $\mathcal{D}_b$ is the derivative defined by the connection A_a^i, to get,

$$S(e,A) = \int d^4x \, (2\tilde{E}_i^a \mathcal{L}_t A_a^i - 2N^b \tilde{E}_i^a F_{ab}^i - \tfrac{i}{2} \underset{\sim}{N} \epsilon^{ij}{}_k \tilde{E}_i^a \tilde{E}_j^b F_{ab}^k). \tag{23}$$

This action is exactly in the form we want. There is a term of the "$p\dot{q}$" form, $(\tilde{E}_i^a \mathcal{L}_t A_a^i)$, from which we can read off that the variable canonically conjugate to A_a^i is $\tilde{E}_i^a$. The theory also has constraints, given by,

$$\tilde{\mathcal{G}}^i = (\mathcal{D}_a \tilde{E}^a)^i \tag{24}$$

$$\tilde{\mathcal{C}}_a = \tilde{E}_i^a F_{ab}^i \tag{25}$$

$$\underset{\sim}{\tilde{\mathcal{C}}} = \epsilon^{ij}{}_k \tilde{E}_i^a \tilde{E}_j^b F_{ab}^k \tag{26}$$

The last four equations correspond to the usual diffeomorphism and Hamiltonian constraints of canonical General Relativity. The first three equations are extra constraints that stem from our use of triads as fundamental variables. These equations, which have exactly the same form as a Gauss Law of an $SU(2)$ Yang-Mills theory, are the generators of infinitesimal $SU(2)$ transformations. They tells us that the formalism is invariant under triad rotations, as it should be.

Notice that a dramatic simplification of the constraint equations has occurred. In particular the Hamiltonian constraint is a polynomial function of the canonical variables, of quadratic order in each variable. Moreover, the canonical variables, and the phase space of the theory are exactly those of a (complex) $SU(2)$ Yang-Mills theory. The reduced phase space is actually a subspace of the reduced phase space of the Yang-Mills theory (the phase space modulo the Gauss Law), since General Relativity has four more constraints that further reduce its phase space. This resemblance in the formalism to that of a Yang-Mills theory will be the starting point of all the results we will introduce in the last sections of this paper.

---

[1]One can proceed in a more elegant, albeit slightly more complicated, way, see ref. [5]

3.5 The constraint algebra

When one has a theory with constraints, one needs to check that these are consistent with each other, i.e. that by taking Poisson brackets among the constraints one does not generate new constraints. If this were the case, these secondary constraints should also be enforced. Fortunately, the system we have here is first class, i.e. the Poisson bracket of each two constraints is a combination of the other constraints. Actually, in terms of the New Variables, the structure of the constraints is simple enough for the reader to be able to compute the constraint algebra without great effort (this computation can also be carried along with the traditional variables and the results are the same). We only summarize the results here. To express them in a simpler form (and to avoid confusing manipulations of distributions while performing the computations), it is again convenient to smooth out the constraints with arbitrary test fields and to perform some recombinations. We denote,

$$\mathcal{G}(N_i) = \int d^3x N_i (\mathcal{D}_a \tilde{E}^a)^i \tag{27}$$

$$C(\vec{N}) = \int d^3x N^b \tilde{E}^a_i F^i_{ab} - \mathcal{G}(N^a A^i_a) \tag{28}$$

$$C(\utilde{N}) = \int d^3x \utilde{N} \epsilon^{ij}{}_k \tilde{E}^a_i \tilde{E}^b_j F^k_{ab} \tag{29}$$

and as before the notation is unambiguous. The constraint algebra then reads,

$$\{\mathcal{G}(N_i), \mathcal{G}(N_j)\} = \mathcal{G}([N_i, N_j]) \tag{30}$$

$$\{C(\vec{N}), C(\vec{M})\} = C(\mathcal{L}_{\vec{M}} \vec{N}) \tag{31}$$

$$\{C(\vec{N}), \mathcal{G}(N_i)\} = \mathcal{G}(\mathcal{L}_{\vec{N}} N_i) \tag{32}$$

$$\{C(\vec{N}), C(\utilde{M})\} = C(\mathcal{L}_{\vec{N}} \utilde{M}) \tag{33}$$

$$\{C(N_i), C(\utilde{N})\} = 0 \tag{34}$$

$$\{C(\utilde{N}), C(\utilde{M})\} = C(\vec{K}) - \mathcal{G}(A^i_a K^a), \tag{35}$$

where the vector $\vec{K}$ is defined by $K^a = 2\tilde{E}^a_i \tilde{E}^b_i (\utilde{N} \partial_a \utilde{M} - \utilde{M} \partial_a \utilde{N})$. Here we clearly see that the constraints are first class. The reader should notice, however, that the algebra is not a true Lie algebra, since one of the structure constants (the one defined by the last equation), is not a constant but depends on the fields $\tilde{E}^a_i$ (through the definition of the vector $\vec{K}$).

3.6 The evolution equations and the reality conditions

Up to now we have been concerned only with instantaneous relations among the fields (the constraints). However, if one wants to evolve the fields in time, one needs the evolution equations, which are simply obtained taking the Poisson bracket of the fields with the Hamiltonian.

These equations give the "time derivative" of the triad and the connection.

$$\dot{\tilde{E}}_i^a = \{\tilde{E}_i^a, H(\underset{\sim}{N})\} = -i\sqrt{2}\mathcal{D}_b(\underset{\sim}{N}\tilde{E}^{[b}\tilde{E}^{a]})_i \tag{36}$$

$$\dot{A}_a^i = \{A_a^i, H(\underset{\sim}{N})\} = -\tfrac{i}{\sqrt{2}}[\underset{\sim}{N}\tilde{E}^b, F_{ab}]^i. \tag{37}$$

From here one can straightforwardly derive the equations of motion for the traditional variables, the metric and the extrinsic curvature.

We now turn our attention to the issue of "reality conditions". As was mentioned before, the formalism we are dealing with describes complex General Relativity. In fact, the action we are using is complex! If we want to recover the classical theory we must take a "section" of the phase space that corresponds to the dynamics of real Relativity. This can be done. One gives data on the initial surface that corresponds to a real spacetime and the evolution equations will keep these data real through the evolution. Now, strictly speaking, this procedure is not really canonical, since we are imposing these conditions by hand at the end. That does not mean it is not useful[2]. In fact, one can eliminate the reality conditions and have a canonical theory. However, many of the beauties of the new formulation are lost, in particular the structure of the resulting constraints is basically that of the traditional formalism.

The issue of the reality conditions acquires a different dimension in the quantum theory. A point of view that is strongly advocated, and may turn out to be successful, is the following. Start by considering the complex theory and apply the steps towards canonical quantization that we discussed subsection (2.4). After the space of physical states has been found, when one decides to find an inner product, the reality conditions are used in order to choose an inner product that implements them. That is, the reality conditions can be a guideline to find the appropriate inner product of the theory. One simply requires that the quantities that have to be real according to the reality conditions of the classical theory, become self-adjoint operators under the chosen inner product. This solves two difficulties at once, since it allows us to recover the real quantum theory and the appropriate inner product at the same time. This point of view is strictly speaking a deviation from standard Dirac quantization, and works successfully for several model problems [11]. The success or failure in Quantum Gravity of this approach is yet to be tested and is one of the most intriguing and attractive features of the formalism. (For a critical viewpoint, see [12]).

What are, therefore, the reality conditions? They can be written in several ways, depending on which variables one chooses to express them in. One could simply write them in terms of the three-metric,

$$\tilde{\tilde{q}}^{ab} = (\tilde{\tilde{q}}^{ab})^* \tag{38}$$

$$\dot{\tilde{\tilde{q}}}^{ab} = (\dot{\tilde{\tilde{q}}}^{ab})^* \tag{39}$$

---

[2] A nontrivial example where it can be worked out to the end is the Bianchi II cosmology [10]

Using the expressions we presented above for the time derivatives, we could easily express these equations in terms of the triad and the connection.

How one writes the reality conditions depends largely on what one wants to accomplish. If one, for instance, is interested in pursuing the quantization program outlined above, and using the reality conditions to fix the inner product, one does not necessarily want them written in the above form. The reason for this is that if one wants to write them as conditions of hermiticity under an inner product, one would want to write them in terms of quantities that are observables of the theory, so that taking their expectation value makes sense. Since we do not know any observable for the theory, we cannot write the reality conditions in this way at present.

Let us finish by making a small digression on the issue of observables. The quantities that one wants to observe in a system are quantities that are invariant under the symmetries of the theory. Any other kind of quantity will be gauge dependent and therefore of no physical relevance. In the canonical theory it is easy to define which kind of quantities are observables. Since the constraints are the infinitesimal generators of the symmetries of the theory, any quantity that has vanishing Poisson brackets with the constraints is invariant under the symmetries of the theory. Therefore, if one wanted to find an observable for General Relativity, one has to look for a quantity with vanishing Poisson brackets with the diffeomorphism and Hamiltonian constraint of the theory. The trouble is that we do not know any single such quantity. This can be a circumstantial problem, merely reflecting our ignorance, or it could be fundamental. There are suggestions that maybe no such quantity exists for General Relativity [13] [3]. This could be related to the fact that the theory displays chaotic behaviour [15, 16]. The issue of observables and their relation to quantization (some people argue that since the theory may have no observables, the whole program of quantization we are pursuing here is doomed) exceeds the scope and is not in line with the emphasis of these talks so we will not discuss it here. We just want to make the reader aware that there is potential for a problem with this issue. For a recent discussion see [17]. From a practical point of view one could argue that even if the full theory has no exact observable, one could find some approximation in which the theory has observables (after all, we live in such an approximation and measure things all the time!). Again, this exceeds the scope of this treatment. See [18] for more details on this point of view.

---

[3] All these remarks refer to the cases of compact spacetimes. For the asymptotically flat case observables as the four momentum and angular momentum are well known. Some non-analytic observables for the compact case may also be written [14]

4 Quantum Theory: The Connection Representation

4.1 Formulation

Let us suppose we now decide to proceed and apply the canonical quantization program to the theory as we have it up to now.

We start by picking a polarization. One has many choices. However, let us remember that our canonical variables are basically similar to those of a Yang-Mills $SU(2)$ theory. When one quantizes Yang-Mills (and Maxwell) theories a usual choice for the polarization is to pick wavefunctionals of the connection $\Psi(A)$. We will pursue in this subsection this treatment for General Relativity. Notice that this is potentially *very* different from what one does with the traditional variables, where the more commonly considered polarization is that in where one takes wavefunctionals of the three metric $\Psi(q)$. In terms of our variables, this polarization would be closer to choosing wavefunctionals of the triad. We see that the use of these new variables leads us to a new perspective even at this level.

A representation for the Poisson algebra of the canonical variables considered can be simply achieved by representing the connection as a multiplicative operator and the triad as a functional derivative,

$$\hat{A}^i_a \Psi(A) = A^i_a \Psi(A), \tag{40}$$

$$\hat{\tilde{E}}^a_i \Psi(A) = \frac{\delta}{\delta A^i_a} \Psi(A). \tag{41}$$

If we now want to promote the constraint equations to operatorial equations, we need first to pick a factor ordering. Two factor orderings have been explored, one with the triads to the right (we will call it II) and one with the triads to the left (I) [4] (see [19] for alternatives). Let us stress that all these calculations are only formal until a regularization is introduced.

4.2 Factor ordering II and the role of Wilson Loops

If one orders the triads to the right, the constraints become,

$$\hat{\tilde{\mathcal{G}}}^i = D_a \frac{\delta}{\delta A^i_a} \tag{42}$$

$$\hat{\tilde{\mathcal{C}}}_a = F^i_{ab} \frac{\delta}{\delta A^i_b} \tag{43}$$

$$\hat{\tilde{\tilde{\mathcal{H}}}} = \epsilon^{ijk} F^i_{ab} \frac{\delta}{\delta A^j_a} \frac{\delta}{\delta A^k_b} \tag{44}$$

An attractive feature of this ordering is that the Gauss Law becomes the infinitesimal generator of $SU(2)$ gauge transformations for the wavefunctions and the

---

[4]The reason for this reversal of notation is to keep it in line with the literature [25].

diffeomorphism constraint becomes the infinitesimal generator of diffeomorphisms on the wavefunctions. This, among other features, attracted the attention of Jacobson and Smolin [20] to this ordering. There is a potential awkwardness when one considers the algebra of constraints. Remember that it is not a true algebra, but as we discussed, the commutator of two Hamiltonians has a structure "constant" that depends on one of the canonical variables, the triad. This means that in this ordering such "constant" would have to appear to the right of the resulting commutator, which is not expected, and when one constructs solutions one has to check the consistency of the constraints [20].

Jacobson and Smolin set out to find solutions to the constraint equations in this formalism. If one starts by considering the Gauss Law, one would like the wavefunctionals to be invariant under $SU(2)$ gauge transformations. A well known infinite parameter family of gauge invariant functionals of a connection are the Wilson Loops,

$$W(A,\gamma) = \text{Tr}\left(\text{Pexp} \oint ds \dot{\gamma}^a(s) A_a(\gamma(s))\right), \quad (45)$$

defined by the trace of the path-ordered exponential of the line integral along a loop γ (parametrized by s) of the connection. In fact, up to some extent *any* Gauge invariant function of a connection can be expressed as a combination of Wilson Loops [21, 22]. In view of this, one can consider Wilson loops as an infinite family of wavefunctions in the connection representation parametrized by a loop $\Psi_\gamma(A) = W(\gamma, A)$ that forms an (overcomplete) basis of solutions to the quantum Gauss Law constraint. So, we have managed to find solutions to the first set of constraints, even perhaps a basis of solutions.

What happens to the diffeomorphism constraint? Evidently Wilson loops are not solutions. When a diffeomorphism acts on a Wilson loop, it gives as a result a Wilson loop with the loop displaced by the diffeomorphism performed. Therefore they are not annihilated by the diffeomorphism constraint and cannot become candidates for physical states of Quantum Gravity. In spite of that, they are worthwhile exploring a bit more. Remember they form an overcomplete basis in terms of which any physical state should be expandable (since any physical state has to be Gauge invariant). We will therefore explore what happens when we act with the Hamiltonian constraint on them. To perform this calculation we only need the formula for the action of a triad on a Wilson Loop,

$$\hat{\tilde{E}}_i^a(x)\Psi_\gamma(A) = \frac{\delta}{\delta A_a^i(x)}\Psi_\gamma(A) = \oint ds\, \delta^3(x-\gamma(s))\dot{\gamma}^a(s)\text{Tr}(U(0,s)\tau^i U(s,1)) \quad (46)$$

where we denote by $U(s_1, s_2) = \int_{s_1}^{s_2} dt\, \dot{\gamma}^a(t) A_a(\gamma(t))$ the holonomy from the point $\gamma(s_1)$ to the point $\gamma(s_2)$; τ^i denotes a Pauli matrix. Note that this expression involves a line integral of a three-dimensional delta function. Using some notational latitude, we can rewrite it as,

$$\hat{\tilde{E}}_i^a(x)\Psi_\gamma(A) = \dot{\gamma}^a(x)\text{Tr}(U(0,s(x))\tau^i U(s(x),1)). \quad (47)$$

The notational latitude consists in the fact that we have "cancelled" a three dimensional Dirac Delta with a one dimensional integral (which hides a regularization problem) and we have denoted by $s(x)$ the parameter value s for which the loop is at the point x (one has to be careful with this notation if the loop multiply traverses such a point, as in the case of an intersection). The point for this notational deviation is to make more transparent the following result (which actually goes through even regularizing with some care [20], up to the extent that that is possible in this context!). Let us evaluate the action of the Hamiltonian constraint on a Wilson Loop. Using the previous formulae we get,

$$\hat{H}(x)\Psi(A) = \epsilon^{ijk} F^i_{ab}(x) \frac{\delta}{\delta A^j_a(x)} \frac{\delta}{\delta A^k_b(x)} \Psi(A) = \qquad (48)$$
$$= F^i_{ab} \dot{\gamma}^a(x) \dot{\gamma}^b(x) \text{Tr}(U(0,s(x))\tau^i U(s(x),1))$$

Notice that in this expression we have an antisymmetric tensor, $F^i_{ab}(x)$, contracted with a symmetric tensor $\dot{\gamma}^a(x)\dot{\gamma}^b(x)$. Therefore, the expression vanishes! We have just proved that a Wilson loop formed with the Ashtekar connection is a solution of the Hamiltonian constraint of Quantum Gravity. This is a remarkable fact. Notice that up to this discover *no* solution of this constraint was known in a general case (without making minisuperspace approximations). Historically, this discovery fostered the interest for loops in this context and led to the use of the loop representation.

There are some drawbacks to this result. An obvious one is that although we solved the Hamiltonian constraint and the Gauss Law, we did not solve the diffeomorphism constraint, therefore these wavefunctions are not states of Quantum Gravity. The second point is that for the above result to hold, we need the tensor $\dot{\gamma}^a(x)\dot{\gamma}^b(x)$ to be symmetric. This is true if we have a smooth loop. If the loop has kinks or intersections, this is not any longer true and the Wilson loops stop to be solutions (at this naive level). Why care about loops with intersections? Why not just restrict ourselves to smooth loops? The problem appears when we try to get some sort of understanding of what these wavefunctionals are. The first question that comes to mind (of a prejudiced relativist at least) is what is the metric for such a state. This in principle is a meaningless question, since the metric is not an observable, but let us ask it anyway to see where it leads. The metric acting on one of these states, gives,

$$\hat{\tilde{q}}^{ab}(x)\Psi_\gamma(A) = \frac{\delta}{\delta A^i_a} \frac{\delta}{\delta A^i_b} \Psi_\gamma(A) = \dot{\gamma}^a(x)\dot{\gamma}^b(x)\Psi_\gamma(A) \qquad (49)$$

So Wilson Loops constructed with smooth loops are eigenstates of the metric operator. First, notice that the metric only has support distributionally along the loop (our sloppy notation may not make this totally transparent, but notice that the quantity $\dot{\gamma}^a(x)$ only can be nonvanishing along the loop). Then, notice that the metric has only one nonvanishing component, the one along the loop. Therefore it is a degenerate metric. Now, this statement is still meaningless in a diffeomorphism invariant context, but it actually can be given a rigorous meaning with a little elaboration. Consider the quantum operator obtained by computing the (square root of the) determinant of the three metric. In terms of the new variables it is given by $\epsilon^{ijk}\epsilon_{abc}\tilde{E}_i^a\tilde{E}_j^b\tilde{E}_k^c$. It is very easy to see, as is expected for a degenerate metric, that this operator vanishes when applied to a Wilson Loop. The problem with this is made clear when we consider General Relativity with a cosmological constant. The only thing that changes in the canonical formalism is that the Hamiltonian constraint gains an extra term,

$$H_\Lambda = H_0 + \Lambda \det q \qquad (50)$$

where H_0 is the vacuum Hamiltonian constraint and Λ is the cosmological constant. The extra term is given by the determinant of the three metric. Now consider our Wilson Loop state. Since it is annihilated by the vacuum Hamiltonian constraint *and* the determinant of the three metric, this means it is a state for an arbitrary value of the cosmological constant! That spells serious trouble. General Relativity with and without a cosmological constant are totally different theories, and one does not expect them to share a common set of states, except for special situations, as for degenerate metrics.

It turns out one can improve the situation a little using intersections. One can find some solutions to the Hamiltonian constraint even for the intersecting case by combining holonomies in such a way that the contributions at the intersection cancel [20, 23, 24]. However, unexpectedly, this is not enough to construct nondegenerate solutions. All the solutions constructed in this fashion, if they satisfy the Hamiltonian constraint, are also annihilated by the determinant of the metric [24]. This, plus the fact that they do not satisfy the diffeomorphism constraint, shows that these solutions are of little physical use in this context. They were, however, very important historically as motivational objects for the study of loops. We will show later on how, when one works in the loop representation, it is possible to generate solutions to all the constraints that, although still based on loops, do not have this degeneracy problem.

4.3 Factor ordering I: The role of the Chern-Simons form

If one orders the constraints with the triads to the left, there is potential for a problem: as we said, apparently in this factor ordering the diffeomorphism constraint fails to generate diffeomorphisms on the wavefunctions. For many of us, this would be a reason to abandon this ordering altogether. However, by considering a very generic regularized calculation one can prove that the diffeo constraint actually generates diffeomorphisms, so this is not a problem [25]. Besides, there is the advantage that when one considers the constraint algebra, one obtains (these are only formal unregulated results) the correct closure [3].

In this ordering, Wilson Loops do not solve the Hamiltonian constraint anymore. However, there is a very interesting and rich solution one can construct. Consider the following state, function of the Chern-Simons form built with the Ashtekar connection,

$$\Psi_\Lambda[A] = \exp(-\tfrac{6}{\Lambda} \int \tilde{\epsilon}^{abc} Tr[A_a \partial_b A_c + \tfrac{2}{3} A_a A_b A_c]) \tag{51}$$

This functional has the property that the triad (you can view it as an electric field) equals the magnetic field formed from the Ashtekar connection.

$$\frac{\delta}{\delta A_a^i}\Psi_\Lambda[A] = \tfrac{6}{\Lambda}\tilde{\eta}^{abc} F_{bc}^i \Psi_\Lambda[A] \tag{52}$$

Besides, it is well known that this functional is invariant under (small) gauge transformations and diffeomorphisms. One can check that it is annihilated by the corresponding constraints. What may come as a surprise is that it is actually annihilated by the Hamiltonian constraint with a cosmological constant. This is easy to see, simply consider the constraint,

$$\hat{\mathcal{H}} = \epsilon_{ijk}\frac{\delta}{\delta A_a^i}\frac{\delta}{\delta A_b^j}F_{ab}^k - \frac{\Lambda}{6}\epsilon_{ijk}\tilde{\epsilon}^{abc}\frac{\delta}{\delta A_a^i}\frac{\delta}{\delta A_b^j}\frac{\delta}{\delta A_c^k} \tag{53}$$

and notice that the rightmost derivative of the determinant of the metric simply reproduces the term on the left when acting on the wavefunction. Notice that the result holds without even considering the action of the other derivatives, and therefore is very robust vis a vis regularization. This result was independently noticed by Ashtekar [26] and Kodama [27]. A nice feature of this result is that the metric is nondegenerate in the sense that we discussed in the previous section. The metric is just given by the trace of the product of two magnetic fields. Such property holds classically for spaces of constant curvature. This has lead some authors to suggest this wavefunction as a "ground state" for a DeSitter geometry [28].

Now, this does not seem a very impressive feat. First of all, it is only one state. Secondly a similar state is present in Yang-Mills theory (this is easy to see, since the Hamiltonian is E^2+B^2 and adjusting constants one gets for the corresponding state $E = iB$) and is known to be nonphysical since it is nonnormalizable. This is true, but it is also true that the nature of a theory defined on a fixed background as Yang Mills theory is expected to be radically different from that of a theory invariant under diffeomorphisms, as General Relativity. Therefore normalizability under the inner product of one theory does not necessarily imply or rule out normalizability under the inner product of the other. It is remarkable that the Chern-Simons form, which is playing such a prominent role in particle physics nowadays, should have such a singular role in General Relativity. It is the only state in the connection representation that we know that may have something to do with a nondegenerate geometry!!

There are more things one could say about the connection representation. There is the compelling work of Ashtekar, Balachandran and Jo [9] concerning the CP violation problem and the partial success (in the linearized theory) of Ashtekar [29] in addressing the issue of time. We do not have space here to make justice to these pieces of work and we refer the reader to the relevant literature.

5 The Loop Representation

5.1 Motivation

A rigorous formulation of a loop representation for a quantum field theory is elaborate and involves many delicate details. However, one can get a very simple intuitive grasp of the idea if one ignores subtleties, and that is exactly what we will do here. We will mention some of the problems, but the reader who wants to get an idea of the subtleties is encouraged to see references [30, 31].

The way to easily view the concept of a Loop Representation is to compare it to the Momentum Representation of ordinary Quantum Mechanics. In the latter, one starts from a wavefunction in the Position Representation $\Psi(\vec{x})$, and convolutes it with an (infinite) basis of functions parametrized by a continuous parameter $\vec{k}$, $\exp(i\vec{k} \cdot \vec{x})$. The result depends on the continuous parameter $\vec{k}$ and we call it a wavefunction in the momentum representation,

$$\Psi(\vec{k}) = \int d^3x \, \exp(i\vec{k} \cdot \vec{x})\Psi(\vec{x}). \tag{54}$$

Now, in the Loop Representation, we start from a wavefunctional in the connection representation $\Psi(A)$. We convolute it with an infinite basis of (gauge invariant) wavefunctions parametrized by a continuous parameter, a loop γ, the Wilson Loops $W_\gamma(A)$. The result depends on the parameter γ and is a wavefunction in the Loop Representation,

$$\Psi(\gamma) = \int \text{``}dA\text{''} W(A, \gamma)\Psi(A). \tag{55}$$

The reader may have several reservations at this point. First of all, what is the measure to perform this integral? We do not know and we denoted that by the quotation marks. Secondly, up to what extent is this transform complete or how faithfully can it represent the wavefunction space. Again we do not quite know the answer. These technical mathematical issues have been addressed up to some extent in the literature and we will not discuss them here. For us to work, it suffices to notice that this transform works for several model problems, as 2+1 gravity [32], Maxwell theory [33] and Chern-Simons theory [34]. Better yet, it works for Yang-Mills on a lattice [35, 36, 37], which includes the phase space of Gravity in terms of the New Variables. For the case of real nonabelian connections, the transform can be given rigorous meaning [30].

Historically, the first construction of a loop representation for Quantum Gravity based on Ashtekar's new variables is due to Rovelli and Smolin [38, 39], immediately following the results with Wilson Loops in the connection representation by Jacobson and Smolin [20]. In the context of gauge theories, and even gravity in terms of traditional variables, loop representation had been considered before by Gambini and Trias [40].

Another point that may disquiet the reader is how can one know or at least guess that the quantity of information conveyed in a connection on a three manifold is equal to that present in the set of all possible loops on the manifold. Leaving aside technicalities, it turns out that the loop basis is way overcomplete. The practical inconvenience of this fact can be clearly seen, for example, in investigations done with the loop representation on the lattice. Formulating exactly what are the "free" degrees of freedom in the loop representation is an open and difficult problem [41]. There are lots of hidden identities among states in the loop representation. They can all be derived from three basic identities among Wilson loops:

- It is easy to see that reversing the orientation of the loop leaves the Wilson loop invariant. Therefore wavefunctions in the Loop Representation are invariant under reversal of orientation of the loop $\Psi(\gamma) = \Psi(\gamma^{-1})$.

- Due to the fact that the Wilson Loop is the trace of a holonomy and that the traces are cyclic, $\Psi(\gamma_1 \circ \gamma_2) = \Psi(\gamma_2 \circ \gamma_1)$.

- There is a nontrivial identity satisfied between traces of $SU(2)$ matrices (other groups have similar identities, although there are different from the specific one for $SU(2)$), $\text{Tr}(A)\text{Tr}(B) = \text{Tr}(A \cdot B) + \text{Tr}(A \cdot B^{-1})$, where A, B are $SU(2)$ matrices. This is usually called the Mandelstam identity. In terms of the Loop Representation, this means we can express any wavefunctional of two loops $\Psi(\gamma_1, \gamma_2)$ in terms of wavefunctionals of one loop by $\Psi(\gamma_1, \gamma_2) = \Psi(\gamma_1 \circ \gamma_2) + \Psi(\gamma_1 \circ \gamma_2^{-1})$. From now on therefore, we need only concentrate on wavefunctionals of a single loop, the multiple loop cases always being reducible to the single loop instance[5].

Unfortunately, the story does not end here. Many nontrivial combinations of these identities are possible, for example the following identity holds for three loops,

$$\Psi(\gamma_1 \circ \gamma_2 \circ \gamma_3) + \Psi(\gamma_1 \circ \gamma_2 \circ \gamma_3^{-1}) = \Psi(\gamma_2 \circ \gamma_1 \circ \gamma_3) + \Psi(\gamma_2 \circ \gamma_1 \circ \gamma_3^{-1}) \qquad (56)$$

(and usually it takes some time to figure out exactly how to derive it from the above identities). The situation only gets worse for higher numbers of loops. An important point, however, is that any wavefunction on loop space that one wants to define (any wavefunction must have some value for a given number of loops) should be consistent with these identities.

---

[5]The Mandelstam identity obviously holds only for a pair of intersecting loops γ_1, γ_2. We are considering loops that share a common basepoint, that is why we can always reduce a wavefunction to that of one loop

There is a very important aspect of the Loop Representation for Quantum Gravity. Since the theory is invariant under diffeomorphisms, this means that the wavefunctionals will be invariant under deformations of the loops. This means the wavefunctionals will be what in the mathematical literature is known as *knot invariants* [38]. Knot theory is the branch of mathematics that studies properties of knot invariants and it has enjoyed a great deal of activity recently. It is really exciting that this newly flourishing branch of mathematics seems to have something to do with Quantum Gravity, since it opens the opportunity for new insights into the field. We will return to these issues at length in the next chapter.

5.2 A more rigorous approach

We said that the introduction of a loop representation via a transform is only a motivational approach since one actually does not know how to perform the functional integral appearing in the transform. It turns out there is a more rigorous way of introducing a loop representation, and this is by quantizing a noncanonical algebra of loop-dependent quantities. This approach was followed for gravity by Rovelli and Smolin in their original article [38]. We will not discuss it in great detail here for reasons of space. The main idea is the following:

- Define a set of classical quantities in the phase space of General Relativity based on loops, called generically T variables. $T_\gamma^0[A]$ is the Wilson Loop, T^1 is a vector density obtained by inserting in the expression of the Wilson loop a triad at a given point $T^a(x)_\gamma[A] = Tr(U(0, s(x))\tilde{E}^a U(s(x), 1))$, T^2 is a two-vector density obtained by inserting in a Wilson loop triads at two points and so on for higher order T's. It is possible, by shrinking the loops to points, to represent any classical quantity in terms of these variables. In particular, one can write the constraints. These quantities close a noncanonical algebra under Poisson Brackets.

- Promote the T variables to a set of operators acting on a space of wavefunctions. The action is obtained by mimicking the action under Poisson brackets of the classical T variables with the Wilson loop. For example, $\hat{T}_\gamma^0 \Psi(\eta) = \Psi(\gamma \circ \eta)$ and so on. The quantum operators close an algebra under commutators that reproduces, in the limit $\hbar \to 0$ the classical T algebra.

- Promote any quantity one is interested in (for example the constraints) to quantum operators simply by writing its classical expression in terms of the T variables and then promoting them to operators by the rules given above.

There are subtleties in the definition of the quantum algebra of T variables that have led to the consideration of strips instead of loops to avoid some regularization issues. See Rovelli [42] for details.

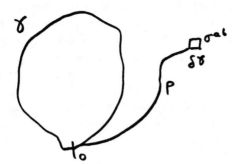

Figure 2: The infinitesimal loop that gives rise to the Loop Derivative

5.3 Differential operators in loop space

If one accepts that a representation of Quantum Gravity exists in which wavefunctions are functionals of loops, the next question is how to represent the constraint equations. To this end we will introduce a differential operator in Loop Space, the loop derivative. Given a functional of a loop $\Psi(\gamma)$, the loop derivative is defined by,

$$\Psi(\gamma \circ \delta\gamma) = (1 + \sigma^{ab}\Delta_{ab}(P))\Psi(\gamma) \tag{57}$$

where $\delta\gamma$ is an infinitesimal loop added at the end of the path P as indicated in the figure 2. σ^{ab} is the area of the infinitesimal loop. The prescription to compute it is, take the wavefunction evaluated on a loop constructing by appending to the original loop an infinitesimal loop $\delta\gamma$ at the end of a path P starting from the basepoint; subtract the wavefunction evaluated on the original loop, and divide the result by the area element σ^{ab}. The result is the loop derivative of the wavefunction. The loop derivative is present in the work of Mandelstam [43], Polyakov [44] and Makeenko and Migdal [45]. The work of Gambini and Trias really brought to the forefront the role of this operator in connection with gauge theories [40].

A fundamental property that can be immediately checked applying the above prescription is,

$$\begin{aligned}\Delta_{ab}(\gamma_0^x)W(\gamma, A) &= Tr(U(0, s(x))F_{ab}(x)U(s(x), 1)) \\ &= F_{ab}^i Tr(U(0, s(x))\tau^i U(s(x), 1)).\end{aligned} \tag{58}$$

As usual, one should be careful if such a point lies at an intersection. However, in most practical situations, intersections are regulated and derivatives act away from them and are evaluated at the intersection only in the limit.

5.4 The diffeomorphism constraint

In terms of the loop derivative it is straightforward to write the generator of infinitesimal diffeomorphisms. Although we could proceed from geometric considerations, let us derive it from the New Variable formulation. Let us consider the generator of diffeomorphisms acting on both sides of the transform (55),

$$\hat{C}(\vec{N})\Psi(\gamma) = \int dA\, W(\gamma, A) \int d^3x N^a(x) \frac{\delta}{\delta A^i_b(x)} F^i_{ab}(x)\Psi(A) \qquad (59)$$

we now integrate by parts and apply the constraint on the Wilson Loop,

$$\hat{C}(\vec{N})\Psi(\gamma) = \int dA \int d^3x N^a(x) F^i_{ab}(x) \frac{\delta}{\delta A^i_b(x)} W(\gamma, A)\Psi(A) \qquad (60)$$

We now functionally differentiate the Wilson Loop and use the equations (46) and (58),

$$F^i_{ab}(x) \frac{\delta}{\delta A^i_b(x)} W(\gamma, A) = F^i_{ab}(x)\dot{\gamma}^b(s(x)) Tr(U(0, s(x))\tau^i U(s(x), 1)) = \qquad (61)$$

$$= \dot{\gamma}^b(x) Tr(F_{ab}(x) U(x, x)) = \dot{\gamma}^b(x) \Delta_{ab}(x) W(\gamma, A) \qquad (62)$$

Therefore we can now replace this in the expression of the constraint to find,

$$\hat{C}(\vec{N}) = \int d^3x N^a(x) \oint ds \delta(x - \gamma(s))\dot{\gamma}^b(s) \Delta_{ab}(x) \qquad (63)$$

Whenever we drop the dependence in the path of the loop derivative, we assume the path goes from the basepoint of the loop to the point of interest. The expression we arrived to is known to be the generator of infinitesimal deformations of the loops. What we see above is what we anticipated at the end of the last subsection, that the wavefunctions will have to be invariant under smooth deformations of the loops, they have to be *knot invariants*.

An important point if we are really to consider the above expression as a diffeomorphism constraint is if it reproduces the constraint algebra of diffeomorphisms. The calculation is possible, though we will not perform it here for reasons of space. The interested reader can consult [40, 46].

5.5 The Hamiltonian constraint

We can now use the same kind of reasoning to determine the action of the Hamiltonian constraint. Naturally, it will be more complicated. Again for reasons of space we will not perform a derivation here. Careful derivations can be found in [47, 48]. The resulting expression is,

$$\hat{H}(x)\Psi(\gamma) = \oint_\gamma ds \oint_\gamma dt\, \delta(x - \gamma(s)) f_\epsilon(\gamma(s), \gamma(t)) \times \qquad (64)$$

$$\times \dot{\gamma}^a(s)\dot{\gamma}^b(t)\Big(\Delta_{ab}(s)\Psi(\gamma_{st} \circ \gamma_{s0t}) + \Delta_{ab}(s)\Psi(\gamma_{ts} \circ \gamma_{t0s})\Big) \qquad (65)$$

This expression requires some explanation. Conceptually it can be viewed as "$\dot{\gamma}^a \dot{\gamma}^b \Delta_{ab}$". However, several details should be taken into account. First of all,

166 Knot Theory and Quantum Gravity

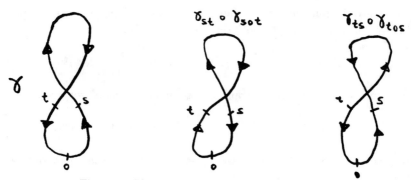

Figure 3: The loops γ, $\gamma_{st} \circ \gamma_{s0t}$ and $\gamma_{ts} \circ \gamma_{t0s}$.

the expression is regulated with a regulator satisfying $f_\epsilon(x,y) \to \delta(x-y)$ when $\epsilon \to 0$. This means the two tangent vectors effectively are evaluated at the same point. Since they are contracted with the loop derivative, which is antisymmetric, this means that the only possibility for the constraint to be nonvanishing is if it acts at an intersection. This fact we already encountered in the connection representation. The parameter values s and t can therefore be thought as slightly displaced from an intersection (which lies at x), but tend to it as we remove the regulator $\epsilon \to 0$. There is a rerouting in the loops, indicated in the wavefunctions. For instance, this should be read in the following way; $\Psi(\gamma_{st} \circ \gamma_{t0s})$ should be understood as the loop composed by traversing the original loop from s to t first and then from t to s, but going through the basepoint 0. This is depicted in the figure 3. The reader may find this expression technically difficult. However, it should be kept in mind that this expression embodies all the dynamical information of the General Theory of Relativity at a quantum level in loop space. When viewed in this way the above expression appears as remarkably simple!

At this point the reader may be puzzled about the kind of derivations we have performed for the constraints. We have made use of the loop transform (55) as if it were an actually well defined expression, instead of the motivational tool we claimed it to be in the introduction. There actually is a way of deriving the form of the constraints in the loop representation without making use at all of the transform (55). One takes the second viewpoint we described in connection with the construction of the loop representation in section 5.2: its elaboration based on a quantization of a noncanonical set of loop-dependent classical quantities (the Rovelli-Smolin T operators). The idea is to express the constraints classically in terms of the T's. Then, since one has a well defined way of promoting the T operators to quantum operators on a space of loop functionals, one has also a prescription for promoting the constraints to differential operators in loop space. It is quite reassuring that proceeding in this way and proceeding via the transform as we detailed above, actually leads to the same results for both the Hamiltonian and diffeomorphism constraints

[48].

5.6 Coordinates on loop space

We have established analytic expressions for the constraints of Quantum Gravity in the Loop Representation. We now develop technology that will allow us to write wavefunctions. We will discuss specifically the construction of wavefunctions in the next chapter. Here we will just set up the framework that will be used.

Let us return for a minute to the connection representation. We are interested in wavefunctions that are $SU(2)$ invariant. Modulo technicalities, all such functions can be expressed as combinations of Wilson Loops [79, 22]. Therefore we can just concentrate on these latter ones. Let us write the expression for a Wilson Loop explicitly,

$$W(\gamma, A) = 2 \qquad (66)$$
$$+ \sum_{i=1}^{\infty} \text{Tr}[\oint ds_1 \int_0^{s_1} ds_2 ... \int_0^{s_{n-1}} ds_n \dot{\gamma}^{a_1}(s_1)...\dot{\gamma}^{a_n}(s_n) A_{a_1}(\gamma(s_1))...A_{a_n}(\gamma(s_n))]$$

We now rearrange this expression in the following way,

$$W(\gamma, A) = 2 + \sum_{i=1}^{\infty} \int d^3x_1 ... \int d^3x_n \text{Tr}(A_{a_1}(x_1)...A_{a_n}(x_n)) \times \qquad (67)$$
$$\times \oint ds_1 ... \oint ds_n \Theta(s_1,...,s_n) \delta^3(x_1 - \gamma(s_1))...\delta^3(x_n - \gamma(s_n)) \dot{\gamma}^{a_1}(s_1)...\dot{\gamma}^{a_n}(s_n) \qquad (68)$$

where $\Theta(s_1,...s_n) = 1$ if $s_1 < ... < s_n$ is a generalized Heaviside Function. The purpose of this rearrangement is to separate the contributions of the loop and the connection to the Wilson Loop. This allows us to write,

$$W(\gamma, A) = 2 + \sum_{n=1}^{\infty} \text{Tr}(A_{a_1 x_1}...A_{a_n x_n}) X^{a_1 x_1 ... a_n x_n}(\gamma) \qquad (69)$$

where,

$$X^{a_1 x_1 ... a_n x_n}(\gamma) = \oint ds_1 ... \oint ds_n \, \dot{\gamma}^{a_1}(s_1)...\dot{\gamma}^{a_n}(s_n) \times$$
$$\times \Theta(s_1,...,s_n) \delta^3(x_1 - \gamma(s_1))...\delta^3(x_n - \gamma(s_n)) \qquad (70)$$

where we have assumed a "generalized Einstein convention" meaning repeated x_i coordinates are integrated over and we treat them as indices. We have isolated all the loop dependence in the quantities X. These quantities behave like multi vector densities (really they are distributions) on the three manifold at the points x_i. They can be viewed as vector densities that have support along the loops associated with the value of the tangent vector to the loop at the point of interest.

The whole point of this construction is that these quantities embody all the information of the loops that is needed to build *any* wavefunction of interest. This

168 Knot Theory and Quantum Gravity

transcends the connection representation and means we can use them in the loop representation. If we are to write functions of the X's as candidates for physical states of gravity, we will need to study the action of the constraints on the X's. We do not have space here to give a detailed account of this, but we will work out an example just to show how this works. The reader should be able to generalize to the other needed cases.

Let us consider the multitangent of order one, X^{ay}, and let us study its loop derivative. We start by the definition (57), appending an infinitesimal parallelogram loop (at the point z) with edges du^a, dv^a to the definition of the X of order one (70),

$$X^{ay}(\gamma \circ \delta\gamma_z) = \int_0^z ds\, \dot\gamma^a(s)\delta(\gamma(s)-x) + du^a\delta(z-x) + dv^a\delta(z+du-x) -$$

$$-du^a\delta(z+du+dv-x) - dv^a\delta(z+dv-x) + \int_z^1 ds\, \dot\gamma^a(s)\delta(\gamma(s)-x) \quad (71)$$

where we now expand to first order assuming du and dv are infinitesimal (for instance, $\delta(z+du-x) = \delta(z-x) + du^a\partial_a\delta(z-x)$, and we also recombine the first and last terms to give a loop integral and get,

$$X^{ax}(\gamma \circ \delta\gamma_z) = X^{ax}(\gamma) + (du^a dv^b - dv^a du^b)\partial_b\delta(x-z). \quad (72)$$

We now read off from the definition of the loop derivative (57), taking into account that the area element of the loop in question is given by $d\sigma^{ab} = du^a dv^b - dv^a du^b$,

$$\Delta_{cb}(z)X^{ax}(\gamma) = \delta^a_{[c}\partial_{b]}\delta(x-z) \quad (73)$$

Similar computations can be carried along for the higher order X's, for instance for a second order one,

$$\Delta_{cd}(z)X^{axby} = X^{by}\delta^a_{[c}\partial_{d]}\delta(x-z) + X^{ax}\delta^b_{[c}\partial_{d]}\delta(y-z) + \delta^a_{[c}\delta^b_{d]}\delta(x-z)\delta(x-y) \quad (74)$$

Generic formulae for an X of any order are given in [49].

The reader may be surprised by the fact that we use the word "coordinates" to describe the multitangents. A "coordinate" should be an object that one prescribes freely. How do we know we can freely prescribe the X's to any order? Actually we cannot. The X's satisfy a series of identities, both algebraic and differential. Examples of these identities are,

$$\partial_a X^{ax} = 0 \quad (75)$$

$$\partial_{ax}X^{ax\,by} = \delta(x-y)X^{by} \quad (76)$$

$$X^{ax\,by} + X^{by\,ax} = X^{ax}X^{by} \quad (77)$$

The algebraic identities, like the last one, stem from identities of the generalized Heaviside Function. The differential identities ensure that the resulting Wilson loop is gauge invariant. Therefore the X's are not coordinates, since they cannot be freely specified.

A remarkable fact, however, is that one can actually solve the aforementioned identities for the "freely specifiable part" of the loop coordinates (in reference [49], they are referred to as Y's). These objects really work as coordinates in loop space. The importance of this last fact cannot be overstressed. Unfortunately, for reasons of space we are unable to present a thorough account of this construction here. The reader is referred to the literature for more details [49]. We will keep on to loosely refer to the X's as loop coordinates.

Another important point connected with the loop coordinates is that they suggest a way to obtain a quantum representation that goes beyond the loop representation. Consider a generic multivector density on a three manifold, X, not necessarily associated with any particular loop. If this quantity satisfies the family of identities mentioned above, one could use it in expression (69) and obtain as a result, not a Wilson loop anymore, but a gauge invariant function of the connection parametrized by the vector density given. This would allow, for example, to use completely smooth vector densities, that have potential for removing several of the regularization difficulties associated with the loop representation. We call the resulting representation, where wavefunctions are functionals of the coordinates,

$$\Psi(X^1, X^2, ...), \qquad (78)$$

"coordinate representation" (some authors call it "form factor representation"). At least for the Maxwell case it has been proven that this provides a viable quantum representation [33]. For the nonabelian case, work is in progress to assess the feasibility of this representation.

6 Knot Theory

6.1 Knots

Knot theory studies the properties of knots in three dimensions that are invariant under smooth deformations of the knots (diffeomorphisms). It was largely set up by the attempts in the last century by P. G. Tait [50] and others to explain the properties of atoms using knotted vortices of "aether" (this was before special Relativity or Quantum Mechanics were invented. A discussion of several appealing aspects of this theory of atomic spectra can be found in the book by Atiyah [51]). We will use the term "knot" loosely to refer either to a single knotted curve or to several curves linked and/or knotted. A central issue in knot theory is to distinguish and classify inequivalent knots. A useful tool for accomplishing this is the use of knot invariants, that is, numbers associated with knots which are invariant under deformations of the knots. At the moment, however, there is no "canonical" or definite set of invariants that would allow us to completely classify knots. The simplest example of

170 Knot Theory and Quantum Gravity

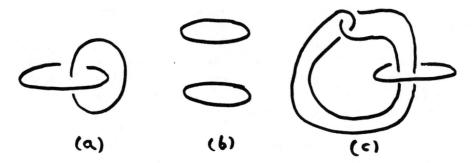

Figure 4: The knots in (b) and (c) (The Whitehead Link) have vanishing linking number.

Figure 5: Upper and under crossings for the definition of the Gauss Linking number

such an invariant is the linking number, first considered by Gauss. Consider two curves as in figure 4, the linking number is one if they are linked (4a) and zero if they are not linked (4b). It is evidently (at least for these cases) invariant under diffeomorphisms. However, how does one compute it in general, say for the curves in 4c. There is a procedure for these cases. One should traverse one of the curves and add 1/2 for every crossing of the type shown in figure 5a and -1/2 for the type 5b. This in particular shows that the curves in figure 4c have vanishing linking number in spite of being obviously linked (this is called the Whitehead link). This clearly shows that we will need other, more complicated, invariants to be able to classify linkings.

It should be apparent to the reader that these concepts are both elegant and clearly related to the issues of Quantum Gravity we discussed in the previous chapter. However, they do not seem quite geared to a direct application. Say we wanted to consider the Gauss linking number as a candidate for a wavefunction of the gravitational field in the loop representation (this actually fails since it is not consistent with the Mandelstam identity). We meet an important requirement in the fact that it is diffeomorphism invariant. But in the form we have casted it up to now, we cannot compute much further. For instance, we do not know how to take an area derivative of this quantity[6].

---

[6] Actually another technical problem appears here, in the fact that the area derivative of dif-

It would be convenient to have an analytic expression for the knot invariants. For the Gauss Linking number there actually is one. It is easy to see (for a demonstration see Maxwell's 1873 Treatise on Electricity and Magnetism! Volume II page 419-423) that the following expression,

$$\mathrm{GL}(\gamma_1,\gamma_2) = \tfrac{1}{4\pi} \oint_{\gamma_1} ds \oint_{\gamma_2} dt\ \dot{\gamma}_1{}^a(s)\dot{\gamma}_2{}^b(t)\epsilon_{abc}\frac{(\gamma_1^c(s) - \gamma_2^c(t))}{|\gamma_1(s) - \gamma_2(t)|^3} \qquad (79)$$

gives an analytic expression for the Linking number. A very exciting fact of the recent work on Chern-Simons theories is that it has provided similar analytic expressions for several other knot invariants.

We can actually write the above invariant in terms of the loop coordinates we introduced in the last chapter. The expression is,

$$\mathrm{GL}(\gamma_1,\gamma_2) = \tfrac{1}{4\pi} X^{ax}(\gamma_1) X^{by}(\gamma_2) \epsilon_{abc} \frac{(x^c - y^c)}{|x - y|^3} \qquad (80)$$

where we again assume that repeated x,y indices are integrated over the whole manifold. In spite of its appearance (it involves a local chart of coordinates and even a distance!) this expression is actually diffeomorphism invariant. The fact that we can express it in terms of the loop coordinates enormously facilitates the application of the constraints to these expressions. For instance, one can easily prove that the diffeomorphism constraint actually annihilates $\mathrm{GL}(\gamma_1,\gamma_2)$, in accordance with the fact that this expression is diffeomorphism invariant. (We will see soon what happens when one applies the Hamiltonian constraint).

There are many other things to be said about knot theory. We will stop here for reasons of space. The reader is encouraged to enjoy the books by Lou Kauffman [52, 53] on the subject for many other amusing and important properties of knots.

6.2 Knot Polynomials

A very important step towards the construction of knot invariants, and indirectly towards the classification problem, has been the invention of the knot polynomials. These are polynomials in an arbitrary variable, usually called t, uniquely associated with a given knot. For each knot there is a given finite order polynomial, although the order may be different for another knot. The important point is that the polynomials are invariant under diffeomorphisms, therefore each coefficient is a knot invariant. Examples of such polynomials are those of Alexander-Conway, The Kauffman Bracket, Jones and HOMFLY. The work on Chern-Simons theories has given rise to even newer polynomials [54] and to previously unknown relationships among the known ones.

---

feomorphism invariant quantities is rather ill defined, being on a similar standing to the derivative of a Heaviside function, since diffeomorphism invariant functions do not change smoothly under addition of an infinitesimal loop

Knot polynomials are usually prescribed by a set of implicit relations, known as skein relations. These relations are enough to find out the particular polynomial associated with a given knot (sometimes the task is not totally trivial!). Let us see an example. Consider the Conway Polynomial $C(\gamma)[t]$, defined by the Skein Relation that appears in figure 6.

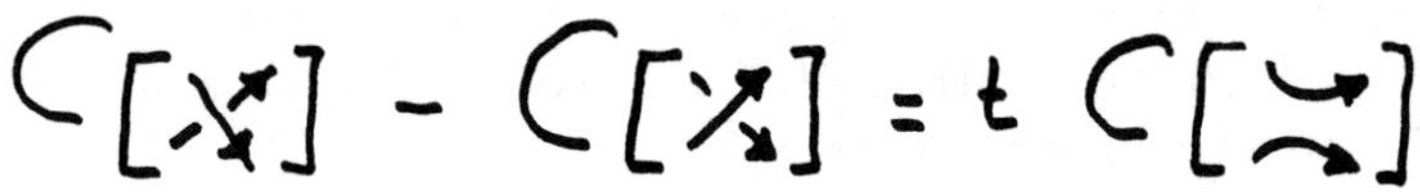

$$C[\times] - C[\times] = t\, C[\smile]$$

Figure 6: Skein relation for the Conway Polynomial

This relation should be interpreted in the following way: for a given knot, project it on a plane and focus on a single crossing. Cut out the crossing and leave four incoming threads. The value of the polynomial for the knot with the crossing drawn glued into the threads left out is related to that of the polynomial with the other crossings drawn via the skein relation. This plus the fact that the polynomial on the unknot is normalized to one is enough for computing the polynomial for an arbitrary knot.

Let us work out an example, evaluating the Conway Polynomial for the trefoil knot given in figure 7. We focus on the crossing encircled by the dotted line and apply the Skein Relation of figure 6. We see in figure 8 that it relates the value of the polynomial evaluated on the knot of interest (our final result) to the value of the polynomial for two other knots. For the left one, one can immediately see that since the knot is homotopic to the unknot, the value of the polynomial on it is 1. For the link on the right we need to do some more work. We again apply the skein relation focusing on the encircled crossing, to evaluate the value of the polynomial on the link of interest. Again we see in figure 8 that one of the polynomials is one and the other is evaluated on two unlinked curves. Applying again the skein relation we can see that the value of the polynomial on such a link is vanishing, Therefore, substituting back, one sees that the value of the Conway Polynomial for the knot of interest is $1 + t^2$.

So we see that the innocent looking skein relations actually contain all the information needed to associate a series of knot invariants to a knot (the reader can verify that by deforming the knot in question and applying the skein relations, one gets the same result).

Again, having a knot polynomial written as a skein relation is not very useful for our purposes. We would like to have a more analytic kind of expression for the knot polynomials to, say, apply the Hamiltonian constraint of Quantum Gravity to it and see if it is a quantum state of the gravitational field. We will see in the

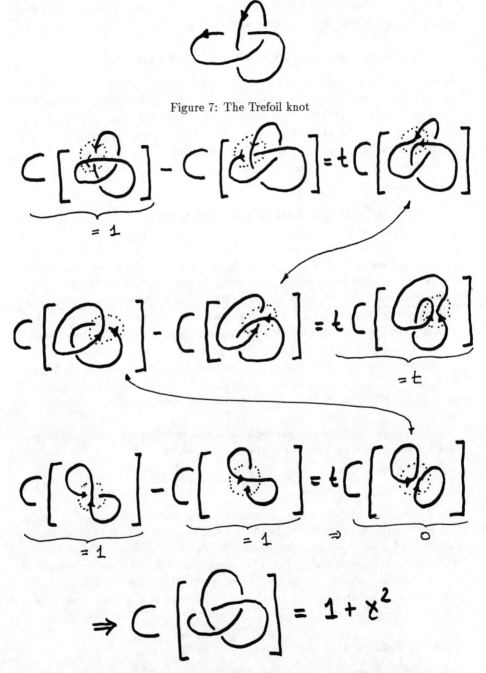

Figure 7: The Trefoil knot

Figure 8: Evaluation of the Conway Polynomial for the trefoil knot

following subsections that Chern-Simons theory has been extremely useful to find such analytic expressions for the knot polynomials.

6.3 Chern-Simons theory and knot polynomials

As the reader must have noticed from previous sections, Chern-Simons theories seem to play a crucial role concerning Knots. Dozens of articles have been written studying various aspects of these theories and this brief subsection can certainly not be anything else but a gross oversimplification. The reader interested in gaining a good understanding is encouraged to explore the relevant papers.

In a nutshell, Chern-Simons theories are gauge theories defined in 2+1 dimensions having as action,

$$S_{CS} = k \int d^3x \; \mathrm{Tr}(A_a \partial_b A_c + \tfrac{2}{3} A_a A_b A_c)\epsilon^{abc}. \tag{81}$$

where the connection A takes value on some group, let us fix for our interests $SU(2)$. k is the coupling constant of the theory. Notice that no use of a spacetime metric was needed to write this action (as one needs, for instance to write the Yang Mills action when one raises or lowers indexes in $F_{ab}F^{ab}$). This is therefore a topological field theory (although what is meant by this is quite context dependent). It is invariant under diffeomorphisms. The classical equations of motion for these theories simply state that the theory is $SU(2)$ invariant and that the connection is flat (F_{ab} constructed out of A_a is zero). Wilson Loops are actually observables in the theory. Since the connection is flat, when one deforms the loop with a diffeomorphism, the value of the Wilson loop does not change. Notice that this does not happen for gravity.

Since the Wilson loop is an observable, one can ask what is its quantum expectation value. In the language of Path Integrals,

$$< W(\gamma) >= \int dA \; e^{iS_{CS}} \; W(A,\gamma) \tag{82}$$

Now this expectation value should be invariant under diffeomorphisms, i.e. it should be a knot invariant, parametrized by the coupling constant k. In the following subsubsections we will explore in some detail what sort of knot invariant this quantity is. We will first derive a skein relation for it and then, taking advantage of the fact that perturbation theory can be used to evaluate (82) (remember this is Chern-Simons theory, not Quantum Gravity!) we will present an analytic expression for the knot invariant.

6.3.1 A skein relation for $< W(\gamma) >$

It turns out that one can prove that $< W(\gamma) >$ satisfies the Skein relations of the knot polynomial known as Kauffmann Bracket, which is intimately related to the

Jones Polynomial. This was an important discovery due to Witten [55]. Witten obtained his result nonperturbatively by recurring to conformal field theories. It turns out that the result can be obtained also in a perturbative fashion with a more modest machinery [56, 57]. The result also holds when the loops have intersections [25]. We here sketch part of the perturbative proof in order to give the reader a flavor of these calculations and also to illustrate the usefulness in this context of some of the techniques we introduced in section 5.

In order to establish a skein relation for the expectation value of the Wilson loop we basically need to relate an under and an upper crossing. This can be done since, as illustrated in figure 9, one can obtain an upper (or under, depending on the orientation of the small loop) crossing by adding a small loop at a given point of the loop. It turns out that we have already developed a technique for evaluating the

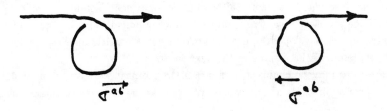

Figure 9: Under and upper crossings created by adding an oriented small loop

change in a function of a loop when one adds a small loop: it is the loop derivative. Therefore what we basically need to compute is the loop derivative of $<W(\gamma)>$.

The change in $<W(\gamma)>$ due to the addition of a small loop of area σ^{ab} is given by formula (57). When combined with formula (58), it gives rise to,

$$\sigma^{ab}\Delta_{ab}(x)\Psi[\gamma] = \int dA \ \sigma^{ab} F^i_{ab}(x) \ \text{Tr}[\tau^i U(\gamma^x_x)] \exp(S_{CS}). \tag{83}$$

Using the relation (52), integrating by parts, and applying (47) one obtains:

$$2k \int dA \ \sigma^{ab} \epsilon_{abc} \int dy^c \ \delta(x-y) \text{Tr}[\tau^i U(\gamma^y_x)\tau^i U(\gamma^x_y)] \exp(S_{CS}) \tag{84}$$

The integral depends on the volume factor

$$\sigma^{ab}\epsilon_{abc}dy^c\delta(x-y) \tag{85}$$

which depending on the relative orientation of the two-surface σ^{ab} and the differential dy^c (which is tangent to γ), can lead to ± 1 or zero. (This expression should really be regularized. We have absorbed appropriate extra factors in the definition of the coupling c constant so to normalize the volume to ±1). Consequently, depending on the value of the volume there are three possibilities:

$$\delta \Psi[\gamma] = 0 \qquad (86)$$
$$\delta \Psi[\gamma] = \pm 2k\Psi[\gamma] \qquad (87)$$

These equations can be diagrammatically interpreted in the following way:

$$\Psi[\hat{L}_\pm] - \Psi[\hat{L}_0] = \pm 2k\Psi[\hat{L}_0] \qquad (88)$$

and coincide with part of the skein relations of a known knot polynomial, the Kauffman Bracket [52, 53]. So we see that $< W(\gamma) >$ is actually an analytic expression for the Kauffman Bracket in the variable k.

It is interesting to notice how helpful, in order to perform this calculations, were the notion of an area derivative and of its properties. The original derivations [57] did not use these concepts (although the treatment is fully equivalent) and therefore the proof we give here is much more economical. This is a good example where techniques developed for gravity are of use in a particle physics problem. We will see more of this happening in the next subsection.

The reader may be confused by figure 9. In the past we have considered these kinds of "curls" in knots as removable. We will see the meaning of this in the next section.

6.3.2 Perturbative calculation of the Kauffman Bracket

Chern-Simons theory being a renormalizable theory, one could compute the expectation value of the Wilson Loop perturbatively. One gets as a result a polynomial in the variable k (the coupling constant of the theory), which should provide analytic expressions for the coefficients of the Kauffman Bracket. In this language therefore a coefficient of the Kauffman Bracket becomes a sum of Feynmann diagrams for the perturbative expansion of $< W(\gamma) >$. We will sketch here the derivation. The reader should be aware that the proofs given here are very schematic. They ignore, for instance, the presence of ghosts (it can actually be seen that ghosts do not contribute to the order of perturbation we are going to discuss). The complete treatment can be seen in [58] and a rigorous mathematical derivation in [59].

In terms of the expression of the Wilson loop written as a function of the Loop Coordinates (69), we can write for the expectation value,

$$< W(\gamma) > = \quad 2 + < A_{ax}A_{by} > X^{ax\,by}(\gamma) + < A_{ax}A_{by}A_{cz} > X^{ax\,by\,cz}(\gamma) +$$
$$+ < A_{ax}A_{by}A_{cz}A_{dw} > X^{ax\,by\,cz\,dw}(\gamma) + \ldots \qquad (89)$$

which corresponds to a diagrammatic expansion given in figure 10, where we have represented the propagator by a wavy line and the loop coordinates of nth order as a circle with n insertions of propagators.

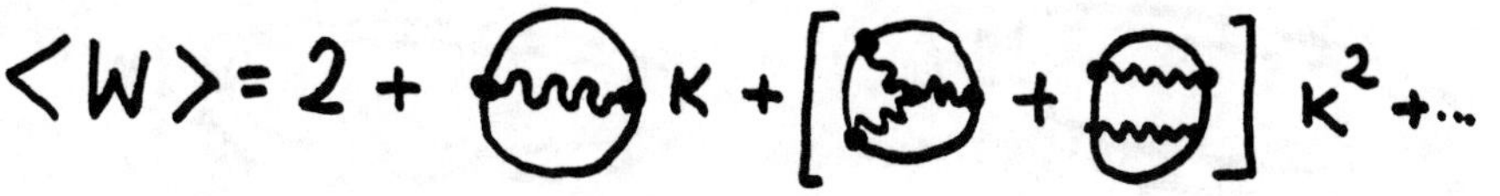

$$\langle W \rangle = 2 + \bigcirc\!\!\!\!\!\sim\!\!\!\sim\!\!\bigcirc \; K + \left[\bigcirc\!\!\!\!\!\Join\!\!\bigcirc + \bigcirc\!\!\!\!\!\bowtie\!\!\bigcirc \right] K^2 + \ldots$$

Figure 10: Diagrammatic expansion of the Wilson Loop in a Chern-Simons theory

The Chern-Simons theory has a propagator given by,

$$g_{axby} = k \, \epsilon_{abc} \frac{(x-y)^c}{|x-y|^3} \tag{90}$$

and a vertex (contracted with three propagators),

$$h_{axbycz} = k^2 \int d^3w \, \epsilon^{def} \, g_{axdw} \, g_{byew} \, g_{czfw} \tag{91}$$

One simply has to contract them with the propagators and vertices to get the expression for the Feynmann diagram.

The Feynmann diagram of first order in k is therefore simply given by,

$$< W(A,\gamma) >^{(1)} = X^{axby} g_{axby} \tag{92}$$

If we expand this out, remembering the expressions for X and g we get,

$$< W(A,\gamma) >^{(1)} = \oint_\gamma ds \oint_\gamma dt \dot{\gamma}^a(s) \dot{\gamma}^b(t) \epsilon_{abc} \frac{(\gamma^c(s) - \gamma^c(t))}{|\gamma(s) - \gamma(t)|^3} = \text{GSL}(\gamma) \tag{93}$$

The reader may recognize in this expression the Gauss Linking number, except for the fact that instead of having two knots, we have only one. We are actually computing the linking of the knot with itself! This sometimes is called "Gauss self-linking number" and we denote it as GSL(γ). This presents a small difficulty. The expression seems to be ill defined when $s = t$ since the denominator vanishes. This is actually not true since the numerator also vanishes, and faster. But there is a problem with this expression, if one computes it carefully, one finds out it not only depends on the loop but on the definition of an arbitrary normal vector to the loop [60, 55, 57]. This means that in the calculation diffeomorphism invariance has been broken (mildly). A simple solution for it is to "frame" the loop, i.e. convert it to a ribbon and compute the self-linking number as the linking number of the two loops on the sides of the ribbon. This however is not a univocous prescription since one can add twists to the ribbon and change its value. Moreover, as we see in the

Figure 11: Two different framings for a given loop. In one case the self-linking number (computed as the linking number of the two curves defined by the framing) is zero and in the other +1.

figure 11, two equivalent loops may yield inequivalent ribbons, so diffeomorphism invariance in the loop sense is lost when one generalizes to ribbons. We will see how one can deal with this problem of the loss of diffeomorphism invariance later on.

The order k^2 Feynmann diagram is the sum of two terms. These terms can be rearranged into three contributions. One of them is the square of the self-linking number. The other two terms together give an analytic representation for another well known knot invariant. Details can be seen in reference [57].

$$< W(\gamma) >^{(2)} = k^2 (\text{GSL}(\gamma)^2 + A_2(\gamma)) \qquad (94)$$

where A_2 is related to the second coefficient of the Conway Polynomial (it is actually $(A_2 + \frac{1}{12})$ and is also related to the second coefficient in an expansion of the Jones Polynomial. From now on we will loosely refer to it as "the second coefficient of the Jones Polynomial". Notice that this invariant is truly diffeomorphism invariant, i.e. it is framing-independent.

This construction goes on at higher orders, each coefficient of the Kauffman Bracket breaks up into a framing dependent portion function of the coefficients of lower order plus a new knot invariant, which is framing independent. The third order expression in k is [61],

$$< W(\gamma) >^{(3)} = k^3 (\text{GSL}(\gamma)^3 + \text{GSL}(\gamma) A_2(\gamma) + A_3(\gamma)) \qquad (95)$$

where $A_3(\gamma)$ is another framing independent knot invariant related to the third coefficient in the expansion of the Jones Polynomial. We do not have space to discuss it here, but crucial in the recombination of the terms to forming expressions of the type (95) is the use of the free part of the loop coordinates, discussed in section 5.5. If one uses the free parts of the loop coordinates it is immediate to find expressions for the framing independent knot invariants that appear at each order. Although this decomposition can be done "by hand" (and that was the way it was done in ref. [62]), it is much more economical to perform it using the loop coordinates. At order three and higher, the use of loop coordinates is almost mandatory due to the complexity of the expressions involved, and actually the first calculation of the third order terms was done in that way. The reader is referred to references [49, 61] for more details on the use of loop coordinates to generate the expansions discussed.

Therefore, as a consequence of this analysis we have the following analytic expression for the Kauffman Bracket,

$$\text{Kauffman Bracket}(\gamma)[k] = 1(\gamma) + \text{GSL}(\gamma)k + (\text{GSL}(\gamma)^2 + A_2(\gamma))k^2 + \quad (96)$$
$$+ (\text{GSL}(\gamma)^3 + \text{GSL}(\gamma)A_2(\gamma) + A_3(\gamma))k^3 + ...$$

What can we do about the framing problem? Clearly we cannot use the Kauffman Bracket as a candidate for a state of gravity because it is not invariant under deformations of the loops (remember figure 11). However, we notice that "buried" inside each coefficient of the Kauffman Bracket is present a framing independent knot invariant (a quantity that is really invariant under deformations of the loops). What we are seeing is the perturbative emergence of the exact relation, known to mathematicians,

$$\text{Kauffman Bracket}(\gamma)[k] = e^{(k\text{GSL}(\gamma))} \text{JonesPolynomial}(\gamma)[k] \quad (97)$$

so we see that all the framing dependence can be concentrated in the "phase factor" $\exp(k\text{GSL}(\gamma))$. We will see that this result allows us to construct real diffeomorphism invariant states of gravity in the next section.

7 Knot theory and quantum states of gravity

The reader may seem surprised by the rather lengthy detour into Chern-Simons theory in the last section. The reason for this will become apparent here.

In section 4.3 we saw that there existed an exact solution to all the constraints of Quantum Gravity in the connection representation (with cosmological constant) given by,

$$\Psi_\Lambda^{CS}(A) = \exp(-\tfrac{6}{\Lambda} \int \tilde{\epsilon}^{abc} Tr[A_a \partial_b A_c + \tfrac{2}{3} A_a A_b A_c]). \quad (98)$$

An interesting question would be: what is the counterpart of this state in the loop representation? In general such a question goes unanswered, since we do not how to perform the integral in the loop transform,

$$\Psi_\Lambda^{CS}(\gamma) = \int \text{``}dA\text{''} W(A,\gamma) \Psi_\Lambda^{CS}(A). \quad (99)$$

However, the reader may notice that if one replaces in (99) the value for $\Psi_\Lambda^{CS}(A)$ given by (98), one gets back the expression for the expectation value of a Wilson loop in a Chern Simons theory we derived last section!,

$$<W(\gamma)> = \int dA \, e^{iS_{CS}} W(A,\gamma) \qquad (100)$$

where the role of the coupling constant of the theory k of last section is now played by $\frac{6}{\Lambda}$. So we see that for the particular wavefunction (98) we can actually compute the transform into the loop representation, and we already know the answer, it is the Kauffman Bracket!. Again we should stress that we cannot consider this knot polynomial strictly as a state of Quantum Gravity since it is not diffeomorphism invariant due to the issue of framing. It is still remarkable that we can find an analogue of state (98) in the loop representation.

If all the formalism works, the Kauffman bracket should be a solution of the Hamiltonian constraint of Quantum Gravity with a cosmological constant in the loop representation. Can we check this fact? We actually can. That is what all the technology of the loop representation we developed in section 5 is good for. We have a wavefunction in the loop representation (the Kauffman Bracket), we can write it in terms of the loop coordinates, (as we saw in the last section) and therefore we can apply to it the constraints of Quantum Gravity to see if it is a solution.

Let us therefore apply the Hamiltonian constraint of Quantum Gravity (with a cosmological constant) in the Loop Representation $\hat{H}_\Lambda$, to the Kauffman Bracket $\Psi_\Lambda^{CS}(\gamma)$,

$$\hat{H}_\Lambda \Psi_\Lambda^{CS}(\gamma) = (\hat{H}_0 + \Lambda \widehat{\det q})(1(\gamma) + \Lambda \, \text{GSL}(\gamma) + \Lambda^2(\text{GSL}(\gamma) + A_2(\gamma)) + \ldots) \quad (101)$$

where $\hat{H}_0$ is the vacuum Hamiltonian constraint.

The result of this calculation is a polynomial in Λ. If it is to vanish, it should do so order by order in Λ. This leaves us with the following equations,

$$\Lambda^0: \quad \hat{H}_0 \, 1(\gamma) = 0 \qquad (102)$$

$$\Lambda^1: \quad \widehat{\det q} \, 1(\gamma) + \hat{H}_0 \, \text{GSL}(\gamma) = 0 \qquad (103)$$

$$\Lambda^2: \quad \widehat{\det q} \, \text{GSL}(\gamma) + \hat{H}_0 \, \text{GSL}(\gamma)^2 + \hat{H}_0 \, A_2(\gamma) = 0 \qquad (104)$$

and so on for higher orders.

Equation (102) trivially holds, since the area derivative in the expression of $\hat{H}_0$ (65) annihilates the constant. Equation (103) is a bit harder to check, but one can actually show that it holds with minor effort.

Something really interesting happens at order Λ^2, since the terms $\widehat{\det q} \, \text{GSL}(\gamma) + \hat{H}_0 \, \text{GSL}(\gamma)^2$ cancel among themselves (this calculation is rather lengthy). That means that for the equation to hold at this order it must happen that,

$$\hat{H}_0 \, A_2(\gamma) = 0 \qquad (105)$$

That is, the second coefficient of the Jones Polynomial has to be a solution of the vacuum Hamiltonian constraint of Quantum Gravity! Therefore, in order to prove that the Kauffman Bracket is a solution of the Hamiltonian constraint with cosmological constant, it must happen that the second coefficient of the Jones Polynomial has to be a solution of the Hamiltonian constraint without cosmological constant.

Historically, eqs. (103) and (105) were shown to hold previous to this discovery [63, 64]. Actually eq. (105) was quite involved to prove, requiring the use of a complicated computer algebra code. Whereas here we find a very natural argument why it should hold.

What happens to higher orders? At each order a similar decomposition occurs for the coefficients of the Kauffman Bracket. That led us to conjecture [65] that maybe at all orders the same occurred. That is, maybe at all orders "nested" inside the Kauffman Bracket was a state of vacuum Quantum Gravity. Even better, since at each order the portion that is a candidate to be a state of vacuum gravity is a coefficient of an expansion of the Jones Polynomial, we could conjecture that,

$$\hat{H}_0 \text{ Jones}_\Lambda(\gamma) = 0??? \tag{106}$$

Unfortunately, computations to try to prove it get more and more involved for higher orders. There is preliminary evidence of a possible proof to all orders involving heavy use of the loop coordinates, but we are unprepared to report about it here [66].

Notice that this state we have found for the vacuum Hamiltonian constraint is *framing independent*, that is, it is a true knot invariant, and therefore a true state of Quantum Gravity. Therefore the use of framing dependent objects before can be seen as an intermediate artifact in the calculation, as if one used a non-diffeomorphism invariant proof to show that a diffeomorphism identity holds. This would be true for all the conjectured states.

How confident can one be of this result? To put this issue in perspective, we should list the potential points where our argumentation has been weak.

- Framing dependence. As we mentioned, when using the result from Chern-Simons theory to obtain the "loop transform" of the Chern-Simons state, one introduces a framing dependence. Crudely put, this means the "loop transform" of a wavefunction that was invariant under diffeomorphisms fails to be invariant. We do not know how to improve this situation. The framing dependence of the Chern-Simons result is well established and is related to spin-statistics in three dimensions. We can just argue that this result was used as an intermediate result and the final result, that the Jones Polynomial coefficient solves the vacuum constraint, is framing independent. Another possible way out of the framing difficulty would be to abandon loops and work directly in the Coordinate Representation mentioned at the end of section 5.6. One would then lose the connection with knot theory but all the expressions involved could be written as well defined functions of smooth vector densities. Unfortunately, many details have to be worked out before we can really claim this is a solution to the problem.

- Regularization. The Hamiltonian constraint we are using involves a regularization and when we claim that something is annihilated by the constraint we really mean it is annihilated at leading order when the regulator is removed. A more careful study of regularization is in order.

- The measure. When we use the Chern-Simons result for the expectation value of the Wilson loop, implicitly we are assuming that the measure used in Chern-Simons theory to perform the path integral is the same as the one to be used in Quantum Gravity. This is by no means obvious. The measure in the Loop Transform should in the end be related to the reality conditions of the formalism and it is clear that the one in Chern-Simons theory is, prima facie, not taking into account this fact. This is related to the next point.

- Reality. In Chern-Simons theory the connection is real, whereas in Quantum Gravity it is complex. This affects our calculation of the expectation value of the Wilson Loop. Clearly if one allows the connection to be complex, formulae like (55) cease to make sense. The integrals basically fail to converge. Even proofs like the one of the skein relation should be taken with care in the complex case since the quantities involved in the skein relations diverge. At the moment, lacking any control about the reality conditions in the loop representation for Quantum Gravity, there is very little we can say about this point. The only hope is that the correct measure, reflecting the reality conditions, could somehow be analytically connected with the Chern-Simons measure, and therefore the results in Chern-Simons theory could be taken as analytic continuations to the purely real case of the gravitational ones. It is evident that this is just a hope and that we cannot say anything else at present.

Given these reservations about our result, do we have any hope that it is correct? We believe there are some supportive elements, that although far from offering a proof, give some reassurance that our result may hold. They are schematically shown in figure 12, and they can be summarized as follows.

- Constraints in connection representation. These constraints were shown to generate the correct diffeomorphism symmetry of the theory [25] and to formally close the commutator algebra [3].

- Equivalence between constraints in both representations. It was shown both using the transform [47] and based on the T operators of Rovelli and Smolin [48].

- Equivalence between wavefunctions. It was proven using perturbative techniques in loop space [57, 56] even for the intersecting case [25]. It was also proven nonperturbatively [55] and using Feynmann diagrammatics [58].

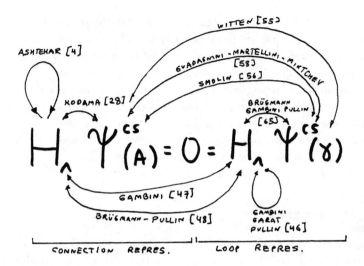

Figure 12: Redundancies in the calculation offer hope that it may be correct

- Constraints in the Loop Representation. Their consistency has been partially proven at the formal level and studies are been done taking into account regularization [46].

All this means that if the result is wrong one or more of the previous results should also be wrong. This could well be. For instance, we are implicitly using the same measure to perform the transform of the constraints and of the states. However, even if the result is wrong, one would learn an important lesson about various aspects of the formalism.

This was the main result we wanted to highlight in these lectures, that using this new formalism for Canonical Quantum Gravity one could find for the first time some nondegenerate physical states of the theory, maybe an infinite family of them. Moreover a new branch of mathematics has been brought into contact with Quantum Gravity, Knot Theory, both at a kinematical level as was emphasized by Rovelli and Smolin[38] but now also at a dynamical level, due to the role of the Jones Polynomial as a state. It is a remarkable fact that there is a connection between General Relativity and Knot Theory at a dynamical level. After all, the Jones Polynomial was developed without taking into account at all the Einstein Equations. This may just be a coincidence or it may mean that the notions of Knot Theory are deeply intertwined with gravity in a way we do not know at present. Will this mean that the Jones Polynomial is a state of *any* theory of gravity one proposes? At present we can just offer this as a conjecture.

Assuming the Jones Polynomial is a state, as conjectured, how general a state can it be? Mathematicians seem to agree that the Jones Polynomial is not enough to solve the main problem of knot theory, the classification of inequivalent knots. That means other invariants are to be found in the future that are more powerful. In this view, one would also expect to find states of Quantum Gravity among them, and therefore the conjectured present family of states would be incomplete. A recent trend in mathematics is to consider Vassiliev invariants as more general invariants to classify knots. It is remarkable that these invariants are defined for loops with intersections, exactly the kind of loops that are relevant for gravity. It seems that our generalization of the Jones Polynomial for loops with intersections [25] does *not* coincide with that used in the theory of Vassiliev invariants. However, a more careful study of this aspect is in order.

8 Final Remarks

Due to space limitations these lecture notes can only be taken as a "tourist brochure" of the subject in question. Many oversimplifications have been introduced that allow the reader to quickly view several important results, but may also obscure a detailed understanding of the topics. We urge the readers who want more than a lax overview to consult the appropriate references. We acknowledge that chronologically this may be the first complete account of these findings that sees the light through publication. We urge the interested readers to pay attention, since in the immediate future more detailed accounts of these topics will be published. We would like to finish by referring to some topics that were not even discussed in the text and making some final remarks on the present status and prospect of the subject. Even here we will have to be unfair and leave unmentioned important topics.

In these lectures we have reviewed basically three things.

- The Ashtekar reformulation of General Relativity.

- Some attempts to construct canonical quantizations using these variables.

- The relation of some results from Chern-Simons theories to the Loop Representation of Quantum Gravity.

The first item is clearly of great importance. The Ashtekar variables are finding new applications in Classical Relativity every day and will certainly become a standard tool of analysis for Relativists *even if attempts to quantize the theory using them fail*. Of the many aspects not even mentioned in these notes concerning this subject, we would like to point out to the reader the following: a) The Capovilla-Dell-Jacobson Lagrangian reformulation of the theory *purely in terms of a connection* [67]. This was a long-cherished dream of many relativists. The work also presents a novel way of solving the constraints which may have implications for the issue of free data of the theory. b) The work of Samuel [68] and Torre [69] that showed how instantons can be transferred from Yang-Mills theory to General Relativity and their stability analysis showing they can actually be a countable number in some cases. c) The application of the variables to Bianchi cosmologies [27, 28, 71], offering a new picture of the classical (and maybe quantum) dynamics of these systems. d) The Newman-Rovelli [70] method for solving the constraint equations using a Hamilton-Jacobi reformulation. d) The possibility that topology change [8] and negative energy [72] may occur in the theory. Summarizing, this is a healthy area of research in which many new and important developments will be studied.

In the second item we count the connection and loop representations. These may or may not succeed in providing a basis for a quantum theory of gravity. Even in the case of failure, it is clear that many lessons have been learnt from their use. We list some important pieces of work on these areas not mentioned in these notes a) The application to two [73] and one [74] killing vector spacetimes, allowing in some cases to find observables in the systems. Recent work also shows that the quantization scheme for one polarization two Killing vector fields may coincide with the usual quantization based on the equivalence with a scalar field [75]. b) The work on 2+1 gravity, which shows for a model system how the connection and loop representations and the loop transform can actually be given a rigorous meaning [32, 76]. b) The application of loop techniques to Gauge theories on the lattice [36, 37], linearized gravity [77] and Maxwell theory [33], again offering test cases where the quantization program works to the end. c) The work on C-P violation [9]. d) The discussion of the issue of time in the linearized theory [29].

An aspect that cannot be overstressed is the development in the loop representation, of techniques for writing differential operators in loop space and to write wavefunctions and knot invariants in analytic form [49]. These findings transcend the area of Quantum Gravity and have immediate application in the quantization of gauge theories (in the continuum and lattice). In fact, we have seen some examples of their application in the sections on Chern-Simons theory. They can also become standard tools of analysis for knot theorists. In fact, understanding in this area is just beginning and we may see even more progress in the near future. Of particular interest is the Coordinate Representation of section 5.2 that could allow for the quantization of diffeomorphism invariant theories without the framing ambiguities of the Loop Representation.

On the final item, we can just say that it is work in progress and that in the end technical difficulties may hamper further development or even disprove the present results. An interesting point seems to be that some of the results seem to survive the inclusion of matter [78]. Another result is that some notions of Knot theory seem to be useful to select an inner product for the theory at least in some toy subsectors [79]. Moreover, the first hint of what a semiclassical interpretation may look like in this context is starting to emerge [18].

Notice that our treatment has evaded the "big questions" of Quantum Gravity, as what is the inner product, the issue of observables and the issue of time. The only comment we can make is that the fact of being able to explore (tentatively) the space of states of the theory may provide a better framework in which to address these problems in the future.

We do not know what will the outcome of this –at the moment– happy marriage of Knot Theory and Quantum Gravity be. As in any other case where a new mathematical technique is introduced into an area of Physics there is potential for striking new results and also for a lot of red herrings. Only time and much more effort will decide which of these two situations we are actually creating with our work.

Acknowledgments

I am most grateful to the organizers, especially Jose Luis Lucio and Octavio Obregón for giving me the opportunity to present my views on these topics. I am also grateful to the faithful audience that endured my lectures in Guanajuato. Most of the original research presented in sections 4 and subsequents was done in collaboration with Rodolfo Gambini (Montevideo) and Bernd Brügmann (Syracuse). Collaborating with them has been a great pleasure and has enriched a lot my view of the subject. I am also indebted by their careful reading of this manuscript. I am also thankful to John Baez, Alejandro Jakubi and Charlie Torre for pointing out mistakes in previous versions of this paper. My understanding (or lack thereof) of the New Variables and the quantization program has been shaped to a great extent through interactions with Abhay Ashtekar, John Baez, Daniel Boyanovsky, Louis Crane, Alan Daughton, Mario Díaz, Cayetano Di Bartolo, Alcides Garat, Joshua Goldberg, Gabriela González, Jorge Griego, Viqar Husain, Ted Jacobson, Lou Kauffman, Karel Kuchař, Renate Loll, Jorma Louko, Pablo Mora, Ted Newman, Octavio Obregón, Richard Price, Joe Romano, Carlo Rovelli, Joseph Samuel, Lee Smolin, Rafael Sorkin, Ranjeet Tate, Peter Thomi, Paul Tod, Charles Torre, Claes Uggla, Madhavan Varadarajan, Enric Verdaguer, Eric Woolgar, Yong-Shi Wu, and many visitors to the Syracuse and Utah groups. To all of them I am indebted for the insights they offered on various topics. This work was supported in part by grants NSF PHY92-07225, NSF PHY93-96246 and by research funds of the University of Utah and The Pennsylvania State University.

References

[1] E. Witten, Nuc. Phys. **B311**, 46 (1988).

[2] S. Deser, J. McCarthy, Z. Yang, Phys. Lett. **B222**, 61 (1989).

[3] A. Ashtekar, Phys. Rev. Lett. **57** 2244, (1986); Phys. Rev. **D36**, 1587 (1987).

[4] A. Ashtekar "New perspectives in canonical gravity" (with invited contributions), Bibliopolis, Naples 1988.

[5] A. Ashtekar "Lectures on non-perturbative canonical quantum gravity" (prepared in collaboration with R. Tate), World Scientific Advanced Series in Astrophysics and Cosmology Vol. 6, Singapore, 1992.

[6] B. DeWitt, Phys. Rev. **150**, 1113 (1967).

[7] Palatini, Rend. Circ. Mat. Palermo **43**, 203 (1917).

[8] G. Horowitz, Class. Quan. Grav. **8**, 587 (1991).

[9] A. Ashtekar, A. P. Balachandran, S. Jo, Int. J. Mod. Phys. **A4**, 1493 (1989).

[10] G. González, R. Tate, (in preparation).

[11] R. Tate, Ph. D. Thesis, Syracuse University 1992.

[12] K. Kuchař, in "General Relativity and Gravitation 1992", R. J. Gleiser, C. N. Kozameh, O. M. Moreschi, editors, University of Utah PreprintInstitute of Physics Publishing (1993).

[13] I. Anderson, C. Torre, Phys. Rev. Lett. **70**, 3525 (1993).

[14] J. Goldberg, J. Lewandowski, C. Stornaiolo, Commun. Math. Phys. **148**, 377 (1992).

[15] J. Frauendiener, E. T. Newman, Ann. Isr. Phys. Soc. **9**, 52 (1990) (Published as a book "Developments in General Relativity, Astrophysics and Quantum Theory", a Jubilee Volume in Honour of Nathan Rosen, F. Cooperstock, L. Horwitz, J. Rosen editors, IOP Publishing, Bristol 1990).

[16] J. Pullin in "Relativity and Gravitation: Classical and Quantum (Silarg VII)", C. D'Olivo, E. Nahmad-Achar, M. Rosembaum, M. Ryan, F. Zertuche, editors, World Scientific, Singapore (1992), p. 189.

[17] A. Anderson, Preprint Imperial/TP/92-93-09, to appear in the Dieter Brill Festschrift.

[18] A. Ashtekar, C. Rovelli, L. Smolin, Phys. Rev. Lett. **69**, 237 (1992).

[19] L. N. Chang, C. Soo, in "Proceedings of the XXth DGM", Twentieth International Meeting on Differential Geometric Methods in Theoretical Physics, S. Catto, A. Rocha, eds., World Scientific, Singapore, (1991), p. 946.

[20] T. Jacobson, L. Smolin, Nuc. Phys. **B299**, 295 (1988).

[21] J. Barrett, Int. J. Theor. Phys. **30**, 1171 (1991).

[22] R. Giles, Phys. Rev. **D24**, 2160 (1981).

[23] V. Husain, Nuc. Phys. **B313**, 711 (1989).

[24] B. Brügmann, J. Pullin, Nuc. Phys. **B363**, 221 (1991).

[25] B. Brügmann, R. Gambini, J. Pullin, Nuc. Phys. **B385**, 581 (1992).

[26] A. Ashtekar, Lectures delivered at The University of Poona (unpublished).

[27] H. Kodama, Phys. Rev. **D42**, 2548 (1990).

[28] H. Kodama, Int. J. Mod. Phys. **D** (to appear).

[29] A. Ashtekar, Proceedings of the Osgood Hill meeting on Conceptual problems in Quantum Gravity, A. Ashtekar, J. Stachel, editors, Birkhauser, Boston, (1991).

[30] A. Ashtekar, C. Isham, Class. Quan. Grav. **9**, 1433 (1992).

[31] J. Lewandowski, Syracuse University Preprint 1992.

[32] A. Ashtekar, V. Husain, C. Rovelli, J. Samuel, L. Smolin, Class. Quan. Grav. **6**, L185 (1989).

[33] R. Gambini, A. Trias, Phys. Rev. **D22**, 1380 (1980); C. Di Bartolo, F. Nori, R. Gambini, L. Leal, A. Trias, Lett. Nuo. Cim. **38**, 497 (1983); A. Ashtekar, C. Rovelli, Class. Quan. Grav. **9**, 1121 (1992)

[34] Miao Li, Nuo. Cim. **B105**, 1113 (1990).

[35] R. Gambini, L. Leal, A. Trias, Phys. Rev. **D39**, 3127 (1989).

[36] J. Aroca Farrerons, Ph. D. Thesis, Universitat Autònoma de Barcelona 1990.

[37] B. Brügmann, Phys. Rev. **D43**, 566 (1991).

[38] C. Rovelli, L. Smolin, Phys. Rev. Lett. **61**, 1155 (1988).

[39] C. Rovelli, L. Smolin, Nuc. Phys. **B331**, 80 (1990).

[40] R. Gambini, A. Trias, Nuc. Phys. **B278**, 436 (1986).

[41] R. Loll, Nuc. Phys. **B350**, 831 (1991).

[42] C. Rovelli, Class. Quan. Grav. **8**, 1613 (1991).

[43] S. Mandelstam, Ann. Phys. (NY) **19**, 1 (1962).

[44] A. M. Polyakov, Nuc. Phys. **B164**, 171 (1979).

[45] Yu. M. Makeenko, A. A. Migdal, Phys. Lett. **B88**, 135 (1979).

[46] R. Gambini, A. Garat, J. Pullin (in preparation).

[47] R. Gambini, Phys. Lett. **B255**, 180 (1991).

[48] B. Brügmann, J. Pullin, Nuc. Phys. **B390**,399 (1993).

[49] C. Di Bartolo, R. Gambini, J. Griego, L. Leal, Preprint IFFI (1991) (Montevideo).

[50] P. G. Tait, "Scientific Papers", Cambridge (1898).

[51] M. Atiyah, "The Geometry and Physics of Knots", Cambridge (1990).

[52] L. Kauffman, "On Knots", Annals of Mathematics Studies 115, Princeton University Press (1987).

[53] L. Kauffman, "Knots and Physics", World Scientific, (1992).

[54] E. Guadagnini, Int. J. Mod. Phys. **A7**, 877 (1992).

[55] E. Witten, Commun. Math. Phys. **121**, 351 (1989).

[56] L. Smolin, Mod. Phys. Lett. **A4**, 1091, (1989).

[57] P. Cotta-Ramusino, E. Guadagnini, M. Martellini, M. Mintchev Nuc. Phys. **B330**, 557 (1990).

[58] E. Guadagnini, M. Martellini, M. Mintchev, Phys. Lett. **B227**, 111 (1989).

[59] D. Bar-Natan, Harvard Preprint (1991).

[60] G. Calugareanu, Rev. Math. Pures Appl. (Bucarest) 4, 5 (1959).

[61] C. Di Bartolo, J. Griego, R. Gambini (in preparation).

[62] E. Guadagnini, M. Martellini, M. Mintchev Nuc. Phys. Phys. **B330**, 575 (1990).

[63] R. Gambini, B. Brügmann, J. Pullin, "Knot invariants as nondegenerate states of four dimensional quantum gravity",in "Proceedings of the XXth DGM", Twentieth International Meeting on Differential Geometric Methods in Theoretical Physics, S. Catto, A. Rocha, eds., World Scientific, Singapore (1992) p. 784.

[64] B. Brügmann, R. Gambini, J. Pullin, Phys. Rev. Lett. **68**, 431 (1992).

[65] B. Brügmann, R. Gambini, J. Pullin, Gen. Rel. Grav. **25**, 1 (1993).

[66] C. Di Bartolo, R. Gambini, J. Griego, J. Pullin (in preparation).

[67] R. Capovilla, J. Dell, T. Jacobson, Phys. Rev. Lett. **63**, 2325 (1989).

[68] J. Samuel, Class. Quan. Grav. **5**, L123 (1988).

[69] C. Torre, Phys. Rev. **D41**, 3620 (1990).

[70] T. Newman, C. Rovelli, Phys. Rev. Lett. **69**, 1300 (1992).

[71] A. Ashtekar, J. Pullin, Ann. Isr. Phys. Soc. **9**, 65 (1990) (Published as a book "Developments in General Relativity, Astrophysics and Quantum Theory", a Jubilee Volume in Honour of Nathan Rosen, F. Cooperstock, L. Horwitz, J. Rosen editors, IOP Publishing, Bristol 1990).

[72] M. Varadarajan, Class. Quan. Grav. **8**, 235 (1991).

[73] V. Husain, L. Smolin, Nuc. Phys. **B327**, 205 (1989).

[74] V. Husain, J. Pullin, Mod. Phys. Lett. **A5**, 733 (1990).

[75] A. Ashtekar, M. Varadarajan (in preparation).

[76] A. Ashtekar, "Lessons from 2+1 gravity" in Strings '90, R. Arnowitt et al. editors, World Scientific, Singapore (1991).

[77] A. Ashtekar, C. Rovelli, L. Smolin, Phys. Rev. **D44**, 1740 (1991).

[78] R. Gambini, J. Pullin, Phys. Rev. **D47**, R5214 (1993).

[79] J. Baez, Class. Quan. Grav. **10**, 673 (1993).

QUANTUM GROUPS

M. Ruiz–Altaba

Dépt. Physique Théorique, Université de Genève, CH–1211 Genève 4

ABSTRACT

These notes correspond rather accurately to the translation of the lectures given at the Fifth Mexican School of Particles and Fields, held in Guanajuato, Gto., in December 1992. They constitute a brief and elementary introduction to quantum symmetries from a physical point of view, along the lines of the forthcoming book by C. Gómez, G. Sierra and myself.

1. Introduction

2. Factorizable S–matrices

3. Bethe's diagonalization of spin chain hamiltonians

4. Integrable vertex models: the six–vertex model

5. The Yang–Baxter algebra

9. Physical spectrum of the Heisenberg spin chain

6. Yang–Baxter algebras and braid groups

7. Yang–Baxter algebras and quantum groups

8. Affine quantum groups

10. Hopf algebras

11. The quantum group $U_q(\mathcal{G})$

12. Comments

13. References

1 Introduction

Since the advent of physics as a pleasure for the human mind, those in our trade have played with idealizations of reality which preserve the essence of the phenomenon and yet are simple enough to be modelled by the mathematical tools available. These "toy models", from the point–like friction-less particle in newtonian mechanics to scalar quantum electrodynamics, constitute most of the syllabus of a physicist's education. At the very worst, the theories and models we shall discuss in these lectures can be taken as paradigmatic toy models; indeed most of two–dimensional field theory was invented as a theoretical laboratory for confinement, dimensional transmutation, instantons, and other such niceties of the real world. Yet something deep remains hidden in the guts of two–dimensional physics. The historic success of string theory in unifying gauge symmetries and general relativity has given the two-dimensional world a new and fruitful life of its own: from a stringy point of view, enough remains to be learned about the fundamental world–sheet that we need not bother for a while about other dimensions. Quite surprisingly, condensed matter physicists have also come to appreciate the interest of low–dimensional field theories, motivated by the planar character of the quantum

© 1994 American Institute of Physics

hall effect and high–temperature superconductivity or the technological interest in thin plastic and silicon chips, among other noteworthy phenomena.

These lectures are meant to illustrate some of the beautiful tricks that can be applied to understanding two–dimensional physics. Were we to think of these lectures as a meal, a menu would read more or less as follows. For appetizers, some relativistic dynamics in one spatial dimension. The reader should then sit down to a light salad of Bethe ansatz, followed by a hot soup of integrable vertex models on the plane. The main course consists of Yang–Baxter algebras, with a variety of sauces of various mathematical origins. A few words about the possible generalization of these treats to higher dimensions are left for desert. All the wines and liquors come from the quantum group vintage, distilled at Kyoto and Kharkov from the well-known and now extinct Leningrad stock. To avoid indigestion, the beautiful example of two–dimensional integrability provided by conformal field theories is not presented; Professor Weyers's lectures in this same volume cover that.

Quantum groups have been discovered relatively recently by physicists and mathematicians concerned with integrable two–dimensional systems. An integrable system has as many integrals of motion (constants) as co-ordinates or, equivalently, momenta; accordingly, in an integrable theory we know that the phase space is spanned by the so–called action–angle variables, essentially a bunch of conserved quantities (hamiltonians) and their conjugate variables (times). Classical two–dimensional statistical physics is equivalent to quantum field theory in one (spatial) dimension. In these lectures we shall investigate two-dimensional systems with an infinite number of degrees of freedom: we shall be concerned with the thermodynamic limit of classical two–dimensional statistical models.

It is perhaps more intuitive to start with quantum field theories in one space and one time dimension, that is quantum field theories on the real line. Integrable field theories in such a small dimension is essentially equivalent to the description of solitons. Indeed, a soliton is a non–dispersive classical solution to the classical equations of motion which survives quantization and acquires a particle–like interpretation. Feynman rules are, in general, rather useless in the description of solitons. Collective phenomena of this sort are not perturbative at all, and in low dimensions we might as well attack the problem directly to find the n–point solitonic Green's functions. In this endeavor, integrability comes in quite handy: we shall see shortly that the existence of an infinity of conserved charges is equivalent to a deceivingly simple cubic equation in the $2 \rightarrow 2$ scattering amplitudes, the celebrated Yang–Baxter equation. In order to get a real theory, we shall in addition impose unitarity and crossing symmetry. Unitarity just means that the probability is conserved, so that nothing comes of nothing and something comes of just as much. Crossing symmetry is a more subtle requirement, familiar from string theories, which can be viewed somehow as a strong relativistic invariance, whereby particles moving forward or backward in time can be traded off (suitably) by other particles moving forward or backward in time.

2 Factorizable S–matrices

We shall be interested in integrable field theoretical systems, *i.e.* systems with an infinite number of mutually commuting conserved charges. One of these charges will be called the hamiltonian, an operator which defines the time evolution of the system. To each conserved charge one can associate a different time evolution: what we call the hamiltonian is a matter of interpretation.

Consider the scattering of relativistic massive particles in a $(1+1)$-dimensional spacetime. There is only one spatial dimension and therefore the ordering of the particles is well–defined, from left to right, say. In more spatial dimensions we should not expect that the interesting features depending strongly on the ordering of the particles remain valid.

Introduce the rapidity θ:

$$p^0 = m \cosh\theta \quad , \quad p^1 = m \sinh\theta \tag{1}$$

This parametrization ensures the on–shell condition $\vec{p}^2 = (p^0)^2 - (p^1)^2 = m^2$.

Alternatively, we could use the light-cone momenta p and $\bar{p}$,

$$p = p^0 + p^1 = m\,e^\theta \quad , \quad \bar{p} = p^0 - p^1 = m\,e^{-\theta} \tag{2}$$

which transform under a Lorentz boost $L_\alpha : \theta \to \theta + \alpha$ as

$$p \to p e^\alpha \quad , \quad \bar{p} \to \bar{p} e^{-\alpha} \tag{3}$$

Quite generally, an irreducible tensor Q_s of the Lorentz group in $1+1$ dimensions is labelled by its spin s according to the rule $L_\alpha : Q_s \to e^{s\alpha} Q_s$, so that p is of spin 1 and its parity conjugate $\bar{p}$ is of spin -1.

If Q_s is a local conserved quantity of integer spin $s > 0$, then in a scattering process involving n particles

$$\sum_{i \in \{\text{in}\}} p_i^s = \sum_{f \in \{\text{out}\}} p_f^s \tag{4}$$

Similarly, if Q_{-s} is conserved, then

$$\sum_{i \in \{\text{in}\}} \bar{p}_i^s = \sum_{f \in \{\text{out}\}} \bar{p}_f^s \tag{5}$$

Setting $s = 1$ in (4) and (5), we recover the usual energy and momentum conservation laws of a relativistic theory.

The physical behavior of integrable systems is quite remarkable. For instance, if (4) and (5) hold for an infinity of different spins s, it follows immediately that the incoming and outgoing momenta must be the same. This means that no particle production or annihilation may ever occur. Also, particles with equal mass may reshuffle their momenta among themselves in the scattering, but particles with different masses may not. Equivalently, we may say that the momenta are conserved individually and that particles of equal mass may interchange additional internal quantum numbers. If all the incoming particles have different masses, then the only effect of the scattering is a time delay (a phase shift) in the outgoing state with respect to the incoming one.

All scattering processes can be understood and pictured as a sequence of two–particle scatterings. This property is called factorizability.

By relativistic invariance, the scattering amplitude between two particles A_i and A_j may only depend on the scalar

$$p_i^\mu p_j^\nu \eta_{\mu\nu} = m_i m_j \cosh(\theta_i - \theta_j) \tag{6}$$

so that, in fact, it may depend only on the rapidity difference $\theta_{ij} = \theta_i - \theta_j$. The general form of the basic two–particle S–matrix is

$$|A_i(\theta_1), A_j(\theta_2)\rangle_{\text{in}} \longrightarrow \sum_{k,\ell} S_{ij}^{k\ell}(\theta_{12}) |A_k(\theta_2), A_\ell(\theta_1)\rangle_{\text{out}} \tag{7}$$

In this notation, $|A_i(\theta_1), A_j(\theta_2)\rangle_{\text{in(out)}}$ stands for the initial (respectively, final) state of two incoming (respectively, outgoing) particles of kinds A_i and A_j and rapidities θ_1 and θ_2.

The second crucial feature of a factorizable S–matrix theory, from which such models get their name, is the property of factorizability: the N–particle S–matrix can always be written as the product of $\binom{N}{2}$ two–particle S–matrices.

We choose an initial state of N particles with rapidities $\theta_1 > \theta_2 > \cdots > \theta_N$ arranged in the infinite past in the opposite order, i.e. $x_1 < x_2 < \cdots < x_N$. This presumes simply that no scatterings may have occured before we study the process, i.e. that we have been looking long before any particles meet. After the $N(N-1)/2$ pair collisions, the particles reach the infinite future ordered along the spatial direction in increasing rapidity. Thus we write

$$S\,|A_{i_1}(\theta_1),\ldots,A_{i_N}(\theta_N)\rangle_{\text{in}} =$$
$$= \sum_{j_1,\ldots,j_N} S^{j_1\cdots j_N}_{i_1\cdots i_N}\,|A_{j_1}(\theta_N),\ldots,A_{j_N}(\theta_1)\rangle_{\text{out}} \tag{8}$$

Factorization means that this process can be interpreted as a set of independent and consecutive two–particle scattering processes.

The spacetime picture of this multi-particle factorized scattering is obtained by associating with each particle a line whose slope is the particle's rapidity. The scattering process is thus represented by a planar diagram with N straight world–lines, such that no three ever coincide at the same point. Any world–line will therefore intersect, in general, all the other ones. The complete scattering amplitude associated to any such diagram is given by the (matrix) product of two–particle S–matrices. For instance, in a four–particle scattering we could get

$$S^{j_1 j_2 j_3 j_4}_{i_1 i_2 i_3 i_4}(\theta_1,\theta_2,\theta_3,\theta_4) = \sum_{\substack{k,\ell,m,n,\\p,q,r,u}} S^{k\ell}_{i_1 i_2}(\theta_{12}) S^{mn}_{\ell i_3}(\theta_{13}) \times$$
$$\times S^{pq}_{km}(\theta_{23}) S^{rj_4}_{ni_4}(\theta_{14}) S^{uj_3}_{qr}(\theta_{24}) S^{j_1 j_2}_{pu}(\theta_{34}) \tag{9}$$

The kinematical data (the rapidities of all the particles) does not fix a diagram uniquely. In fact, for the same rapidites we have a whole family of diagrams, differing from each other by the parallel shift of some of the straight world–lines. The parallel shift of any one line can (and should) be interpreted as a symmetry transformation. It corresponds to the translation of the (asymptotic in– and out–) x co-ordinates of the particle associated to the line. Requiring the factorizability condition is equivalent to imposing that the scattering amplitudes of diagrams differing by such parallel shifts should be the same.

For the simple case of three particles, the condition that the factorization be independent of parallel shifts of the world–lines amounts to the following noteworthy factorization equation, which is the necessary and sufficient condition for any two diagrams differing by parallel shifts to have equal associated amplitudes:

$$\sum_{p_1,p_2,p_3} S^{p_1 p_2}_{i_1 i_2}(\theta_{12}) S^{p_3 j_3}_{p_2 i_3}(\theta_{13}) S^{j_1 j_2}_{p_1 p_3}(\theta_{23}) =$$
$$= \sum_{p_1,p_2,p_3} S^{p_2 p_3}_{i_2 i_3}(\theta_{23}) S^{j_1 p_1}_{i_1 p_2}(\theta_{13}) S^{j_2 j_3}_{p_1 p_3}(\theta_{12}) \tag{10}$$

This is the famous Yang–Baxter equation.

To formalize this a bit more, consider a set of operators $\{A_i(\theta)\}$ ($i = 1, \ldots, n$) associated to each particle i with rapidity θ, obeying the following commutation relations:

$$A_i(\theta_1)A_j(\theta_2) = \sum_{k,\ell} S_{ij}^{k\ell}(\theta_{12}) A_k(\theta_2) A_\ell(\theta_1) \tag{11}$$

This equation encodes the two–particle scattering process (7), where "collision" has been replaced by "commutation". Furthermore, the relation between (7) and (11) becomes evident if we interpret $A_i(\theta)$ as an operator (Zamolodchikov operator) which creates the particle $|A_i(\theta)\rangle$ when it acts on the Hilbert space vacuum $|0\rangle$:

$$A_i(\theta)|0\rangle = |A_i(\theta)\rangle \tag{12}$$

The factorization equation (10) emerges in this context as a "generalized Jacobi identity" of the algebra (11), assumed associative.

The following conditions are needed to guarantee the physical consistency of the Zamolodchikov algebra (11):

i) Normalization:

$$\lim_{\theta \to 0} S_{ij}^{k\ell}(\theta) = \delta_i^k \delta_j^\ell \quad \Longleftrightarrow \quad \lim_{\theta \to 0} S(\theta) = 1 \tag{13}$$

This condition is obtained by setting $\theta_1 = \theta_2$ in (11). In physical terms, it means that no scattering takes place if the relative velocity of the two particles vanishes, i.e. if the two world–lines are parallel.

ii) Unitarity:

$$\sum_{j_1,j_2} S_{j_1 j_2}^{i_1 i_2}(\theta) S_{k_1 k_2}^{j_1 j_2}(-\theta) = \delta_{k_1}^{i_1} \delta_{k_2}^{i_2} \quad \Longleftrightarrow \quad S(\theta)S(-\theta) = 1 \tag{14}$$

This follows from applying (11) twice.

iii) Real analyticity:

$$S^\dagger(\theta) = S(-\theta^*) \tag{15}$$

which together with (14) implies the physical unitarity condition $S^\dagger S = 1$.

iv) Crossing symmetry:

$$S_{ij}^{k\ell}(\theta) = S_{j\bar{\ell}}^{\bar{i}k}(i\pi - \theta) \tag{16}$$

where $\bar{j}$ and $\bar{k}$ denote the antiparticles of j and k, respectively.

As an example, let us consider a theory with only one kind of particle A and its antiparticle $\overline{A}$. Due to CPT invariance, there exist only three different amplitudes (we also assume conservation of particle number, i.e. $\mathbb{Z}_2$ invariance). The scattering amplitude between identical particles (or antiparticles) is denoted S_I, whereas S_T and S_R denote the transmission and reflection amplitudes, respectively:

$$\begin{aligned}
A(\theta_1)A(\theta_2) &= S_I(\theta_{12})A(\theta_2)A(\theta_1) \\
A(\theta_1)\overline{A}(\theta_2) &= S_T(\theta_{12})\overline{A}(\theta_2)A(\theta_1) + S_R(\theta_{12})A(\theta_2)\overline{A}(\theta_1) \\
\overline{A}(\theta_1)A(\theta_2) &= S_T(\theta_{12})A(\theta_2)\overline{A}(\theta_1) + S_R(\theta_{12})\overline{A}(\theta_2)A(\theta_1) \\
\overline{A}(\theta_1)\overline{A}(\theta_2) &= S_I(\theta_{12})\overline{A}(\theta_2)\overline{A}(\theta_1)
\end{aligned} \tag{17}$$

It is not hard to check that the factorization equations for this algebra read as

$$S_I S'_R S''_I = S_T S'_R S''_T + S_R S'_I S''_R$$
$$S_I S'_T S''_R = S_T S'_I S''_R + S_R S'_R S''_T \qquad (18)$$
$$S_R S'_T S''_I = S_R S'_I S''_T + S_T S'_R S''_R$$

where we have set $S_a = S_a(\theta_{12})$, $S'_a = S_a(\theta_{13})$, $S''_a = S_a(\theta_{23})$ for $a \in \{I, T, R\}$ to lighten the notation.

The normalization conditions read

$$S_I(0) = 1 \quad , \quad S_T(0) = 0 \quad , \quad S_R(0) = 1 \qquad (19)$$

whereas unitarity requires

$$S_T(\theta)S_T(-\theta) + S_R(\theta)S_R(-\theta) = 1$$
$$S_T(\theta)S_R(-\theta) + S_R(\theta)S_T(-\theta) = 0 \qquad (20)$$

and the crossing symmetry implies

$$S_I(\theta) = S_T(i\pi - \theta) \quad , \quad S_R(\theta) = S_R(i\pi - \theta) \qquad (21)$$

The equations (18) imply that the quantity

$$\Delta = \frac{S_I(\theta)^2 + S_T(\theta)^2 - S_R(\theta)^2}{2S_I(\theta)S_T(\theta)} \qquad (22)$$

is independent of the rapidity θ. An interesting factorized S-matrix is provided by the sine–Gordon theory, where the states A and $\overline{A}$ of (17) are identified with the soliton and antisoliton

3 Bethe's diagonalization of spin chain hamiltonians

Consider now a periodic one–dimensional regular lattice (a periodic chain) with L sites. At each site, the spin variable may be either up or down, so that the Hilbert space of the spin chain is simply $\mathcal{H}^{(L)} = \otimes^L V^{\frac{1}{2}}$ where $V^{\frac{1}{2}}$ is the spin-$\frac{1}{2}$ irreducible representation of $SU(2)$ with basis $\{|\uparrow\rangle, |\downarrow\rangle\}$. By simple combinatorics, the dimension of the Hilbert space is $\dim \mathcal{H}^{(L)} = 2^L$. On $\mathcal{H}^{(L)}$, we consider a very general hamiltonian H, subject to three constraints.

First, we assume that the interaction is of short range, for example only among nearest neighbors. Next, we impose that the hamiltonian H be translationally invariant. Letting e^{iP} denote the operator which shifts the states of the chain by one lattice unit to the right, then this requirement reads as $\left[e^{iP}, H\right] = 0$. From periodicity of the closed chain, we must have $e^{iPL} = 1$ Finally, we demand that the hamiltonian preserve the third component of the spin:

$$[H, S^z_{\text{total}}] = \left[H, \sum_{i=1}^{L} S^z_i\right] = 0 \qquad (23)$$

This requirement allows us to divide the Hilbert space of states into different sectors, each labelled by the third component of the spin or, equivalently, by the total number of spins down. We shall denote by $\mathcal{H}^{(L)}_M$ the subspace of $\mathcal{H}^{(L)}$ with M spins down. Obviously, $\dim \mathcal{H}^{(L)}_M = \binom{L}{M}$, so that $\dim \mathcal{H}^{(L)} = \sum_{M=0}^{L} \dim \mathcal{H}^{(L)}_M$.

We wish to study the eigenstates and spectrum of H. The zero-th sector $\mathcal{H}_0^{(L)}$ contains only one state, the "Bethe reference state" with all spins up. The most natural ansatz for the eigenvectors of H in the other sectors is some superposition of "spin waves" with different velocities. For the first sector, *i.e.* the subspace of states with all spins up except one down, the ansatz for the eigenvector is thus of the form $|\Psi_1\rangle = \sum_{x=1}^{L} f(x) |x\rangle$ where $|x\rangle$ represents the state with all spins up but for the one at lattice site x ($1 \leq x \leq L$). The unknown wavefunction $f(x)$ determines the probability that the single spin down is precisely at site x.

From the complete translational invariance due to periodic boundary conditions, it is reasonable to assume that $f(x)$ is just the wavefunction for a plane wave

$$f(x) = e^{ikx} \tag{24}$$

with some particular momentum k to be fixed by the boundary condition $f(x+L) = f(x)$. Thus $k = 2\pi I/L$, with $I = 0, 1, \ldots, L-1$. Hence the eigenvectors of H with one spin down span indeed a basis of the Hilbert space $\mathcal{H}_1^{(L)}$, by dimensionality counting.

The wavefunction solving the eigenvalue problem for the sector with two spins down, $H|\Psi_2\rangle = E_2|\Psi_2\rangle$, is of the form $|\Psi_2\rangle = \sum_{x_1,x_2} f(x_1,x_2)|x_1,x_2\rangle$, where $|x_1,x_2\rangle$ stands for the state with ll spins up except two spins down at positions x_1 and x_2.

The periodicity condition reads now $f(x_1, x_2) = f(x_2, x_1 + L)$. The most naive ansatz for $f(x_1, x_2)$ generalizing the plane wave is $f(x_1, x_2) = A_{12} \exp i(k_1 x_1 + k_2 x_2)$. This ansatz is inappropriate, however, because it violates the periodicity condition. Physically, we have forgotten to include the scattering of the two "spin waves" with "quasi-momenta" k_1 and k_2. The solution to this problem was found by Bethe, who wrote the useful ansatz

$$f(x_1, x_2) = A_{12} e^{i(k_1 x_1 + k_2 x_2)} + A_{21} e^{i(k_1 x_2 + k_2 x_1)} \tag{25}$$

which does satisfy the periodicity condition provided

$$A_{12} = A_{21} e^{ik_1 L} \quad , \quad A_{21} = A_{12} e^{ik_2 L} \tag{26}$$

Note that these two conditions imply, in particular, that $\exp i(k_1 + k_2)L = 1$, which reflects the invariance of the wavefunction under a full turn around the chain, *i.e.* under the shift of L units of lattice space:

$$f(x_1 + L, x_2 + L) = f(x_1, x_2) \iff e^{i(k_1+k_2)L} = 1 \tag{27}$$

This equation must hold if the wavefunction is to be single valued.

The ansatz (25) already assumes that the S-matrix for two spin waves is purely elastic. In fact, the only dynamics allowed is the permutation of the quasi-momenta.

To capture the physical meaning behind equations (26), let us introduce the "scattering amplitudes for spin waves"

$$\hat{S}_{12} = \frac{A_{21}}{A_{12}} \quad , \quad \hat{S}_{21} = \frac{A_{12}}{A_{21}} \tag{28}$$

in terms of which (26) read as

$$e^{ik_1 L} \hat{S}_{12}(k_1, k_2) = 1 \quad , \quad e^{ik_2 L} \hat{S}_{21}(k_2, k_1) = 1 \tag{29}$$

These equations tell us that the total phase shift undergone by a spin wave after travelling all the way around the closed chain is one. This phase shift receives two contributions; one is purely kinematic ($e^{ik_1 L}$ or $e^{ik_2 L}$) and depends only on the quasi-momentum of the spin waves, while the other reflects the phase shift produced by the interchange of the two spin waves.

Summarizing the previous discussion, we have found that the Bethe ansatz for the eigenvector of the hamiltonian H in the sector $M = 2$ is

$$f(x_1, x_2) = A_{12} \left(e^{i(k_1 x_1 + k_2 x_2)} + \hat{S}_{12}(k_1, k_2) \, e^{i(k_1 x_2 + k_2 x_1)} \right) \tag{30}$$

The generic form of a state $|\Psi_M\rangle \in \mathcal{H}_M^{(L)}$ in the sector with $M > 2$ spins down is

$$|\Psi_M\rangle = \sum_{1 \leq x_1 < x_2 < \cdots < x_M \leq L} f(x_1, \ldots, x_M) \, |x_1, \ldots, x_M\rangle \tag{31}$$

The Bethe ansatz is now

$$f(x_1, \ldots, x_M) = \sum_{p \in S_M} A_p \, e^{i(k_{p(1)} x_1 + \cdots + k_{p(M)} x_M)} \tag{32}$$

where the sum runs over the $M!$ permutations p of the labels of the quasi-momenta k_i. The periodicity condition is now

$$f(x_1, x_2, \ldots, x_M) = f(x_2, \ldots, x_M, x_1 + L) \tag{33}$$

When $M = 3$, we get the following six equations:

$$e^{i k_1 L} = \frac{A_{123}}{A_{231}} = \frac{A_{132}}{A_{321}} \quad , \quad e^{i k_2 L} = \frac{A_{231}}{A_{312}} = \frac{A_{213}}{A_{132}} \quad , \quad e^{i k_3 L} = \frac{A_{312}}{A_{123}} = \frac{A_{321}}{A_{213}} \tag{34}$$

Thus, in addition to the relations among the quasi-momenta k_i and the amplitudes A_p, there exist additional constraints among the amplitudes of three quasi-particles, which were absent in the simpler case with $M = 2$. These relations tell us that the interchange of two particles is independent of the position of the third particle. Locality of the interactions is thus equivalent to the factorization property of the S-matrix, according to which the scattering amplitude of M quasi-particles factorizes into a product of $\binom{M}{2}$ two-point S-matrices.

The Yang–Baxter content of the Bethe ansatz for $M = 3$ is illustrated with the following equalities:

$$A_{321} = \begin{cases} \hat{S}_{12} \, A_{312} = \hat{S}_{12} \hat{S}_{13} \, A_{132} = \hat{S}_{12} \hat{S}_{13} \hat{S}_{23} \, A_{123} \\ \hat{S}_{23} \, A_{231} = \hat{S}_{23} \hat{S}_{13} \, A_{213} = \hat{S}_{23} \hat{S}_{13} \hat{S}_{12} \, A_{123} \end{cases} \tag{35}$$

We thus arrive to the all-important "Bethe ansatz equations"

$$e^{i k_i L} = \prod_{\substack{j=1 \\ j \neq i}}^{M} \hat{S}_{ji}(k_j, k_i) \qquad \text{for } i = 1, \ldots, M \tag{36}$$

written in general for a sector with arbitrary M. The actual solution to these equations far transcends the framework of these lectures. Suffice it to say that a variety of methods have been devised to attack them.

The spin wave scattering amplitude $\hat{S}_{12}$ depends of course on the detailed form of the hamiltonian, and it can be computed by solving the $M = 2$ eigenvalue equation, which reads more explicitly as

$$E_2 \, f(x_1, x_2) = \sum_{1 \leq y_1 < y_2 \leq L} \langle x_1, x_2 | H | y_1, y_2 \rangle \, f(y_1, y_2) \tag{37}$$

Using (30) in (37), we would find $\hat{S}_{12}$ as a function of k_1, k_2 and the matrix elements of H.

Unfortunately, there does not exist a simple criterion to decide when a spin chain hamiltonian is integrable, *i.e.* when it allows the Bethe construction. As we have shown, however, the Bethe ansatz will work whenever the spin wave S-matrix satisfies the integrability condition and factorization. Let us stress that the diagonalization of a hamiltonian with the help of the Bethe ansatz does not even work for any translationally invariant and short range hamiltonian preserving the total spin. Only a very special class of such hamiltonians can be diagonalized via the Bethe procedure, namely those which describe integrable models. An important spin chain model, to which the Bethe ansatz technique is applicable, is the XXZ model

$$H_{XXZ} = J \sum_{i=1}^{L} \left(\sigma_i^x \sigma_{i+1}^x + \sigma_i^y \sigma_{i+1}^y + \Delta \sigma_i^z \sigma_{i+1}^z \right) \tag{38}$$

4 Integrable vertex models: the six–vertex model

Let us now turn to classical statistical systems in two spatial dimensions (in equilibrium, so no time dimension) on a lattice. Whereas in the previous section the main problem consisted in diagonalizing a one–dimensional hamiltonian, in this section we address the computation of the partition function of the lattice system.

A vertex model is a statistical model defined on a lattice $\mathcal{L}$, taken regular and rectangular for simplicity. We shall thus consider an $L \times L'$ lattice with L vertical lines (columns) and L' horizontal lines (rows).

A physical state on this lattice is defined by the assignment to each lattice edge of a state variable, characterized by some labels; allowing for two possible states on each link suffices for our purposes. These two possiblities may be interpreted as spins up or down. Alternatively, if we imagine the lattice links as electric wires with a current of constant intensity running through them, then the two states are associated with the direction of the current.

The dynamics of the model is characterized by the interactions among the lattice variables, which take place at the vertices, whence the name vertex model. The energy ε_V associated with a vertex V depends only on the four states on the edges meeting at that vertex (locality). This is also true for the Boltzmann weights $W_V = \exp(-\varepsilon_V/k_B T)$ which measure the probability of each local configuration. Quite generally, it is convenient to represent Boltzmann weights as $W\begin{pmatrix} \beta & \nu \\ \mu & \alpha \end{pmatrix}$, where μ and ν are the horizontal edge state labels, and α and β the vertical ones:

$$W\begin{pmatrix} \beta & \nu \\ \mu & \alpha \end{pmatrix} \quad = \quad \begin{array}{c} \beta \\ \mu \!\!-\!\!\!\!+\!\!\!\!-\!\! \nu \\ \alpha \end{array} \tag{39}$$

If we impose that the interaction conserve the total spin or the local current, then all but six Boltzmann weights must vanish. In addition to the spin, particle number or current conservation, we may also impose the $\mathbb{Z}_2$ reversal symmetry under $|\!\Uparrow\rangle \leftrightarrow |\!\Downarrow\rangle$. Under this condition, the independent Boltzmann weights are reduced to three, which we shall call just a, b and c. These weights define the symmetric or zero–field six–vertex model, which we shall call the six–vertex model for short.

The six–vertex model is characterized by link variables $\in \mathbb{Z}_2 = \{0,1\}$ with Boltzmann weights subject to current conservation

$$W\begin{pmatrix} \beta & \nu \\ \mu & \alpha \end{pmatrix} = 0 \quad \text{unless} \quad \mu + \alpha = \nu + \beta \tag{40}$$

and reflection symmetry ($\bar{x} = 1 - x$)

$$W\begin{pmatrix} \beta & \nu \\ \mu & \alpha \end{pmatrix} = W\begin{pmatrix} \bar{\beta} & \bar{\nu} \\ \bar{\mu} & \bar{\alpha} \end{pmatrix} \tag{41}$$

A compact way to write the weights is

$$W\begin{pmatrix} \beta & \nu \\ \mu & \alpha \end{pmatrix} = b\,\delta_{\mu\nu}\,\delta_{\alpha\beta} + c\,\delta_{\mu\beta}\,\delta_{\nu\alpha} + (a-b-c)\,\delta_{\mu\alpha}\,\delta_{\nu\beta} \tag{42}$$

The partition function is

$$Z_{L \times L'}(a,b,c) = \sum_{\mathcal{C}} e^{-E(\mathcal{C})/k_B T} = \sum_{\mathcal{C}} \prod_V W_V \tag{43}$$

where the sum runs over all possible configurations $\mathcal{C}$, of which there are $2^{LL'}$. In the thermodynamic limit, when L and L' tend to infinity, the computation of the sum (43) becomes a rather formidable and apparently insurmountable problem. Lieb's breakthrough to compute the partition function (43) of the six–vertex model relies basically on rephrasing the problem as the diagonalization of the anisotropic spin–$\frac{1}{2}$ chain, which had been solved already by the Bethe ansatz. First, let us perform the sum over the horizontal variables, which involves only the Boltzmann weights on the same row of the lattice, and then carry out the sum over the vertical variables. The double sum (43) can thus be rearranged as follows:

$$Z_{L \times L'}(a,b,c) = \sum_{\substack{\text{vertical rows} \\ \text{states}}} \prod \left(\sum_{\substack{\text{horizontal} \\ \text{states}}} \prod_{V \in \text{row}} W_V \right) \tag{44}$$

The quantity in parenthesis depends on the two sets of vertical states above and below the row of horizontal variables: it is the (row to row) transfer matrix of the model. For conceptual clarity, it is convenient to introduce the "fixed time states" as the set of vertical link variables on the same row:

$$|\alpha\rangle = \begin{array}{ccccccc} \alpha_1 & \alpha_2 & \alpha_3 & & \alpha_{L-1} & \alpha_L \\ \uparrow & \uparrow & \uparrow & \cdots & \uparrow & \uparrow \end{array} \tag{45}$$

The transfer matrix element $\langle \beta | t | \alpha \rangle$ can then be understood as the transition probability for the state $|\alpha\rangle$ to project on the state $|\beta\rangle$ after a unit of time. We are thinking now of the horizontal direction as space, and the vertical one as time:

$$\langle \beta | t(a,b,c) | \alpha \rangle = \sum_{\mu_i} W\begin{pmatrix} \beta_1 & \mu_2 \\ \mu_1 & \alpha_1 \end{pmatrix} W\begin{pmatrix} \beta_2 & \mu_3 \\ \mu_2 & \alpha_2 \end{pmatrix} \cdots$$

$$\cdots W\begin{pmatrix} \beta_{L-1} & \mu_L \\ \mu_{L-1} & \alpha_{L-1} \end{pmatrix} W\begin{pmatrix} \beta_L & \mu_1 \\ \mu_L & \alpha_L \end{pmatrix} \tag{46}$$

We agree with the Chinese who think that a picture is better than a formula:

$$\langle\beta|t|\alpha\rangle \;=\; \sum_{\mu_i} \begin{array}{c} \beta_1 \;\; \beta_2 \qquad\qquad \beta_{L-1} \;\; \beta_L \\ |\;\;\;|\;\;\;\;\;\cdots\;\;\;\;|\;\;\;| \\ \mu_1\,|\,\mu_2\,|\,\mu_3\;\;\;\;\;\mu_{L-1}\,|\,\mu_L\,|\,\mu_1 \\ |\;\;\;|\;\;\;\;\;\cdots\;\;\;\;|\;\;\;| \\ \alpha_1 \;\; \alpha_2 \qquad\qquad \alpha_{L-1} \;\; \alpha_L \end{array} \qquad (47)$$

The transfer matrix $t(a,b,c)$ plays the role of a discrete evolution operator acting on the Hilbert space $\mathcal{H}^{(L)}$ spanned by the row states $|\alpha\rangle$ (dim $\mathcal{H}^{(L)} = 2^L$), isomorphic to the one considered above in the diagonalization of the spin-$\frac{1}{2}$ hamiltonian. The full partition function reads thus $Z_{L\times L'}(a,b,c) = \mathrm{tr}_{\mathcal{H}^{(L)}}\left(t(a,b,c)\right)^{L'}$. The trace on $\mathcal{H}^{(L)}$ implements periodic boundary conditions in the "time" direction. This expression is just the hamiltonian formulation of the partition function (43). Thus evaluating the partition function is in fact equivalent to finding the eigenvalues of the transfer matrix. We are led, therefore, to essentially the same problem considered in the previous section, namely the diagonalization of an operator on $\mathcal{H}^{(L)}$.

First of all, the local conservation law (40) translates into $\langle\beta|\,t\,|\alpha\rangle = 0$ unless the total spin is equal for both $|\alpha\rangle$ and $|\beta\rangle$, $\sum_{i=1}^{L}\alpha_i = \sum_{i=1}^{L}\beta_i$. More technically, the number operator $M = \sum_{i=1}^{L}\alpha_i$ commutes with the transfer matrix:

$$[t(a,b,c), M] = 0 \qquad (48)$$

This is the analog of equation (23) and the relation between the total spin S^z and M is simply $S^z = \frac{L}{2} - M$. Once again, the Hilbert space $\mathcal{H}^{(L)}$ can be broken down into sectors $\mathcal{H}_M^{(L)}$ labelled by $M \in \{0, 1, \ldots, L\}$. In each of these sectors, the transfer matrix can be diagonalized independently, $t(a,b,c)|\Psi_M\rangle = \Lambda_M(a,b,c)|\Psi_M\rangle$. The states $|x_1,\ldots,x_M\rangle$ with 1's at the positions $x_1,\ldots,x_M$ and 0's elsewhere form a basis of $\mathcal{H}_M^{(L)}$. Expanding $|\Psi_M\rangle$ in this basis,

$$|\Psi_M\rangle = \sum_{1\le x_1 < x_2 < \cdots < x_M \le L} f(x_1,\ldots,x_M)\,|x_1,\ldots,x_M\rangle \qquad (49)$$

we find the equation for the eigenfunctions $f(x_1,\ldots,x_M)$:

$$\sum_{1\le y_1 < y_2 < \cdots < y_M \le L} \langle x_1,\ldots,x_M|\,t(a,b,c)\,|y_1,\ldots,y_M\rangle\, f(y_1,\ldots,y_M)$$
$$= \Lambda_M(a,b,c) f(x_1,\ldots,x_M) \qquad (50)$$

The transfer matrix connects states with the same number M of down spins, whose locations may change. The eigenvalue problem (50) can be solved with the help of the Bethe ansatz technique.

The sector $M = 0$ contains only one state $|\Omega\rangle = |00\ldots0\rangle$ which is the Bethe reference state. This state plays the role of a vacuum in the construction of the other states, but it need not coincide with the ground state of the model: the physical vacuum minimizes the free energy and may have nothing to do with $|\Omega\rangle$. From (46) and (50), we obtain

$$\Lambda_0 = \langle\Omega|\,t\,|\Omega\rangle = \sum_{\mu=0,1} \left[W\begin{pmatrix}0 & \mu \\ \mu & 0\end{pmatrix}\right]^L = a^L + b^L \qquad (51)$$

In the sector $M = 1$, we choose $f(x) = e^{ikx}$ and some elementary algebra yields (with the assumption of periodic boundary conditions)

$$\Lambda_1(k) = a^L P(k) + b^L Q(k) \tag{52}$$

with

$$P(k) = \frac{ab + (c^2 - b^2)\,e^{-ik}}{a^2 - ab\,e^{-ik}}, \quad Q(k) = \frac{a^2 - c^2 - ab\,e^{-ik}}{ab - b^2\,e^{-ik}} \tag{53}$$

The sector with $M = 2$ excitations contains more structure. The Bethe ansatz reads in this case

$$f(x_1, x_2) = A_{12}\,e^{i(k_1 x_1 + k_2 x_2)} + A_{21}\,e^{i(k_2 x_1 + k_1 x_2)} \tag{54}$$

subject to the periodic boundary conditions, which yields the equations (28) and (29). The eigenvalue of (54) is given by

$$\Lambda_2 = a^L P_1 P_2 + b^L Q_1 Q_2 \tag{55}$$

where $P_i = P(k_i)$ and $Q_i = Q(k_i)$. The ratio of the amplitudes A_{12} and A_{21}, which again may be interpreted as a spin wave scattering matrix, is thus

$$\hat{S}_{12} = \frac{A_{21}}{A_{12}} = -\frac{1 - 2\frac{a^2+b^2-c^2}{2ab}\,e^{ik_2} + e^{i(k_1+k_2)}}{1 - 2\frac{a^2+b^2-c^2}{2ab}\,e^{ik_1} + e^{i(k_1+k_2)}} \tag{56}$$

For later convenience, we define the anisotropy parameter Δ as

$$\Delta = \frac{a^2 + b^2 - c^2}{2ab} \tag{57}$$

Notice the strong similarity between (57) and (22).

The generalization of the above results to sectors with more than two excitations proceeds through the factorization properties of the higher order Bethe amplitudes $A_{1 \cdots M}$ [see equations (35)]. The general formula for the eigenvalue Λ_M of a vector of the form (32) is

$$\Lambda_M = a^L \prod_{i=1}^{M} P(k_i) + b^L \prod_{i=1}^{M} Q(k_i) \tag{58}$$

and the quasi-momenta k_i ($i = 1, \ldots, M$) must satisfy the Bethe equations (36) which follow from the periodicity (33) of the wave functions and the factorization properties of the Bethe amplitudes. In this case, they read explicitly as

$$e^{ik_i L} = (-1)^{M-1} \prod_{\substack{j=1 \\ j \neq i}}^{M} \frac{1 - 2\Delta\,e^{ik_i} + e^{i(k_i+k_j)}}{1 - 2\Delta\,e^{ik_j} + e^{i(k_i+k_j)}} \tag{59}$$

The final step in the computation of the eigenvalues of the transfer matrix and, ultimately, of the partition function, hinges upon the solution of the Bethe equations (59). It is very important that the Bethe equations associated with the six–vertex model depend on the Boltzmann weights a, b and c only through the combination yielding the anisotropy Δ in (57). This is the key for understanding the integrability of the six–vertex model. The first immediate consequence from this observation is that two different transfer matrices $t(a, b, c)$ and $t(a', b', c')$ sharing the same value for Δ have the same eigenvectors and thus they commute:

$$[\,t(a,b,c),\,t(a',b',c')\,] = 0 \quad \Longleftrightarrow \quad \Delta(a,b,c) = \Delta(a',b',c') \tag{60}$$

Therefore, given a value of Δ, through the Bethe procedure we diagonalize not just a transfer matrix but a whole continuous family of mutually commuting transfer matrices. Since each transfer matrix defines a different time evolution, to each transfer matrix with the same parameter Δ is associated a conserved quantity. So the six–vertex model has a large number of conserved quantities, in fact an infinity of them in the thermodynamic limit.

Let us sketch now the relation between the one–dimensional hamiltonian $H_{XXZ}(\Delta) = H_\Delta$ of the anisotropic Heisenberg chain (38) and the two–dimensional six–vertex model. From the identification of the Bethe ansatz eigenvectors under (57), we see that

$$\left[H_{\Delta(a,b,c)}, t(a,b,c) \right] = 0 \tag{61}$$

Comparison of this expression with (60) leads us to suspect that the hamiltonian H_Δ must already be contained somehow in the transfer matrix $t(a, b, c)$, i.e. it should be one of the conserved quantities in the system. The same argument applies also to the translation operator e^{iP} which commutes both with the hamiltonian and with the transfer matrix.

To make these suggestive connections explicit, let us start with the momentum operator e^{-iP}. Suppose we make the following choice of Boltzmann weights:

$$a = c = c_0 \quad , \quad b = 0 \tag{62}$$

which is consistent with any value of Δ. Then from (42) we get

$$W \begin{pmatrix} \beta & \nu \\ \mu & \alpha \end{pmatrix} \bigg|_{\substack{a=c=c_0 \\ b=0}} = c_0\, \delta_{\mu\beta}\, \delta_{\nu\alpha} \tag{63}$$

which can be imagined as an operator which multiplies by c_0 the incoming state $\{\mu, \alpha\}$ but otherwise leaves it untouched: the horizontal state on the left becomes the vertical state on top, and the vertical state below becomes the horizontal state to the right. Thus the transfer matrix from these weights behaves as the shift operator e^{-iP}:

$$\begin{aligned} t_0 \ket{\alpha} &= t(c_0, 0, c_0) \ket{\alpha_1, \alpha_2, \ldots, \alpha_L} \\ &= c_0^L \quad \ket{\alpha_L, \alpha_1, \ldots, \alpha_{L-1}} \end{aligned} \tag{64}$$

and the momentum operator P is identified with

$$P = i \log \left(\frac{t_0}{c_0^L} \right) \tag{65}$$

This identification is easily checked on one–particle states. From (64), we see that $t_0 \ket{x} = \ket{x+1}$, and thus in the Fourier transformed states $\ket{k}$ we find

$$\begin{aligned} t_0 \ket{k} &= t_0 \sum_{x=1}^{L} e^{ikx} \ket{x} = \sum_{x=1}^{L} e^{ikx} \ket{x+1} \\ &= \sum_{x=1}^{L} e^{ik(x+1)} e^{-ikx} \ket{x+1} = e^{-ik} \ket{k} \end{aligned} \tag{66}$$

Similarly, the hamiltonian H_Δ can be obtained by expanding the transfer matrix in the vicinity of the parameter point (62), keeping the value of Δ constant. Note that this amounts to expanding the transfer matrix about t_0, *i.e.* about a matrix proportional to the shift operator. So we fix

$$\Delta = \frac{\delta a - \delta c}{\delta b} \qquad (67)$$

and obtain

$$t_0^{-1}\,\delta t = \frac{\delta b}{2c_0}\sum_{i=1}^{L}\left\{\frac{\delta a + \delta c}{\delta b}\,1 + \sigma_i^x\sigma_{i+1}^x + \sigma_i^y\sigma_{i+1}^y + \Delta\sigma_i^z\sigma_{i+1}^z\right\} \qquad (68)$$

Thus the hamiltonian H_Δ [see (38)] appears in the expansion of the logarithm of the transfer matrix about the shift operator,

$$H_\Delta = i\frac{\partial}{\partial u}\log\frac{t(u)}{c_0^L}\bigg|_{u=0} \qquad (69)$$

We urge the reader to carry out the above simple and most instructive calculation explicitly.

Expanding $\log t(u)$ in powers of u we get a whole set of local conserved quantities involving in general interactions over a finite range, not just among nearest neighbors. In this sense, the transfer matrix is the generating functional for a large class of commuting conserved quantities: this follows from the integrability equation (60). Recall that we arrived at that equation after a long analysis involving the Bethe ansatz. Instead, we could have taken equation (60) as the starting point to define a vertex model, and asked ourselves under what conditions the Boltzmann weights of such vertex model lead to integrability. The answer to this question is that, in order that the vertex model be integrable, the Boltzmann weights must satisfy the justly celebrated Yang–Baxter equation:

$$\sum_{\mu',\nu',\gamma} W\begin{pmatrix}\gamma & \mu'\\ \mu & \alpha\end{pmatrix} W'\begin{pmatrix}\beta & \nu'\\ \nu & \gamma\end{pmatrix} W''\begin{pmatrix}\nu'' & \mu''\\ \nu' & \mu'\end{pmatrix} =$$

$$= \sum_{\mu',\nu',\gamma} W''\begin{pmatrix}\nu' & \mu'\\ \nu & \mu\end{pmatrix} W'\begin{pmatrix}\gamma & \mu''\\ \mu' & \alpha\end{pmatrix} W\begin{pmatrix}\beta & \nu''\\ \nu' & \gamma\end{pmatrix} \qquad (70)$$

where W, W' and W'' are three different sets of Boltzmann weights.

Equation (70) is the Yang–Baxter equation for vertex models. Note that in the two problems worked out so far, the diagonalization of spin chain hamiltonians and the diagonalization of the transfer matrix for vertex models, integrability is encoded in the same mathematical structure, namely the factorization of the spin wave S-matrix (36) and the vertex Yang–Baxter equation (70). There exist several other formulations of the Yang–Baxter equation, which is always a cubic equality; it first appeared under the name of star–triangle relation in Onsager's solution to the two–dimensional Ising model.

A side comment: in statistical mechanics, the local energies must be real and thus the Boltzmann weights must **be real and** positive. It is nevertheless useful to allow the Boltzmann weights to be complex in general. This freedom is useful from the technical viewpoint, but it is also physically meaningful. In particular, the region of parameter space in which the one–dimensional spin chain hamiltonian is hermitian need not coincide with that in which the two–dimensional Boltzmann weights are real and positive. Thus the physical spin chain hamiltonian is indeed an analytic extension of the hamiltonian derived from realistic Boltzmann weights.

5 The Yang–Baxter algebra

Let us analyze in more detail the structure of the Yang–Baxter equation (70). Our goal is to capture in a general algebraic framework the integrability properties of the models studied above. The transfer matrix is an endomorphism

$$t(a,b,c) : V_1 \otimes \cdots \otimes V_L \to V_1 \otimes \cdots \otimes V_L \tag{71}$$

where V_i stands for the spin-$\frac{1}{2}$ representation space at the i-th position of the lattice. This operator is built up by multiplying Boltzmann weights on the same row and summing over the horizontal states connecting them, while keeping the vertical states above and below them fixed [see eq. (46)]. To make this distinction even clearer, we shall refer to the space of horizontal states as auxiliary, and denote it by V_a. The space of vertical states on which the transfer matrix acts we shall call quantum and denote it as $\mathcal{H}^{(L)} = V_1 \otimes \cdots \otimes V_L$, as in (71). For the six–vertex model, V_a is also a spin-$\frac{1}{2}$ representation space, like V_i ($i = 1, \ldots, L$).

According to these definitions, it is natural to interpret the Boltzmann weights associated to the i-th vertex as an operator $\mathcal{R}_{ai}$:

$$\mathcal{R}_{ai} : V_a \otimes V_i \to V_a \otimes V_i \tag{72}$$

where the subindices in $\mathcal{R}$ label the vector spaces it acts upon. The operator $\mathcal{R}_{ai}$ is defined by its matrix elements

$$\begin{array}{c}\beta_i\\ \mu_i \!\!\!\! -\!\!\!\!\!\!-\!\!\!\!\!\!-\!\!\!\! \mu_{i+1}\\ \alpha_i \end{array} = W\begin{pmatrix} \beta_i & \mu_{i+1} \\ \mu_i & \alpha_i \end{pmatrix} \equiv \mathcal{R}^{\mu_{i+1}\beta_i}_{\mu_i\alpha_i} \tag{73}$$

$$= {}_a\langle\mu_{i+1}| \otimes {}_i\langle\beta_i| \, \mathcal{R}_{ai} \, |\mu_i\rangle_a \otimes |\alpha_i\rangle_i$$

Note that if $\mathcal{R}$ appears with two subindices, they label the spaces $\mathcal{R}$ acts upon, whereas if $\mathcal{R}$ appears with two subindices and two superindices, they label the basis vectors of the spaces $\mathcal{R}$ is acting between. Using these notations, the transfer matrix (46) can be written as

$$\langle\beta|t|\alpha\rangle = \sum_{\mu's} \mathcal{R}^{\mu_1\beta_L}_{\mu_L\alpha_L} \mathcal{R}^{\mu_L\beta_{L-1}}_{\mu_{L-1}\alpha_{L-1}} \cdots \mathcal{R}^{\mu_3\beta_2}_{\mu_2\alpha_2} \mathcal{R}^{\mu_2\beta_1}_{\mu_1\alpha_1} \tag{74}$$

We have reversed the order of multiplication of the Boltzmann weights to agree with the conventions for multiplying matrices in the auxiliary space, namely $(XY)^\mu_\nu = X^\mu_\lambda Y^\lambda_\nu$. We thus arrive finally to a label–independent expression for the transfer matrix:

$$t = \mathrm{tr}_a \left(\mathcal{R}_{aL} \mathcal{R}_{aL-1} \cdots \mathcal{R}_{a2} \mathcal{R}_{a1} \right) \tag{75}$$

Here, tr_a denotes the trace over the auxiliary space V_a.

After these preliminaries, we may ask whether two transfer matrices t and t', derived from two sets of Boltzman weights $\mathcal{R}$ and $\mathcal{R}'$, do commute. Of course, t and t' must act on the same quantum space $V_1 \otimes \cdots \otimes V_L$, but the auxiliary spaces for each of them may be different. We multiply t and t' and for clarity we label their respective auxiliary spaces as V_a and V_b, even in the case when these spaces are isomorphic:

$$t\,t' = \mathrm{tr}_{a\times b} \left(\mathcal{R}_{aL} \mathcal{R}'_{bL} \cdots \mathcal{R}_{a1} \mathcal{R}'_{b1} \right) \tag{76}$$

where $\mathrm{tr}_{a\times b}$ denotes the trace on $V_a \otimes V_b$. Similarly, multiplying t' and t we get

$$t'\,t = \mathrm{tr}_{a\times b} \left(\mathcal{R}'_{bL} \mathcal{R}_{aL} \cdots \mathcal{R}'_{b1} \mathcal{R}_{a1} \right) \tag{77}$$

Hence t commutes with t' if and only if there exists an invertible matrix X_{ab} such that
$$\mathcal{R}'_{bi}\mathcal{R}_{ai} = X_{ab}\mathcal{R}_{ai}\mathcal{R}'_{bi}X_{ab}^{-1} \qquad \forall i = 1, \ldots, L \tag{78}$$
Indeed, using the cyclicity of the trace we find
$$t't = \mathrm{tr}_{a\times b}\Big(X_{ab}\mathcal{R}_{aL}\mathcal{R}'_{bL}X_{ab}^{-1}X_{ab}\mathcal{R}_{aL-1}\mathcal{R}'_{bL-1}X_{ab}^{-1}\cdots$$
$$\cdots X_{ab}\mathcal{R}_{a1}\mathcal{R}'_{b1}X_{ab}^{-1}\Big) = tt' \tag{79}$$

Moreover, the matrix X_{ab} may be interpreted as arising from Boltzmann weights on the space $V_a \otimes V_b$: we shall call them $\mathcal{R}''_{ab}$. The integrability condition (78) is the Yang–Baxter equation in operator formalism:
$$\mathcal{R}''_{ab}\mathcal{R}_{ai}\mathcal{R}'_{bi} = \mathcal{R}'_{bi}\mathcal{R}_{ai}\mathcal{R}''_{ab} \tag{80}$$

With some minor changes in notation, equation (80) can be written as
$$\mathcal{R}_{12}\mathcal{R}'_{13}\mathcal{R}''_{23} = \mathcal{R}''_{23}\mathcal{R}'_{13}\mathcal{R}_{12} \tag{81}$$
where $\mathcal{R}_{12}$, $\mathcal{R}'_{13}$ and $\mathcal{R}''_{23}$ are Yang–Baxter matrices acting on the spaces $V_1 \otimes V_2$, $V_1 \otimes V_3$ and $V_2 \otimes V_3$, respectively. In components, the operator Yang–Baxter equation (81) reads as follows:
$$\sum_{j_1,j_2,j_3} \mathcal{R}^{k_1 k_2}_{j_1 j_2}\mathcal{R}'^{j_1 k_3}_{i_1 j_3}\mathcal{R}''^{j_2 j_3}_{i_2 i_3} = \sum_{j_1,j_2,j_3} \mathcal{R}''^{k_2 k_3}_{j_2 j_3}\mathcal{R}'^{k_1 j_3}_{j_1 i_3}\mathcal{R}^{j_1 j_2}_{i_1 i_2} \tag{82}$$

Equation (82) is the most general form of the Yang–Baxter equation for vertex models, in the sense that the spaces V_1, V_2 and V_3 need not be isomorphic. We shall not consider this possibility, but keep all these vector spaces two–dimensional, so that the $\mathcal{R}$ operator is a 4×4 matrix which, in the case of the six–vertex model, is
$$\mathcal{R}^{(6v)}(a,b,c) = \begin{pmatrix} a & & & \\ & b & c & \\ & c & b & \\ & & & a \end{pmatrix} \tag{83}$$

If $\mathcal{R}^{(6v)}$ is to be invertible, then we must require $a \neq 0$ and $b \neq \pm c$.

Taking now three six–vertex $\mathcal{R}$–matrices $\mathcal{R} = \mathcal{R}^{(6v)}(a,b,c)$, $\mathcal{R}' = \mathcal{R}^{(6v)}(a',b',c')$ and $\mathcal{R}'' = \mathcal{R}^{(6v)}(a'',b'',c'')$, then the Yang–Baxter equation holds provided
$$\Delta(a,b,c) = \Delta(a',b',c') = \Delta(a'',b'',c'') \tag{84}$$
in full agreement with equation (60). The Yang–Baxter equation captures completely the integrability of the six–vertex model, encoded in (84).

Expressing the weights a, b and c in terms of u, we find that the Yang–Baxter matrix $\mathcal{R}(u) = \mathcal{R}^{(6v)}(a(u), b(u), c(u))$ satisfies the Yang–Baxter equation (81) in the form
$$\mathcal{R}_{12}(u)\mathcal{R}_{13}(u')\mathcal{R}_{23}(u'') = \mathcal{R}_{23}(u'')\mathcal{R}_{13}(u')\mathcal{R}_{12}(u) \tag{85}$$
with u'' fixed in terms of u and u'. Now, on a sphere all points are equivalent in the sense that any point can be mapped to any other one by a conformal transformation. We may therefore choose the functions $a(u)$, $b(u)$ and $c(u)$ in such a way that u'' is just $u' - u$. Then (85) adopts the usual additive form
$$\mathcal{R}_{12}(u)\mathcal{R}_{13}(u+v)\mathcal{R}_{23}(v) = \mathcal{R}_{23}(v)\mathcal{R}_{13}(u+v)\mathcal{R}_{12}(u) \tag{86}$$
valid for any complex u and v.

The monodromy matrix $T(u)$ is defined in the same manner as the transfer matrix, except that we do not trace over the first (or last, due to periodic boundary conditions) horizontal states in (74), that is to say

$$T(u) = \mathcal{R}_{aL}\mathcal{R}_{aL-1}\cdots\mathcal{R}_{a2}\mathcal{R}_{a1} \qquad (87)$$

The trace of the monodromy matrix on the auxiliary space is just the transfer matrix

$$t(u) = \mathrm{tr}_a T(u) \qquad (88)$$

Using $i, j, \ldots$ as labels in the auxiliary space V_a, we see that $T(u)$ is in fact a matrix $T_i^j(u)$ of operator valued functions which act, in this case, on the Hilbert space $\mathcal{H}^{(L)} = V_1 \otimes \cdots \otimes V_L$. These operators will be represented graphically as

$$T_i^j(u) = \quad i \;\text{—}\; j \qquad (89)$$

with the double line standing for the Hilbert space $\mathcal{H}^{(L)}$. The characteristic feature of these operators is that they satisfy an important set of quadratic relations reflecting their behavior under monodromy:

$$\mathcal{R}_{ab}(u-v)\left(T_a(u)\otimes T_b(v)\right) = \left(T_b(v)\otimes T_a(u)\right)\mathcal{R}_{ab}(u-v) \qquad (90)$$

This equation constitutes the cornerstone of the quantum inverse scattering method; it is also at the origin of the quantum group. The subindices a and b are short-hand for the two auxiliary spaces V_a and V_b on which the operators T and $\mathcal{R}$ act. For extra clarity, we have indulged in a notational redundance indicating the tensor product in (90) which is taken over these auxiliary spaces, while the quantum indices (not shown) are multiplied as ordinary matrix indices. The proof of (90) uses the Yang–Baxter equation (86) repeatedly and elucidates the index interplay:

$$\begin{aligned}\mathcal{R}_{ab}(u-v)\left(T_a(u)\otimes T_b(v)\right) &= \\ &= \mathcal{R}_{ab}(u-v)\mathcal{R}_{aL}(u)\mathcal{R}_{bL}(v)\cdots\mathcal{R}_{a1}(u)\mathcal{R}_{b1}(v) \\ &= \mathcal{R}_{bL}(v)\mathcal{R}_{aL}(u)\cdots\mathcal{R}_{b1}(v)\mathcal{R}_{a1}(u)\mathcal{R}_{ab}(u-v) \\ &= \left(T_b(v)\otimes T_a(u)\right)\mathcal{R}_{ab}(u-v)\end{aligned} \qquad (91)$$

For practical purposes, it is often convenient to write equation (90) in components:

$$\sum_{j_1,j_2}\mathcal{R}_{j_1 j_2}^{k_1 k_2}(u-v)T(u)_{i_1}^{j_1}T(v)_{i_2}^{j_2} = \sum_{j_1,j_2}T(v)_{j_2}^{k_2}T(u)_{j_1}^{k_1}\mathcal{R}_{i_1 i_2}^{j_1 j_2}(u-v) \qquad (92)$$

Given (90), it is an easy task to prove the commutativity of the transfer matrices, i.e. $[\mathrm{tr}\,T_a(u),\mathrm{tr}\,T_b(u)] = 0$.

Equation (92) has been derived for the six–vertex model, but it can be taken as the starting point for the construction of integrable vertex models, at least for those with the difference property. To this end, we shall introduce the formal notion of a Yang–Baxter algebra.

A Yang–Baxter algebra $\mathcal{A}$ consists of a couple $(\mathcal{R},T)$, where $\mathcal{R}$ is an $n^2\times n^2$ invertible matrix and $T_i^j(u)$ $(i,j\in\{1,\ldots,n\}; u\in\mathbf{C})$ are the generators of $\mathcal{A}$. They must satisfy the quadratic relations (92), whose consistency implies the Yang–Baxter equation (86) for $\mathcal{R}(u)$. The entries of the matrix $\mathcal{R}(u)$ play the role of structure constants of the algebra $\mathcal{A}$. This is quite analogous to a Lie algebra, or better yet to its universal enveloping algebra, which is also defined in terms of a set of generators and structure constants. Following this analogy, the Yang–Baxter relation plays the role of the Jacobi identity: they both reflect the associativity of the corresponding algebras.

An important property of Yang–Baxter algebras is their "addition law", called co-product or co-multiplication Δ, which maps the algebra $\mathcal{A}$ into the tensor product $\mathcal{A} \otimes \mathcal{A}$ while preserving the algebraic relations of $\mathcal{A}$:

$$\Delta : \mathcal{A} \to \mathcal{A} \otimes \mathcal{A}$$
$$T_i^j(u) \mapsto \sum_k T_i^k(u) \otimes T_k^j(u) \tag{93}$$

The diagrammatic representation of the co-product follows from (89):

$$\Delta \left(i \,\text{—}\!\!\!\!=\!\!\!\!\text{—}\, j \right) = \sum_k i \,\text{—}\!\!\!\!=\!\!\!\!\text{—}\, k \,\text{—}\!\!\!\!=\!\!\!\!\text{—}\, j \tag{94}$$

It is left as an exercise for the reader to check that ΔT_i^j satisfy the same relations as T_i^j in (92). The algebra $\mathcal{A}$ has thus both a multiplication and a co-multiplication; $\mathcal{A}$ is called a bi-algebra.

The definition of a Yang–Baxter algebra just provided is general, and it can be applied to any $\mathcal{R}$-matrix satisfying the Yang–Baxter relation. We confine ourselves once more to the Yang–Baxter algebra constructed from the $\mathcal{R}$-matrix of the six-vertex model.

We represent the four generators T_i^j of $\mathcal{A}^{(6v)}$ as operators acting on a Hilbert space $\mathcal{H}$. We shall name them as follows, for convenience:

$$\begin{aligned} T_0^0(u) &= A(u) \,, & T_1^0(u) &= B(u) \\ T_0^1(u) &= C(u) \,, & T_1^1(u) &= D(u) \end{aligned} \tag{95}$$

Just as the structure constants of Lie algebras provide a representation of the algebra (the adjoint), the $\mathcal{R}^{(6v)}$ matrix provides a representation of $\mathcal{A}^{(6v)}$ of dimension two under the identification

$$\left(T_i^j(u) \right)_\ell^k = \mathcal{R}_{i\ell}^{jk}(u) \tag{96}$$

or explicitly

$$\begin{aligned} A(u) &= \begin{pmatrix} a(u) & 0 \\ 0 & b(u) \end{pmatrix} = \frac{a+b}{2} \mathbf{1} + \frac{a-b}{2} \sigma_3 \\ B(u) &= \begin{pmatrix} 0 & 0 \\ c(u) & 0 \end{pmatrix} = c \sigma^- \\ C(u) &= \begin{pmatrix} 0 & c(u) \\ 0 & 0 \end{pmatrix} = c \sigma^+ \\ D(u) &= \begin{pmatrix} b(u) & 0 \\ 0 & a(u) \end{pmatrix} = \frac{a+b}{2} \mathbf{1} - \frac{a-b}{2} \sigma_3 \end{aligned} \tag{97}$$

Equations (97) yield what we might call the spin-$\frac{1}{2}$ representation of the algebra (92). This nomenclature is appropriate since $C(u)$ and $B(u)$ act as raising and lowering operators, respectively, while $A(u)$ and $D(u)$ span the Cartan subalgebra of $SU(2)$. Using now the bi-algebra structure of $\mathcal{A}^{(6v)}$ defined by the co-multiplication (93), we may obtain a representation of $\mathcal{A}^{(6v)}$ on the space $\mathcal{H}^{(L)} = \otimes^L V_{\frac{1}{2}}$. In particular, for

$L = 2$ we get
$$\Delta(A(u)) = A(u) \otimes A(u) + C(u) \otimes B(u)$$
$$\Delta(B(u)) = B(u) \otimes A(u) + D(u) \otimes B(u)$$
$$\Delta(C(u)) = C(u) \otimes D(u) + A(u) \otimes C(u) \qquad (98)$$
$$\Delta(D(u)) = D(u) \otimes D(u) + B(u) \otimes C(u)$$

It is an easy exercise to check that ΔC annihilates the reference state $|\Omega\rangle \equiv |00\rangle = |\uparrow\uparrow\rangle$:
$$\Delta C(u) |00\rangle = 0 \qquad (99)$$

Under this interpretation of $B(u)$ and $C(u)$ as creation and annihilation operators, it follows that $\Delta B(u)$ acting on the reference state $|\Omega\rangle$ yields a state in the sector with the number of spins down equal to one. We can rewrite this state as
$$\Delta B(u) |\Omega\rangle = |\Psi_1\rangle = \sum_x f(x) |x\rangle = f(1) |10\rangle + f(2) |01\rangle \qquad (100)$$

with
$$f(1) = c(u)a(u) \quad , \quad f(2) = b(u)c(u) \qquad (101)$$

Comparing (101) with (24), we deduce the relation between Boltzmann weights and quasi-momenta:
$$\frac{b(u)}{a(u)} = e^{ik} \qquad (102)$$

This method for lowering spins (*i.e.* creating 1's) from a reference state by means of the B operators can be extended to a lattice with $L > 2$ sites. To do so we recall the definition (95) of the operator B as the entry T_1^0 of the monodromy matrix; thanks to the co-product (93), it can be made to act on the space $\mathcal{H} = \otimes^L V_{\frac{1}{2}}$:
$$B(u) = \Delta^{L-1}\left(T_1^0(u)\right) \qquad (103)$$

where
$$\Delta^{L-1} : \mathcal{A} \to \overbrace{\mathcal{A} \otimes \cdots \otimes \mathcal{A}}^{L \text{ times}} \qquad (104)$$

is the associative generalization of (93), $\Delta^{L-1} = (1 \otimes \Delta)\Delta^{L-2}$ with $L \geq 2$. Hence a state with M spins down can be built as follows:
$$|\Psi_M\rangle = \prod_{i=1}^{M} B(u_i) |00\cdots 0\rangle = \prod_{i=1}^{M} B(u_i) |\Omega\rangle \qquad (105)$$

The states (105) are called algebraic Bethe ansatz states ["algebraic" in contrast with the "co-ordinate" description (31)], and constitute a very good starting point for solving the eigenvalue problem of the transfer matrix. In order to show this, let us work out more explicitly the relations satisfied by the generators of the six-vertex Yang–Baxter algebra.

From (83) and (92) we obtain, for arbitrary u and v,
$$B(u)B(v) = B(v)B(u) \qquad (106a)$$
$$A(u)B(v) = \frac{a(v-u)}{b(v-u)} B(v)A(u) - \frac{c(v-u)}{b(v-u)} B(u)A(v) \qquad (106b)$$

$$D(u)B(v) = \frac{a(u-v)}{b(u-v)} B(v)D(u) - \frac{c(u-v)}{b(u-v)} B(u)D(v) \quad (106c)$$

$$C(u)B(v) - B(v)C(u) = \frac{c(u-v)}{b(u-v)} \left(A(v)D(u) - A(u)D(v) \right) \quad (106d)$$

Equation (106a) implies that the algebraic Bethe ansatz state (105) is independent of the ordering in which the B operators are multiplied.

The transfer matrix of the six–vertex model can be written from equations (88) and (95) as

$$t^{(6v)}(u) = \mathrm{tr}_a T^{(6v)}(u) = A(u) + D(u) \quad (107)$$

Therefore, the problem of diagonalizing the transfer matrix (107) in the algebraic Bethe ansatz basis (105) amounts to finding a choice of the parameters $\{u_i, i = 1, \ldots, M\}$ such that

$$\begin{aligned} t^{(6v)}(u)|\Psi_M\rangle &= [A(u) + B(u)] \prod_{i=1}^{M} B(u_i) |\Omega\rangle \\ &= \Lambda_M(u; \{u_i\}) \prod_{i=1}^{M} B(u_i) |\Omega\rangle \end{aligned} \quad (108)$$

The advantage of using the algebraic Bethe ansatz states is that the whole computation involved in (108) reduces to a systematic use of the commutation relations (106), in addition to the obvious relations

$$A(u)|\Omega\rangle = a(u)^L |\Omega\rangle \quad , \quad D(u)|\Omega\rangle = b(u)^L |\Omega\rangle \quad (109)$$

Indeed, using (106b), (106c) and (109), we find

$$(A(u) + D(u)) \prod_{i=1}^{M} B(u_i) |\Omega\rangle = \text{unwanted terms} + \\ + \left[a^L(u) \prod_{i=1}^{M} \frac{a(u_i - u)}{b(u_i - u)} + b^L(u) \prod_{i=1}^{M} \frac{a(u - u_i)}{b(u - u_i)} \right] \prod_{i=1}^{M} B(u_i) |\Omega\rangle \quad (110)$$

The first term of the right–hand side of this equation gives us the eigenvalue of the transfer matrix:

$$\Lambda_M(u; \{u_i\}) = a^L(u) \prod_{i=1}^{M} \frac{a(u_i - u)}{b(u_i - u)} + b^L(u) \prod_{i=1}^{M} \frac{a(u - u_i)}{b(u - u_i)} \quad (111)$$

From the second summands in (106b) and (106c), however, we also obtain terms which are not of the desirable form $\prod_{i=1}^{M} B(u_i) |\Omega\rangle$. If these terms are present, the algebraic Bethe ansatz does not work. The unwanted terms actually cancel, for the six–vertex model, under a judicious choice of the u_i parameters. The condition that the parameters u_i should satisfy to guarantee the cancellation of the unwanted terms is precisely the Bethe equations, written in the form

$$\left(\frac{a(u_i)}{b(u_i)} \right)^L = \prod_{\substack{j=1 \\ j \neq i}}^{M} \frac{a(u_i - u_j) b(u_j - u_i)}{a(u_j - u_i) b(u_i - u_j)} \quad (112)$$

Equations (111) and (112) are the final outcome of the diagonalization of the transfer matrix through the algebraic Bethe ansatz, which we now can compare with equations (58) and (59). Matching the eigenvalues (58) and (111) yields

$$\frac{a(u_i - u)}{b(u_i - u)} = P(k_i) = \frac{a(u)b(u) + (c^2(u) - b^2(u))\,e^{-ik_i(u_i)}}{a^2(u) - a(u)b(u)\,e^{-ik_i(u_i)}} \qquad (113a)$$

$$\frac{a(u - u_i)}{b(u - u_i)} = Q(k_i) = \frac{a^2(u) - c^2(u) - a(u)b(u)\,e^{-ik_i(u_i)}}{a(u)b(u) - b^2(u)\,e^{-ik_i(u_i)}} \qquad (113b)$$

Choosing $u = 0$ in (113a) and using the fact that $a(0) = c(0) \neq 0$, $b(0) = 0$ (see the explicit parametrization below), we get

$$\frac{b(u_i)}{a(u_i)} = e^{ik_i(u_i)} \qquad (114)$$

Hence the comparison between (112) and (59) ends up producing

$$\hat{S}_{ji} = \frac{a(u_j - u_i)b(u_i - u_j)}{a(u_i - u_j)b(u_j - u_i)} = -\frac{1 - 2\Delta\,e^{ik_i} + e^{i(k_i+k_j)}}{1 - 2\Delta\,e^{ik_j} + e^{i(k_i+k_j)}} \qquad (115)$$

which confirms the result (102). Equations (113), (114) and (115) provide the map between the quasi-momenta k_i and the uniformization variables u_i used in the algebraic Bethe ansatz construction.

Let us use the following uniformization of the Boltzmann weights of the six-vertex model:

$$\begin{aligned} a(u) &= \mathcal{R}_{00}^{00}(u) = \mathcal{R}_{11}^{11}(u) = \sinh(u + i\gamma) \\ b(u) &= \mathcal{R}_{10}^{10}(u) = \mathcal{R}_{01}^{01}(u) = \sinh u \\ c(u) &= \mathcal{R}_{01}^{10}(u) = \mathcal{R}_{10}^{01}(u) = i\sin\gamma \end{aligned} \qquad (116)$$

where the parameter γ is related to the anisotropy Δ by the relation $\Delta = \cos\gamma$. Using now the map

$$e^{ik_j} = \frac{\sinh u_j}{\sinh(u_j + i\gamma)} \qquad (117)$$

it is easy to check that equations (113) and (115) are satisfied with the six-vertex $\mathcal{R}$-matrix (116), and that the Bethe equations can be written as

$$\left(\frac{\sinh(u_j + i\gamma)}{\sinh u_j}\right)^L = \prod_{\substack{k=1 \\ k \neq j}}^{M} \frac{\sinh(u_j - u_k + i\gamma)}{\sinh(u_j - u_k - i\gamma)} \qquad (118)$$

We may calculate the energy of the Bethe ansatz state (105) from the eigenvalue of the transfer matrix (111) if we recall that the hamiltonian is defined as

$$H = i\frac{\partial}{\partial u}\log\left(\frac{t(u)}{a(u)^L}\right)\bigg|_{u=0} \qquad (119)$$

$$= \frac{1}{2\sin\gamma}\sum_{j=1}^{L}\left[\sigma_j^x\sigma_{j+1}^x + \sigma_j^y\sigma_{j+1}^y + \cos\gamma\left(\sigma_j^z\sigma_{j+1}^z - 1\right)\right]$$

212 Quantum Groups

Indeed, carrying out the computation explicitly we find

$$E_M(\{u_j\}) = i\frac{\partial}{\partial u}\log\left(\frac{\Lambda_M(u,\{u_j\})}{a(u)^L}\right)\bigg|_{u=0}$$

$$= -\sum_{j=1}^{M}\frac{\sin\gamma}{\sinh u_j \sinh(u_j+i\gamma)} \qquad (120)$$

Similarly, the total momentum of the same Bethe state is

$$P_M(\{u_j\}) = i\log\left(\frac{t(u)}{a(u)^L}\right)\bigg|_{u=0} = i\sum_{j=1}^{M}\log\frac{\sinh(u_j+i\gamma)}{\sinh u_j} \qquad (121)$$

Equations (113)–(115) explain the integrability content of the Bethe equations of the six–vertex model, which is codified in the Yang–Baxter equation. The factorization properties for S–matrices simply reflect the consistency of the Yang–Baxter algebra. We have thus reinterpreted the factorization properties as integrability of the six–vertex model.

6 Physical spectrum of the Heisenberg spin chain

Let us abandon formalism for a while and come back to the simple system of the one–dimensional antiferromagnetic Heisenberg model (the XXX model)

$$H = J\sum_{i=1}^{L}(\vec{\sigma}_i\cdot\vec{\sigma}_{i+1}-1) \qquad J>0 \qquad (122)$$

Since J is positive, neighboring spins tend to align antiparallel. If J was negative, then it would be favored for all spins to align in the same direction, and we would be in the ferrromagnetic phase.

The first issue we should address about a one–dimensional spin system like (122) is whether a particle interpretation exists. By this, we mean whether it is possible to define a Fock representation of the Hilbert space of the model, such that the vacuum $|0\rangle$ of the Fock space $\mathcal{F}$ corresponds to the ground state and the many–particle states to the low lying excitations. A particle interpretation is readily available if we establish the correspondence

$$\mathcal{H}_\infty^{\ell.\ell.} \to |0\rangle \oplus \mathcal{H}_1 \oplus \mathcal{H}_2 \oplus \cdots = \mathcal{F} \qquad (123)$$

where $\mathcal{H}_\infty^{\ell.\ell.}$ represents the Hilbert space of the low lying excitations of the hamiltonian (122) in the limit $L\to\infty$, and $\mathcal{H}_n$ is the Hilbert space of n elementary excitations. The Fock vacuum $|0\rangle$ represents the antiferromagnetic ground state.

The two crucial questions about the elementary excitations are

i) The dispersion relation $\epsilon(k)$, from which we get information about the existence of a mass gap. If the mass gap is zero, then the theory may correspond in the continuum limit to a massless or critical field theory, *i.e.* to a conformal field theory.

ii) The internal quantum numbers (spin) of the elementary excitations.

To get some flavor for why the answer to these two questions is so difficult, it is necessary to reflect for a moment on the richness of the antiferromagnetic vacuum. Recall that to solve the model (122) we must diagonalize the hamiltonian in the basis of spin waves. A state with M spin waves is gotten by flipping M spins down from the reference state with all spins up. For the trivial case of only one spin wave in a periodic chain, the dispersion relation for the spin wave of the XXX model is given by

$$E(k) = 4J(\cos k - 1) \qquad (124)$$

and thus the different physical behavior of the ferromagnetic phase and the antiferromagnetic one can be easily distinguished.

In the ferromagnetic regime, the coupling constant J is negative and the energy of the spin wave is positive, so the Bethe reference state coincides with the ground state of minimal energy; this is the ordered phase, and spins tend to align. The solution to the physical problem is relatively straightforward.

The antiferromagnetic regime, with a positive coupling $J > 0$, is trickier. The energy of the spin wave is negative, and flipping one spin down is energetically favored over keeping all spins up. The Bethe reference state has nothing to do with the ground state, which we expect to be a singlet of the global $SU(2)$ symmetry, in fact a state with $S^z_{total} = 0$. To get such a state in our picture, we need a "condensate" of spin waves. This physical intuition, together with the fact that the energy of the spin wave for $J > 0$ is negative, can be combined thanks to the concept of a Dirac sea. Identifying the vacuum as the Dirac sea filled up to the Fermi surface, the elementary excitations will be thought of as holes in the Dirac sea. The integrability of the model, which amounts to the factorizability of the scattering matrix for spin waves, allows us to construct the sea starting from the Bethe equations.

For the isotropic Heisenberg model (122) in the antiferromagnetic phase $J > 0$, the two questions above were pretty much solved by Faddeev and Takhtadjan in the early eighties. The dispersion relation for the low lying excitations turns out to be of the form

$$\epsilon(k) = 2\pi J \sin k \qquad 0 \leq k \leq \pi \qquad (125)$$

which means that the system has no mass gap. This is consistent with a continuum limit described by a free massless scalar field. The surprising result has to do with the spin of the low lying excitations. Since an excitation corresponds to flipping one local spin up into a spin down, with a net change of one unit of angular momentum, you might have guessed from the one–particle hamiltonian that the elementary excitations would have spin one. Instead, it turns out that the particle–like excitations over the antiferromagnetic Dirac sea have spin $\frac{1}{2}$. The Fock space of the model is thus, for a chain with an even number of sites,

$$\mathcal{F} = \bigoplus_{n=0}^{\infty} \int_0^{\pi} \cdots \int_0^{\pi} dk_1 \cdots dk_{2n} \otimes^{2n} \mathbf{C}^2 \qquad (126)$$

where the integrations run over the possible values of the momenta and $\mathbf{C}^2$ represents the internal spin-$\frac{1}{2}$ space of dimension two. Proper symmetrization of the states in (126) must be taken into account as well. The excitations come in pairs [whence the $2n$ in (126)], for otherwise the total spin of a chain with an even number of sites would not be an integer. Let us stress that the internal quantum numbers of the elementary excitations are a completely unexpected collective result, impossible to predict *a priori*. This is the motivation for invoking particles with strange statistics (anyons) in attempts to understand high temperature superconductors.

7 Yang–Baxter algebras and braid groups

The basic relation studied two section ago is the Yang–Baxter equation (86). In components, it reads as

$$\sum_{j_1,j_2,j_3} \mathcal{R}^{k_1 k_2}_{j_1 j_2}(u) \mathcal{R}^{j_1 k_3}_{i_1 j_3}(u+v) \mathcal{R}^{j_2 j_3}_{i_2 i_3}(v) =$$
$$= \sum_{j_1,j_2,j_3} \mathcal{R}^{k_2 k_3}_{j_2 j_3}(v) \mathcal{R}^{k_1 j_3}_{j_1 i_3}(u+v) \mathcal{R}^{j_1 j_2}_{i_1 i_2}(u) \quad (127)$$

where all the indices run from 1 to $n = \dim V$ ($n = 2$ for the six–vertex model). An interesting way to write this equation calls for the permuted R matrix,

$$R = P\mathcal{R} : V_1 \otimes V_2 \to V_2 \otimes V_1 \quad (128)$$

where P is the permutation map

$$P : V_1 \otimes V_2 \to V_2 \otimes V_1$$
$$e_i^{(1)} \otimes e_j^{(2)} \mapsto e_j^{(2)} \otimes e_i^{(1)} \quad (129)$$

with $\{e_r^{(i)}, r = 1, \ldots, n\}$ a basis of V_i. The relation between the entries of R and $\mathcal{R}$ is straightforward:

$$R = P\mathcal{R} \iff R^{k\ell}_{ij} = \mathcal{R}^{\ell k}_{ij} \quad (130)$$

With the help of the permuted R–matrix, the Yang–Baxter equation (127) can be written as

$$(1 \otimes R(u))(R(u+v) \otimes 1)(1 \otimes R(v)) =$$
$$= (R(v) \otimes 1)(1 \otimes R(u+v))(R(u) \otimes 1) \quad (131)$$

Every operator in parentheses acts on the space $V \otimes V \otimes V$:

$$(R(u) \otimes 1) e_{i_1} \otimes e_{i_2} \otimes e_{i_3} = R^{j_2 j_1}_{i_1 i_2}(u) e_{j_2} \otimes e_{j_1} \otimes e_{i_3}$$
$$(1 \otimes R(u)) e_{i_1} \otimes e_{i_2} \otimes e_{i_3} = R^{j_3 j_2}_{i_2 i_3}(u) e_{i_1} \otimes e_{j_3} \otimes e_{j_2} \quad (132)$$

Note that R is very close to a factorizable S–matrix.

The reason for writing the Yang–Baxter equation in the form (131) comes from its relation to the braid group B_L on L strands, which is generated by $L-1$ elements σ_i ($i = 1, \ldots, L-1$) subject to the relations

$$\sigma_i \sigma_{i+1} \sigma_i = \sigma_{i+1} \sigma_i \sigma_{i+1} \quad (133a)$$
$$\sigma_i \sigma_j = \sigma_j \sigma_i \quad |i-j| \geq 2 \quad (133b)$$
$$\sigma_i \sigma_i^{-1} = \sigma_i^{-1} \sigma_i = 1 \quad (133c)$$

The generator σ_i braids the i–th strand under the $(i+1)$–th strand, whereas σ_i^{-1} effects the inverse braiding, i.e. it takes the i–th strand over the $(i+1)$–th strand.

Trying to make (131) look more like (133), we define the operators $R_i(u)$ (for $i = 1, \ldots, L-1$) on $\bigotimes_{i=1}^L V_i$, which act on the spaces $V_i \otimes V_{i+1}$ as $R(u)$ and as the identity elsewhere:

$$R_i(u) = 1 \otimes \cdots \otimes 1 \otimes \overset{(i,i+1)}{R(u)} \otimes 1 \otimes \cdots \otimes 1 \quad (134)$$

Then equation (131) becomes
$$R_{i+1}(u)R_i(u+v)R_{i+1}(v) = R_i(v)R_{i+1}(u+v)R_i(u) \tag{135}$$
while obviously
$$R_i(u)R_j(v) = R_j(v)R_i(u) \qquad |i-j| > 1 \tag{136}$$
The identification of the Yang–Baxter equation in the form (135) with the braid group relation (133a) cannot be realized yet due to the presence of the rapidity variable u. This should be no problem, however, in a situation where $u = v = u + v$ which has two solutions:

i) $u = v = 0$,
ii) $u = v$, $|u| = \infty$.

The first solution is trivial, since from (116) and (130) we get merely
$$R(u=0) = i \sin \gamma \, \mathbf{1} \tag{137}$$
And thus $\mathcal{R}$ is just proportional to a permutation P. Solution (ii) is known as the braid limit. Up to constant factors,
$$\lim_{u \to \pm\infty} e^{-|u|} R(u) \sim P \exp\left[(\pm i\gamma/2)\sigma^z \otimes \sigma^z\right] \tag{138}$$
where we have assumed u real. This limit provides us with a representation of the braid group in terms of, essentially, permutations. The permutation group of L elements S_L satisfies the same defining relations as the braid group B_L, except for the crucial difference that the square of a transposition is the identity, and therefore one cannot distinguish overcrossings from undercrossings. A good representation of the braid group should be able to distinguish between σ and σ^{-1}. In the limit $u \to +\infty$, the Boltzmann weights behave as
$$a(u) \to \frac{1}{2} e^u e^{i\gamma} \quad , \quad b(u) \to \frac{1}{2} e^u \quad , \quad c(u) = i \sin \gamma \tag{139}$$
Hence the information contained in the weight $c(u)$ is washed out in the limit, which accounts for the "triviality" of the result (138). In order not to lose information in the limits $u \to \pm\infty$, we perform a u–dependent rescaling of the basis elements, i.e. a u–dependent "diagonal" change of basis $\tilde{e}_r(u) = f_r(u)e_r(u)$ ($r = 1,\ldots,n$). Recalling that the definition of R is
$$R(e_{r_1}(u_1) \otimes e_{r_2}(u_2)) = R_{r_1 r_2}^{r_2' r_1'}(u_1 - u_2) e_{r_2'}(u_2) \otimes e_{r_1'}(u_1) \tag{140}$$
we deduce that the R–matrix in the new basis $\tilde{e}_r$ is given by
$$\tilde{R}_{r_1 r_2}^{r_2' r_1'}(u_1, u_2) = \frac{f_{r_1}(u_1) f_{r_2}(u_2)}{f_{r_1'}(u_1) f_{r_2'}(u_2)} R_{r_1 r_2}^{r_2' r_1'}(u_1 - u_2) \tag{141}$$
The trick is to preserve the **difference property** of the R–matrix under this change of basis, and thereby fix the scaling functions $f_r(u)$. Indeed, if $\tilde{R}(u_1, u_2)$ is to still depend only on the difference $u_1 - u_2$, the functions in the change of basis must be $f_r(u) = e^{\alpha u r}$ Since $\tilde{R}$ conserves the total quantum number, that is $\tilde{R}_{r_1 r_2}^{r_2' r_1'}(u_1, u_2) = 0$ unless $r_1 + r_2 = r_1' + r_2'$, we may write the rescaled $\tilde{R}$ matrix explicitly as
$$\tilde{R}_{r_1 r_2}^{r_2' r_1'}(u_1 - u_2, \alpha) = e^{\alpha(u_1 - u_2)(r_1 - r_1')} R_{r_1 r_2}^{r_2' r_1'}(u_1 - u_2) \tag{142}$$

We take the braid limit of (142) at a special value of α:

$$R \equiv 2\,e^{-i\gamma/2} \lim_{u \to +\infty} e^{-u} \tilde{R}(u, \alpha = 1) \tag{143}$$

obtaining

$$\begin{aligned}
R_{00}^{00} &= R_{11}^{11} = e^{i\gamma/2} \\
R_{10}^{01} &= R_{01}^{10} = e^{-i\gamma/2} \\
R_{10}^{10} &= e^{-i\gamma/2}\left(e^{i\gamma} - e^{-i\gamma}\right) \\
R_{01}^{01} &= 0
\end{aligned} \tag{144}$$

or, in matrix form,

$$R = \begin{array}{c} \\ 00 \\ 01 \\ 10 \\ 11 \end{array} \begin{pmatrix} \overset{00}{q^{\frac{1}{2}}} & \overset{01}{0} & \overset{10}{0} & \overset{11}{0} \\ 0 & 0 & q^{-\frac{1}{2}} & 0 \\ 0 & q^{-\frac{1}{2}} & q^{-\frac{1}{2}}(q - q^{-1}) & 0 \\ 0 & 0 & 0 & q^{\frac{1}{2}} \end{pmatrix} \tag{145}$$

with

$$q = e^{i\gamma} \tag{146}$$

These R-matrices, first derived by Jimbo, satisfy the Yang–Baxter relation (131) without spectral parameter, and appear often in the literature. It can be checked explicitly that $(R)^2 \neq 1$ for $\gamma \neq 0$, so that we have indeed obtained a genuine representation of the braid group. In the isotropic case ($\gamma = 0$) we fall back to the previous result (138). The inverse matrix R^{-1} can be obtained as the other real infinite limit of the same rescaled R:

$$R^{-1} = -2\,e^{i\gamma/2} \lim_{u \to -\infty} e^{u} R(u, \alpha = 1) \tag{147}$$

8 Yang–Baxter algebras and quantum groups

In the new basis $\{\tilde{e}_r\}$, the $\mathcal{R}$-matrix (without the permutation in R) and the monodromy matrix T are related to those in the basis $\{e_r\}$ by the equations

$$\tilde{\mathcal{R}}_{i_1 i_2}^{j_1 j_2}(u) = e^{u(i_1 - j_1)} \mathcal{R}_{i_1 i_2}^{j_1 j_2}(u) \tag{148a}$$

$$\tilde{T}(u)_i^j = e^{u(i-j)} T(u)_i^j \tag{148b}$$

The reader may check that $\tilde{\mathcal{R}}(u)$ and $\tilde{T}(u)$ do satisfy indeed equation (92).

Now just like we took the limit $u \to \pm\infty$ of the matrix $\tilde{\mathcal{R}}(u)$, we may take the limit of the monodromy matrix $\tilde{T}(u)$. To get a feeling for what this limit may yield, let us consider the spin-$\frac{1}{2}$ representation given by equations (97):

$$\begin{aligned}
\lim_{u \to +\infty} \tilde{A}(u) &= \lim_{u \to +\infty} \frac{1}{2} e^{u} e^{i\gamma/2} \begin{pmatrix} e^{i\gamma/2} & \\ & e^{-i\gamma/2} \end{pmatrix} = \frac{1}{2} e^{u} q^{1/2} q^{S^z} \\
\lim_{u \to -\infty} \tilde{A}(u) &= \lim_{u \to -\infty} -\frac{1}{2} e^{-u} e^{-i\gamma/2} \begin{pmatrix} e^{-i\gamma/2} & \\ & e^{i\gamma/2} \end{pmatrix} \\
&= -\frac{1}{2} e^{-u} q^{-1/2} q^{-S^z}
\end{aligned} \tag{149}$$

where $S^z = \frac{1}{2}\sigma^z$ is the Cartan generator in the spin-$\frac{1}{2}$ representation of $SU(2)$ and we recall that $q = e^{i\gamma}$. The limits $u \to \pm\infty$ of $\tilde{B}(u)$, $\tilde{C}(u)$ and $\tilde{D}(u)$ can be evaluated similarly, whereby the braid limits of the monodromy matrix $\tilde{T}(u)$ are

$$T_+ \equiv 2q^{-1/2} \lim_{u \to +\infty} e^{-u} \begin{pmatrix} \tilde{T}_0^0 & \tilde{T}_0^1 \\ \tilde{T}_1^0 & \tilde{T}_1^1 \end{pmatrix}$$
$$= \begin{pmatrix} q^{S^z} & 0 \\ q^{-1/2}(q-q^{-1})S^- & q^{-S^z} \end{pmatrix} \tag{150a}$$

$$T_- \equiv -2q^{1/2} \lim_{u \to -\infty} e^{u} \begin{pmatrix} \tilde{T}_0^0 & \tilde{T}_0^1 \\ \tilde{T}_1^0 & \tilde{T}_1^1 \end{pmatrix}$$
$$= \begin{pmatrix} q^{-S^z} & -q^{1/2}(q-q^{-1})S^+ \\ 0 & q^{S^z} \end{pmatrix} \tag{150b}$$

where $S^\pm = (\sigma^x \pm i\sigma^y)/2$ are the off-diagonal generators of $SU(2)$ in the spin-$\frac{1}{2}$ irrep.

The fun starts when we take the various limits $u \to \pm\infty$, $v \to \pm\infty$ in the $\mathcal{R}TT = TT\mathcal{R}$ equation (90). All the extra factors work out nicely so that the result is

$$\mathcal{R}_{12} (T_+)_1 (T_+)_2 = (T_+)_2 (T_+)_1 \mathcal{R}_{12}$$
$$\mathcal{R}_{12} (T_+)_1 (T_-)_2 = (T_-)_2 (T_+)_1 \mathcal{R}_{12} \tag{151}$$
$$\mathcal{R}_{12}^{-1} (T_-)_1 (T_-)_2 = (T_-)_2 (T_-)_1 \mathcal{R}_{12}^{-1}$$

with $\mathcal{R} = PR$ [R given by (144)] and $T_\pm$ from (150).

This system of equations appears rather complicated at first sight. They are, in fact, equivalent to the following algebraic relations between S^z, S^+ and S^-:

$$\left[S^z, S^\pm\right] = \pm S^\pm \tag{152a}$$

$$\left[S^+, S^-\right] = \frac{q^{2S^z} - q^{-2S^z}}{q - q^{-1}} \tag{152b}$$

These are the defining relations for the quantum group $U_q(s\ell(2))$, which is some kind of deformation of the Lie algebra $s\ell(2)$ with $q = e^{i\gamma}$ acting as deformation parameter.

The notation $U_q(s\ell(2))$ clarifies that the quantum group consists of all the formal powers and linear combinations of S^+, S^- and S^z, subject to the relations (152). Traditionally, $U(s\ell(2))$ denotes the universal enveloping algebra of $s\ell(2)$, that is all the formal powers and linear combinations of $S^\pm$ and S^z modulo the standard Lie algebra relations. In the isotropic limit $\gamma \to 0$, i.e. $q \to 1$, we recover from (152) the usual $s\ell(2)$ algebra. The limit $\gamma \to 0$ is called classical, in the sense that the "quantum" group $U_q(s\ell(2))$ becomes the "classical" universal enveloping algebra $U(s\ell(2))$. From the viewpoint of rigid nomenclature, it is perhaps unfortunate that the classical limit of a quantum group is (the universal enveloping algebra of) a Lie algebra; beware of misled distinctions between quantum groups and quantum algebras!

Finally, we may take the braid limits $u \to \pm\infty$ in the co-multiplication rule (93) to find the co-multiplication for the generators of the quantum group $U_q(s\ell(2))$:

$$\Delta(q^{S^z}) = q^{S^z} \otimes q^{S^z} \tag{153a}$$
$$\Delta(S^\pm) = S^\pm \otimes q^{S^z} + q^{-S^z} \otimes S^\pm \tag{153b}$$

The co-multiplication preserves the algebraic relations (152), as can be checked by using the fact that Δ is a homomorphism, that is to say $\Delta(ab) = \Delta(a)\Delta(b)$. Note that the non–trivial addition rule (153b) is consistent with the non–trivial commutator (152b), and viceversa. Compare also with $\Delta B(u)$ in equation (98).

We have derived an interesting algebraic structure, the quantum group $U_q(s\ell(2))$, by letting the rapidities become infinite in the Yang–Baxter elements $T_i^j(u)$. More precisely, the quadratic relations between the monodromy matrices ($\mathcal{R}TT$ equations of the Yang–Baxter algebra) give us the defining relations of $U_q(s\ell(2))$, while the co-multiplication of $\mathcal{A}^{(6v)}$ implies that of $U_q(s\ell(2))$.

Let us return once again to the Yang–Baxter equation satisfied by the $\mathcal{R}$-matrix:

$$\sum_{j_1,j_2,j_3} \mathcal{R}_{j_1 j_2}^{k_1 k_2}(u-v) \mathcal{R}_{i_1 j_3}^{j_1 k_3}(u) \mathcal{R}_{i_2 i_3}^{j_2 j_3}(v) = \\ = \sum_{j_1,j_2,j_3} \mathcal{R}_{j_2 j_3}^{k_2 k_3}(v) \mathcal{R}_{j_1 i_3}^{k_1 j_3}(u) \mathcal{R}_{i_1 i_2}^{j_1 j_2}(u-v) \quad (154)$$

The $\mathcal{R}TT$ equation is based on the identification of T_i^j in the representation of dimension two with the $\mathcal{R}$-matrix itself:

$$\left(T_i^j(u)\right)_\alpha^\beta = \mathcal{R}_{i\alpha}^{j\beta} = \quad i \!-\!\!\!\!\!\!\!-\!\!\!\!\!\!\!-\!\!\!\!\!\!\!- j \quad (155)$$

where i, j are indices of the auxiliary space (that is, labels for the elements of $\mathcal{A}^{(6v)}$) and α, β are indices of the quantum space (indicating the representation of $\mathcal{A}^{(6v)}$). We wish to emphasize that in all this construction, the $\mathcal{R}$ matrix has played a rather auxiliary role, as indeed its indices in $\mathcal{R}TT = TT\mathcal{R}$ are auxiliary: we would like to see the $\mathcal{R}$-matrix playing a role in quantum space as well.

Let us take advantage of an interesting property satisfied by the $\mathcal{R}$-matrix of the six–vertex model, the parity symmetry:

$$\mathcal{R}_{i_1 i_2}^{j_1 j_2}(u) = \mathcal{R}_{i_2 i_1}^{j_2 j_1}(u) \quad (156)$$

or equivalently

$$P\mathcal{R}(u)P = \mathcal{R}(u) \quad (157)$$

with P the permutation operator (129). With the help of equation (156), we may rewrite (154) as

$$\sum_{j_1,j_2,j_3} \mathcal{R}_{j_1 j_2}^{k_1 k_2}(u-v) \mathcal{R}_{j_3 i_1}^{k_3 j_1}(u) \mathcal{R}_{i_3 i_2}^{j_3 j_2}(v) = \\ = \sum_{j_1,j_2,j_3} \mathcal{R}_{i_3 j_1}^{j_3 k_1}(u) \mathcal{R}_{j_3 j_2}^{k_3 k_2}(v) \mathcal{R}_{i_1 i_2}^{j_1 j_2}(u-v) \quad (158)$$

Note that in (154) the space $V_1 \otimes V_2$ is auxiliary and V_3 quantum, whereas now V_3 is auxiliary but both V_1 and V_2 are quantum. We have thus gotten $\mathcal{R}$ into quantum space. Using (155), we rewrite (158) as

$$\mathcal{R}(u-v)\left(T_j^k(u) \otimes T_i^j(v)\right) = \left(T_i^j(u) \otimes T_j^k(v)\right) \mathcal{R}(u-v) \quad (159)$$

where the tensor product takes place in $V_1 \otimes V_2$ and $\mathcal{R}(u-v) \in \text{End}(V_1 \otimes V_2)$.

What are the consequences of the $\mathcal{R}TT$ equation (159) with $\mathcal{R}$ in quantum space? First of all, for consistency with the parity symmetry, there must exist some function

$\rho(u)$ such that
$$\mathcal{R}(u)\mathcal{R}(-u) = \rho(u)\rho(-u)\mathbf{1} \tag{160}$$

This can be derived by acting on both sides of (159) with the permutation operator P and then using (157). Equation (160) can also be derived from the unitarity condition (16), namely $\mathcal{R}(u)P\mathcal{R}(u)P \sim 1$ for parity invariant $\mathcal{R}$-matrices, i.e. satisfying (159). For the $\mathcal{R}$-matrix of the six-vertex model, $\rho(u) = a(u)$ in (160).

Letting $u = v$ in (159) and knowing that
$$\mathcal{R}(0) = \rho(0)P \tag{161}$$

we obtain
$$T_i^j(u) \otimes T_j^k(u) = P\left(T_j^k(u) \otimes T_i^j(u)\right) P \tag{162}$$

which establishes the equivalence between the co-multiplication (93) and its transpose. More generally, when $u \neq v$, equation (159) establishes an equivalence between the two ways of co-multiplying arbitrary elements of $\mathcal{A}$. Note that u and v are labels of the representation spaces V_1 and V_2, respectively. If, recalling (93), we define
$$\Delta_{u,v}\left(T_i^k\right) = T_i^j(u) \otimes T_j^k(v) \tag{163a}$$
$$\Delta'_{v,u}\left(T_i^k\right) = T_j^k(v) \otimes T_i^j(u) \tag{163b}$$

we can write (159) as
$$\Delta_{u,v}\left(T_i^k\right) = \mathcal{R}(u-v)\Delta'_{v,u}\left(T_i^k\right)\mathcal{R}^{-1}(u-v) \tag{164}$$

This equation means that $\mathcal{R}(u-v)$ intertwines the two possible co-multiplications $\Delta_{u,v}$ and $\Delta'_{v,u}$. In Drinfeld's definition of quantum groups as quasi-triangular Hopf algebras, equation (164) is one of the basic postulates.

To understand how the $\mathcal{R}$-matrix intertwines between a co-product and its transpose, let us take a closer look at the braid limit $u \to \infty$ of equation (159), or rather of its analog for the rescaled $\tilde{\mathcal{R}}(u)$ and $\tilde{T}(u)$ introduced in (148), namely
$$\tilde{\mathcal{R}}(v-u)\left(\tilde{T}_i^j(u) \otimes \tilde{T}_j^k(v)\right) = \left(\tilde{T}_j^k(v) \otimes \tilde{T}_i^j(u)\right)\tilde{\mathcal{R}}(v-u) \tag{165}$$

Using (143) and (150) we find the braid limit of (165), which is the braid limit of (164):
$$\Delta'(g) = R\Delta(g)R^{-1} \qquad g \in \left\{q^{\pm S^z}, S^\pm\right\} \tag{166}$$

Here, the $\mathcal{R}$-matrix is $\mathcal{R} = PR$ with R the Jimbo matrix (144), the co-product $\Delta(g)$ is given by (153), and the transposed co-product Δ' is, explicitly,
$$\begin{aligned}\Delta'(q^{S^z}) &= q^{S^z} \otimes q^{S^z} \\ \Delta'(S^\pm) &= S^\pm \otimes q^{-S^z} + q^{S^z} \otimes S^\pm\end{aligned} \tag{167}$$

Equation (166) should really be written, to avoid confusion, as
$$\Delta'_{\frac{1}{2}\frac{1}{2}}(g) = \mathcal{R}^{\frac{1}{2}\frac{1}{2}}\Delta_{\frac{1}{2}\frac{1}{2}}(g)\left(\mathcal{R}^{\frac{1}{2}\frac{1}{2}}\right)^{-1} \tag{168}$$

where $\Delta_{\frac{1}{2}\frac{1}{2}}$ denotes the restriction of $\Delta(g)$ to the irrep $\frac{1}{2} \otimes \frac{1}{2}$ of $U_q(s\ell(2))$ and the indices on $\mathcal{R}$ remind us that we are using the representation of $\mathcal{R}$ on the vector space $\frac{1}{2} \otimes \frac{1}{2}$. It is nevertheless worth stressing that (166) makes sense even at the level of the quantum group $U_q(s\ell(2))$, prior to the construction of its representations. By this we mean that $\mathcal{R}$ in (167) can be viewed as an element of $U_q(s\ell(2)) \otimes U_q(s\ell(2))$, rather than as a numerical matrix as in (168). For this reason, the matrix $\mathcal{R}$ is called the "universal R–matrix", and its existence guarantees that the co-multiplications Δ and Δ' of $U_q(s\ell(2))$ are equivalent at the purely algebraic level.

9 Affine quantum groups

We have found that the integrability of the six–vertex model in the braid limit (when the dependence of the Boltzmann weights on the rapidity drops out) is encoded in the quantum group $U_q(s\ell(2))$. Motivated by these results, we may ask ourselves whether a mathematical structure similar to the quantum group could be associated with the rapidity–dependent R–matrix solutions to the Yang–Baxter equation.

The $s\ell(2)$ Lie algebra (A_1 in Cartan's classification) has three Chevalley generators E, F and H with the following non–vanishing commutators

$$[E, F] = H$$
$$[H, E] = 2E \qquad (169)$$
$$[H, F] = -2F$$

The usual spin generators are related to these by

$$E = S^+, \quad F = S^-, \quad H = 2S^z \qquad (170)$$

The affine extension of A_1, called $A_1^{(1)}$ in Kac's classification, has six Chevalley generators E_i, F_i and H_i ($i = 0, 1$). Suppose we have an irreducible representation of A_1. It can be affinized, $i.e.$ promoted to an irreducible representation of $A_1^{(1)}$ through the identifications

$$E_0 = e^u F \qquad E_1 = e^u E$$
$$F_0 = e^{-u} E \qquad F_1 = e^{-u} F \qquad (171)$$
$$H_0 = -H \qquad H_1 = H$$

where $x = e^u$ is a complex affinization parameter.

Different irreps of $A_1^{(1)}$ (of zero central extension) may be labelled by the affine parameter e^u and the Casimir of the corresponding representation of A_1. For example, the irreducible ($e^u, \frac{1}{2}$) representation of $A_1^{(1)}$ derives from the usual spin–$\frac{1}{2}$ irrep of A_1:

$$E_0 = \begin{pmatrix} 0 & 0 \\ e^u & 0 \end{pmatrix} \qquad E_1 = \begin{pmatrix} 0 & e^u \\ 0 & 0 \end{pmatrix}$$
$$F_0 = \begin{pmatrix} 0 & e^{-u} \\ 0 & 0 \end{pmatrix} \qquad F_1 = \begin{pmatrix} 0 & 0 \\ e^{-u} & 0 \end{pmatrix} \qquad (172)$$
$$H_0 = \begin{pmatrix} -1 & 0 \\ 0 & 1 \end{pmatrix} \qquad H_1 = \begin{pmatrix} 1 & 0 \\ 0 & -1 \end{pmatrix}$$

Let us turn to the quantum deformations of A_1 and $A_1^{(1)}$, which we shall denote by $U_q(A_1)$ and $U_q(A_1^{(1)})$, respectively. If we define the operator K as

$$K = q^H \qquad (173)$$

we may rewrite equations (152) in terms of E, F and K as

$$KE = q^2 EK$$

$$KF = q^{-2}FK \tag{174}$$

$$[E, F] = \frac{K - K^{-1}}{q - q^{-1}}$$

We shall denote by $U_q(A_1) = U_q(s\ell(2))$ the algebra generated by E, F and H subject to (152).

Affinization of the spin-$\frac{1}{2}$ irrep of $U_q(A_1)$ yields an irrep $(e^u, \frac{1}{2})$ of $U_q(A_1^{(1)})$:

$$E_0 = \begin{pmatrix} 0 & 0 \\ e^u & 0 \end{pmatrix} \qquad E_1 = \begin{pmatrix} 0 & e^u \\ 0 & 0 \end{pmatrix}$$

$$F_0 = \begin{pmatrix} 0 & e^{-u} \\ 0 & 0 \end{pmatrix} \qquad F_1 = \begin{pmatrix} 0 & 0 \\ e^{-u} & 0 \end{pmatrix} \tag{175}$$

$$K_0 = \begin{pmatrix} q^{-1} & 0 \\ 0 & q \end{pmatrix} \qquad K_1 = \begin{pmatrix} q & 0 \\ 0 & q^{-1} \end{pmatrix}$$

which is the same as representation (172) of the classical group $A_1^{(1)}$ provided we take into account relation (173). For the fundamental irrep, as the doublet of $s\ell(2)$, it is always true that the classical and quantum representations coincide.

We expect that the irrep $(e^u, \frac{1}{2})$ of $U_q(A_1^{(1)})$ should be intimately related to the spin-$\frac{1}{2}$ representation (97) of the generators $A(u)$, $B(u)$, $C(u)$ and $D(u)$ of $\mathcal{A}^{(6v)}$. This is so, indeed:

$$A(u) = \frac{1}{2}\left(e^u q^{1/2} K^{1/2} - e^{-u} q^{-1/2} K^{-1/2}\right)$$

$$B(u) = \frac{1}{2}\left(q - q^{-1}\right) q^{-1/2} F K^{1/2}$$

$$C(u) = \frac{1}{2}\left(q - q^{-1}\right) q^{-1/2} E K^{1/2} \tag{176}$$

$$D(u) = \frac{1}{2}\left(e^u q^{1/2} K^{-1/2} - e^{-u} q^{-1/2} K^{1/2}\right)$$

Please check that the algebraic relations (106) follow from those of the quantum group $U_q(A_1)$, equations (174).

The affine quantum group $U_q(A_1^{(1)})$ enjoys also a bi-algebra structure, determined by the co-product

$$\Delta(E_i) = E_i \otimes K_i + 1 \otimes E_i$$
$$\Delta(F_i) = F_i \otimes 1 + K_i^{-1} \otimes F_i \tag{177}$$
$$\Delta(K_i) = K_i \otimes K_i$$

It is thus natural to look for an intertwiner $\mathcal{R}$-matrix for the tensor product of two spin-$\frac{1}{2}$ irreps $(e^{u_1}, \frac{1}{2}) \otimes (e^{u_2}, \frac{1}{2})$ of $U_q(A_1^{(1)})$:

$$\mathcal{R}(e^{u_1}, e^{u_2})\Delta_{e^{u_1}, e^{u_2}}(g) = \Delta'_{e^{u_1}, e^{u_2}}(g)\mathcal{R}(e^{u_1}, e^{u_2}) \qquad \forall g \in U_q(A_1^{(1)}) \tag{178}$$

This is nothing but the affinized version of the intertwiner condition (168) for $U_q(A_1)$. At the risk of offending the reader, we show the transposed co-multiplication Δ' of the generators of $U_q(A_1^{(1)})$:

$$\Delta'(E_i) = E_i \otimes 1 + K_i \otimes E_i$$
$$\Delta'(F_i) = F_i \otimes K_i^{-1} + 1 \otimes F_i \tag{179}$$
$$\Delta'(K_i) = K_i \otimes K_i$$

Comparing (177) with (153), we see that the relation between E, F and S^+, S^- is

$$E = K^{\frac{1}{2}} S^+ \quad , \quad F = K^{-\frac{1}{2}} S^- \tag{180}$$

After some straightforward manipulations, from (178) we get the affine $\mathcal{R}^{\frac{1}{2}\frac{1}{2}}(e^{u_1}, e^{u_2})$ matrix:

$$\frac{\mathcal{R}^{01}_{01}(e^{u_1}, e^{u_2})}{\mathcal{R}^{00}_{00}(e^{u_1}, e^{u_2})} = \frac{e^{u_1-u_2} - e^{u_2-u_1}}{q e^{u_1-u_2} - q^{-1} e^{u_2-u_1}}$$

$$\frac{\mathcal{R}^{10}_{01}(e^{u_1}, e^{u_2})}{\mathcal{R}^{00}_{00}(e^{u_1}, e^{u_2})} = \frac{q - q^{-1}}{q e^{u_1-u_2} - q^{-1} e^{u_2-u_1}} \tag{181}$$

with $\mathcal{R}^{00}_{00} = \mathcal{R}^{11}_{11}$, $\mathcal{R}^{01}_{01} = \mathcal{R}^{10}_{10}$, $\mathcal{R}^{10}_{01} = \mathcal{R}^{01}_{10}$, and all other matrix elements of $\mathcal{R}$ equal to zero.

Identifying now $q = e^{i\gamma}$ and $e^u = e^{u_1 - u_2}$, we see that $\mathcal{R}(e^{u_1}, e^{u_2})$ is just the six-vertex $\mathcal{R}$-matrix (116). With this happy result we conclude our preview the relation between the six-vertex model and the affine quantum group $U_q(A_1^{(1)})$.

10 Hopf algebras

Let $(\mathcal{A}, m, \iota)$ be an algebra whose multiplication $m : \mathcal{A} \otimes \mathcal{A} \to \mathcal{A}$ is associative, that is

$$[m(m \otimes 1)](a \otimes b \otimes c) = [m(1 \otimes m)](a \otimes b \otimes c) \quad \forall a, b, c \in \mathcal{A} \tag{182}$$

We write $ab = m(a \otimes b)$, $\forall a, b \in \mathcal{A}$. If $a \in \mathcal{A}$ and $\lambda \in \mathbf{C}$, to make formal sense of $\lambda a \in \mathcal{A}$ we need the unit map $\iota : \mathbf{C} \to \mathcal{A}$ of $\mathcal{A}$, which is intimately tied to the identity $1 \in \mathcal{A}$:

$$\iota : \lambda \in \mathbf{C} \mapsto \lambda 1 \in \mathcal{A} \tag{183}$$

The unit ι and the multiplication m are compatible in the sense that

$$m(a \otimes \iota(\lambda)) = a\lambda = \lambda a = m(\iota(\lambda) \otimes a) \quad \forall a \in \mathcal{A} \;\; \forall \lambda \in \mathbf{C} \tag{184}$$

Let us consider now a co-multiplication or co-product $\Delta : \mathcal{A} \to \mathcal{A} \otimes \mathcal{A}$, which should be "co-associative":

$$(\Delta \otimes 1)(\Delta(a)) = (1 \otimes \Delta)(\Delta(a)) \quad \forall a \in \mathcal{A} \tag{185}$$

We also need a co-unit map $\epsilon : \mathcal{A} \to \mathbf{C}$ to define a co-algebra $(\mathcal{A}, \Delta, \epsilon)$; it must satisfy

$$(1 \otimes \epsilon)\Delta = (\epsilon \otimes 1)\Delta = 1 \tag{186}$$

A simultaneous algebra and co-algebra $(\mathcal{A}, m, \iota, \Delta, \epsilon)$ is called a bi-algebra if the co-multiplication Δ and the co-unit ϵ are consistent with the multiplication m, that is if they are homomorphisms:

$$\epsilon(ab) = \epsilon(a)\epsilon(b) \quad , \quad \Delta(ab) = \Delta(a)\Delta(b) \tag{187}$$

Actually, the unit ι and co-unit ϵ must also be compatible:

$$\iota(\epsilon(a)) = \epsilon(a)1 \quad (\forall a \in \mathcal{A}) \tag{188}$$

A Hopf algebra is a bi-algebra enjoying an antipode $\gamma : \mathcal{A} \to \mathcal{A}$, which is an antihomomorphism

$$\gamma(ab) = \gamma(b)\gamma(a) \tag{189}$$

satisfying the following condition:

$$m(\gamma \otimes 1)\Delta(a) = m(1 \otimes \gamma)\Delta(a) = \epsilon(a) 1 \quad \forall a \in \mathcal{A} \tag{190}$$

This condition involves all the ingredients of the bi-algebra structure. Hopf algebras are much more interesting than mere bi-algebras.

If the multiplication m is commutative (respectively, not commutative), we call the algebra commutative (respectively, non–commutative). Just like the multiplication, the co-multiplication may or may not be commutative. The Hopf algebra is accordingly co-commutative or non–co-commutative. Of primary interest are those Hopf algebras which are neither commutative nor co-commutative.

Introduce the permutation map

$$\sigma : \mathcal{A} \otimes \mathcal{A} \to \mathcal{A} \otimes \mathcal{A}$$
$$a \otimes b \mapsto b \otimes a \tag{191}$$

which merely interchanges the order of the operands. Commutativity means thus

$$ab \equiv m(a \otimes b) = m(\sigma(a \otimes b)) \equiv m(b \otimes a) \equiv ba \qquad \forall a, b \in \mathcal{A} \tag{192}$$

On the other hand, if the algebra is co-commutative, then

$$\Delta(a) = \sigma \cdot \Delta(a) \equiv \Delta'(a) \qquad \forall a \in \mathcal{A} \tag{193}$$

Given a co-multiplication Δ, it is not hard to check that the operation $\Delta' = \sigma \circ \Delta \in \text{End}\,(\mathcal{A} \otimes \mathcal{A})$ is also a co-multiplication, with modified antipode $\gamma'(a) = [\gamma(a)]^{-1}$, ($\forall a \in \mathcal{A}$). Given a Hopf algebra $\mathcal{A}$, it is called quasi-triangular if there exists a universal $\mathcal{R}$–matrix $\mathcal{R} \in \mathcal{A} \otimes \mathcal{A}$ such that

$$\Delta'(a) = \mathcal{R}\Delta(a)\mathcal{R}^{-1} \qquad \forall a \in \mathcal{A} \tag{194}$$

and

$$(1 \otimes \Delta)\mathcal{R} = \mathcal{R}_{13}\mathcal{R}_{12} = \sum_{i,j} A_i A_j \otimes B_j \otimes B_i$$
$$(\Delta \otimes 1)\mathcal{R} = \mathcal{R}_{13}\mathcal{R}_{23} = \sum_{i,j} A_i \otimes A_j \otimes B_i B_j \tag{195}$$

and

$$(\gamma \otimes 1)\mathcal{R} = (1 \otimes \gamma^{-1})\mathcal{R} = \mathcal{R}^{-1} \tag{196}$$

where $\mathcal{R}$ is called the universal $\mathcal{R}$–matrix. We write $\mathcal{R} = \sum_i A_i \otimes B_i$ and let

$$\mathcal{R}_{12} = \sum_i A_i \otimes B_i \otimes 1$$
$$\mathcal{R}_{13} = \sum_i A_i \otimes 1 \otimes B_i \tag{197}$$
$$\mathcal{R}_{23} = \sum_i 1 \otimes A_i \otimes B_i$$

Essentially, quasi-triangularity means that the co-multiplication Δ and its "transposed" Δ' are related linearly. In some sense, it establishes an equivalence between two different ways of "adding things up".

A co-commutative algebra is trivially quasi-triangular, with $\mathcal{R} = 1 \otimes 1$. A Hopf algebra is called triangular if $\mathcal{R}_{12}\mathcal{R}_{21} = 1 \otimes 1$, where $\mathcal{R}_{21} = \sum B_i \otimes A_i$.

A non–co-commutative quasi-triangular Hopf algebra is called a quantum group.

224 Quantum Groups

The interest in quasi-triangular Hopf algebras is that they produce solutions to the Yang–Baxter equation naturally. Indeed, from (194) and (195) the Yang–Baxter equation in a "universal form" without spectral parameter may be derived:

$$\mathcal{R}_{12}\mathcal{R}_{13}\mathcal{R}_{23} = \mathcal{R}_{23}\mathcal{R}_{13}\mathcal{R}_{12} \tag{198}$$

The proof goes like this:

$$\begin{aligned}[(\sigma \circ \Delta) \otimes 1]\mathcal{R} &= \sum_i \Delta'(A_i) \otimes B_i \\ &= \sum_i \mathcal{R}_{12}\Delta(A_i)\mathcal{R}_{12}^{-1} \otimes B_i \\ &= \mathcal{R}_{12}\left(\sum_i \Delta(A_i) \otimes B_i\right)\mathcal{R}_{12}^{-1} \\ &= \mathcal{R}_{12}\left[(\Delta \otimes 1)\mathcal{R}\right]\mathcal{R}_{12}^{-1} \\ &= \mathcal{R}_{12}\mathcal{R}_{13}\mathcal{R}_{23}\mathcal{R}_{12}^{-1}\end{aligned} \tag{199}$$

On the other hand,

$$\begin{aligned}[(\sigma \circ \Delta) \otimes 1]\mathcal{R} &= \sigma_{12}(\Delta \otimes 1)\mathcal{R} \\ &= \sigma_{12}(\mathcal{R}_{13}\mathcal{R}_{23}) \\ &= \mathcal{R}_{23}\mathcal{R}_{13}\end{aligned} \tag{200}$$

and thus (198) follows.

11 The quantum group $U_q(\mathcal{G})$

In this section, we present the quantum semi-simple algebras due to Drinfeld and Jimbo, generalizing the construction of $U_q(s\ell(2))$ above.

Let $\mathcal{G}$ be a semi-simple Lie algebra with $A = (a_{ij})$ $(i,j = 1,\ldots,n = \text{rank } \mathcal{G})$ the corresponding Cartan matrix and $D = (D_i)$ the vector or diagonal matrix such that $D_i a_{ij} = a_{ij} D_j$.

The quantum group $U_q(\mathcal{G})$ is defined as the algebra of formal power series in q with generators e_i, f_i, k_i $(i=1,\ldots,n = \text{rank } \mathcal{G})$ subject to the following relations:

$$\begin{aligned} k_i k_j &= k_j k_i \\ k_i e_j &= q_i^{a_{ij}} e_j k_i \\ k_i f_j &= q_i^{-a_{ij}} f_j k_i \\ e_i f_j - f_j e_i &= \delta_{ij}\frac{k_i - k_i^{-1}}{q_i - q_i^{-1}} \end{aligned} \tag{201}$$

$$\sum_{\ell=0}^{1-a_{ij}} (-1)^\ell \begin{bmatrix}1-a_{ij} \\ \ell\end{bmatrix}_{q_i} e_i^{1-a_{ij}-\ell} e_j e_i^\ell = 0 \quad i \neq j \tag{202a}$$

$$\sum_{\ell=0}^{1-a_{ij}} (-1)^\ell \begin{bmatrix}1-a_{ij} \\ \ell\end{bmatrix}_{q_i} f_i^{1-a_{ij}-\ell} f_j f_i^\ell = 0 \quad i \neq j \tag{202b}$$

We have used the notations

$$q_i = q^{D_i} \quad,\quad [x]_{q_i} = \frac{q_i^x - q_i^{-x}}{q_i - q_i^{-1}} \tag{203}$$

The equations (202) are called the quantum Serre relations. For the simplest case of $\mathcal{G} = A_1$, there are no Serre relations and (201) coincide with (174).

The generators e_i, f_i and k_i, the index i ranging over the positive simple roots, constitute the Chevalley basis of the algebra. Supplemented with the Serre relations, it is equivalent to the Cartan basis (one raising operator for each positive root). In the quantum case, the Chevalley basis is much more convenient than the Cartan basis, due to the profusion of q-factors.

The co-multiplication of $U_q(\mathcal{G})$ is given by

$$\Delta(k_i) = k_i \otimes k_i$$
$$\Delta(e_i) = e_i \otimes k_i + 1 \otimes e_i \qquad (204)$$
$$\Delta(f_i) = f_i \otimes 1 + k_i^{-1} \otimes f_i$$

and the antipode by

$$\gamma(e_i) = -e_i k_i^{-1} \quad , \quad \gamma(f_i) = -k_i f_i \quad , \quad \gamma(k_i) = k_i^{-1} \qquad (205)$$

To write the universal $\mathcal{R}$-matrix of $U_q(\mathcal{G})$ we need the "logarithm" of k_i,

$$H_i = \frac{1}{2hD_i} \log\left(k_i^2\right) \qquad (206)$$

where we have set $q = \exp h$. Then

$$\mathcal{R} = \exp\left[h \sum_{i,j=1}^{n} \left(B^{-1}\right)_{ij} H_i \otimes H_j\right] \left[1 + \sum_{i=1}^{n} \left(1 - q_i^{-2}\right) e_i \otimes f_i + \cdots\right] \qquad (207)$$

where $B_{ij} = D_i a_{ij}$ is the symmetrized Cartan matrix. This $\mathcal{R}$-matrix follows form Drinfeld's quantum double construction.

In practical applications, we are interested in the R-matrix in some representation. For example, the R-matrix in the fundamental of $U_q(s\ell(n))$, $R_n = P\mathcal{R}^{(n,n)}$, is

$$R_n = q \sum_{i=1}^{n} e_{ii} \otimes e_{ii} + \sum_{i \neq j} e_{ij} \otimes e_{ji} + \left(q - q^{-1}\right) \sum_{i<j} e_{jj} \otimes e_{ii} \qquad (208)$$

where e_{ij} is an $n \times n$ matrix whose only non-zero entry is the (i,j)-th one. In matrix notation,

$$(R_n)_{ij}^{k\ell} = \begin{cases} q & \text{if } i = j = k = \ell \\ 1 & \text{if } i = \ell \neq k = j \\ q - q^{-1} & \text{if } i = k < \ell = j \\ 0 & \text{otherwise} \end{cases} \qquad (209)$$

It is not hard to verify that

$$(R_n)^{-1} = q^{-1} \sum_{i=1}^{n} e_{ii} \otimes e_{ii} + \sum_{i \neq j} e_{ij} \otimes e_{ji} + \left(q^{-1} - q\right) \sum_{i>j} e_{jj} \otimes e_{ii} \qquad (210)$$

and thus

$$R_n - (R_n)^{-1} = \left(q - q^{-1}\right) 1_{V \otimes V} \qquad (211)$$

where V is the n-dimensional space on which $U_q(s\ell(n))$ is represented.

12 Comments

In these lectures, we have tried to give a flavor for some of the main ideas and techniques underlying two–dimensional integrable systems. Much has been left out, notably the thorny eight–vertex model, all the face models (closely linked with conformal field theories), and the funny models arising from quantum groups when q is a root of unit. This last point alone deserves a full course. Also, we have bypassed all the applications of and to knot theory. In a finite amount of time, however, only so much information can be humanly absorbed. The subject of theories with quantum symmetries is under active research, particularly from the point of view of string theory: the emphasis there is on continuum two–dimensional field theories with $N = 2$ supersymmetry.

Continued and passionate discussions with Matías Moreno are gratefully acknowledged. I wish to thank the continued strenuous efforts of José Luis Lucio, the organizer and soul of the School (now ten years old), for succeeding in creating every two years a convivial atmosphere where scientific discussion can be carried out freely and invigoratingly. My thanks also to this year's most worried organizer, Miguel Vargas. I am much indebted to the interesting questions and inquisitive minds of the participants at the School. Last but not least, I am thankful to the Instituto de Física de la Universidad Nacional Autónoma de México (IFUNAM) for bestowing on me the Cátedra Tomás Brody. This work has been partially supported by the Fonds National Suisse pour la Recherche Scientifique.

13 References

It is perhaps more useful to point out a few good review articles and books; references to the original literature can be found easily starting from the works cited below.

The book on which these notes are based:

C. Gómez, M. Ruiz–Altaba and G. Sierra (1993). *Quantum Groups in Two–Dimensional Physics* (Cambridge University Press, Cambridge).

Nice introductions to the theory of factorizable S–matrices:

A.B. Zamolodchikov and Al.B. Zamolodchikov (1979). Factorized S–matrices in two dimensions as the exact solution of certain relativistic quantum field theory models, *Ann. Phys.* **120**, 253.

A.B. Zamolodchikov (1980). Factorized S–matrices in lattice statistical systems, *Sov. Sci. Rev.* **A2**, 1.

The Bethe ansatz:

J.H. Lowenstein (1982). Introduction to Bethe ansatz approach in $(1 + 1)$–dimensional models, in *Proceedings, Les Houches XXXIX* (Elsevier, Amsterdam).

L.D. Faddeev and L.A. Takhtajan (1981). What is the spin of a spin wave?, *Phys. Lett.* **85A**, 375.

Vertex models:

R.J. Baxter (1982). *Exactly Solved Models in Statistical Mechanics* (Academic Press, London).

The Leningrad school:

E.K. Sklyanin (1993). Quantum inverse scattering method: selected topics, in *Proceedings of the Fifth Nankai Workshop,* ed. Mo-Lin Ge (World Scientific, Singapore).

Quantum groups:

V.G. Drinfeld (1987). Quantum groups, in *Proceedings of the 1986 International Congress of Mathematics,* ed. A.M. Gleason (Am. Math. Soc., Berkeley).

M. Jimbo (1991). *Quantum Groups and the Yang–Baxter Equation* (Springer, Tokyo).

L.D. Faddeev, N.Yu. Reshetikhin and L.A. Takhtajan (1990). Quantum Lie groups and Lie algebras, *Leningrad Math. J.* **1**, 193–225.

L. Alvarez-Gaumé, C. Gómez and G. Sierra (1989). Hidden quantum symmetries in rational conformal field theories, *Nucl. Phys.* **B319,** 155.

V. Pasquier and H. Saleur (1990). Common structures between finite systems and conformal field theories through quantum groups, *Nucl. Phys.* **B330,** 523.

New Results from Experiments at the HERA Storage Ring and from ARGUS

D. Wegener
Institut für Physik [1], Universität Dortmund

Abstract

Recent results from the ep storage ring HERA and from the e^+e^- storage ring DORIS II are discussed. Special emphasis is given to the specific layout of the detectors and to the progress in calorimetry achieved in the last few years. The impact of the ARGUS experiment on B- and τ-physics is discussed.

First Results from HERA

In the history of atomic and subatomic physics electron scattering has turned out to be one of the most efficient methods to explore the structure of matter. It starts in atomic physics with the observations of H.Hertz and P.Lenard who showed that electrons can penetrate an atom [1]. It was followed by the studies of J.Frank and G.Hertz [2] who were the first to detect quantum like excitations of atoms in electron scattering. The finite size of nuclei and nucleons was observed by R.Hofstadter [3] in elastic scattering experiments. Finally, the pioneering work of the SLAC–MIT group [4] was one of the milestones on the way to establish quarks as the constituents of matter and gluons as the quanta of strong interaction.

HERA, the first electron–proton storage ring ever built, extends the kinematical region accessible to this important method by more than a factor of 10 in the longitudinal and transverse direction. Hence one hopes to

---

[1] Supported by the Bundesministerium für Forschung und Technologie of the Federal Republic of Germany under contract number 054DO51P

collect important new information on the structure of matter from the experiments just starting. In table 1 the design parameters of the machine are collected. Comparing them with the values achieved in the first year of running demonstrates that still much room is available for improvements. Nevertheless, interesting first results were achieved, which are discussed in chapters 4,5 of this lecture. In its written version the final results of the 1992 runs are presented, while in the lecture only preliminary data based in part of the 1992 statistics were discussed. To motivate the layout of the detectors we start with the discussion of the physics program of the HERA experiments (ch.1). A full survey of the possibilities is given in the proceedings of two recent workshops [5, 6]. In ch.2 electromagnetic and hadronic calorimeters as the heart of all HERA detectors are discussed, followed by a short description (ch.3) of the H1 and ZEUS detectors, their trigger system and the luminosity measurement.

Table 1: HERA design parameters compared to values achieved (brackets)

	e^--ring	p-ring
circumference	6336 m	
energy	30 GeV (26.6 GeV)	820 GeV (820 GeV)
B field	0.164 T	4.68 T
circulating currents	60 (~ 10)mA	160 (~ 10)mA
number of filled bunches	210 (10)	210 (10)
time between bunches	96 ns	
variance of vertex	10 cm ($< 25 cm$)	
luminosity	$1.5 \cdot 10^{31} cm^{-2} s^{-1}$	

1. Physics Program and Detectors

1.1 Deep Inelastic Scattering

Detailed studies of deep inelastic scattering (fig.1a,b) are the major topics of the HERA program. The following Lorentz invariants are used to fix the kinematics:

$$\text{Four momentum transfer } Q^2 = -q^2 = -(p_1 - p_3)^2 \quad (1)$$
$$\text{Center of mass energy} \quad s = (p_1 + p_2)^2. \quad (2)$$

The scaling variables

$$0 < x = \frac{Q^2}{2 p_1 q} \leq 1 \quad (3)$$

$$0 < y = \frac{p_2\, q}{p_1\, p_2} \leq 1 \qquad (4)$$

have a simple interpretation in the quark model: x is the fractional momentum of the struck parton while y, in the proton rest frame, corresponds to the energy of the exchanged boson normalized to the energy of the primary electron. Since in the Breit–frame q^2 is a three vector, the ratio

$$r = \frac{\hbar}{Q} \qquad (5)$$

can be interpreted as a space resolution. At HERA $r = 3 \cdot 10^{-4}\, fm\, (Q^2 = 4 \cdot 10^4\, GeV^2)$ is obtained, hence structures 10 times smaller than in previous experiment can be resolved. Moreover, at HERA fractional longitudinal momenta $x \geq 10^{-4}$ can be measured, which is a factor 100 smaller than in previous fixed target experiments. The interest in data for this new kinematical region is discussed in ch.1.1.2.

1.1.1 High Q^2 Physics

Two aspects of deep inelastic scattering are of special interest. At large $Q^2 \geq 10^4\, GeV^2$ the exchange of $Z^o, W^\pm$–bosons contributes with a strength comparable to photon exchange. Hence HERA is a weak interaction collider in contrast to hadron colliders, which in the same sense are quark– and gluon–storage–rings. Once the design luminosity at HERA is achieved, challenging possibilities to study the dynamics of Z^o– and $W^\pm$–exchange exist. To some extent one can expect from these studies a progress comparable to that achieved in understanding electromagnetism by the work of H.Hertz [7] who made the step from electrostatics to electrodynamics which allowed to check the full structure of Maxwell equations. In a more technical language anomalous $WW\gamma$–coupling and the production of new charged and neutral heavy bosons can be analyzed [8].

The sensitivity of these experiments is strongly enhanced by the fact that the electrons can be polarized longitudinally, hence the measurements are sensitive to right–handed weak currents. Note that such measurements allow to study parity violation at distances 5 orders of magnitude smaller than investigated in nuclear β–decay.

Even more speculative ideas can be taken up. The improved space resolution allows to search for substructures of quarks and leptons. A unique field is the search for leptoquarks and leptogluons (fig.2). These speculative states should show up as a peak in the inclusive cross section [9]. The mass range accessible at HERA is $m_{LQ} \leq 200\, GeV$. Similarly, the production of excited quarks and leptons is possible, the mass range can be extended by a factor of 3 up to $m_{e^*, q^*} \leq 250\, GeV$.

The electroweak studies presume a precise knowledge of the proton structure function, which for $Q^2 \leq 10^4\, GeV^2$ can be extracted from the expression [10]

$$\frac{d^2\sigma}{dx\,dy} = \frac{4\pi\alpha^2 s}{Q^4}\left\{(1-y)F_2(x,Q^2) + y^2\,x\,F_1(x,Q^2)\right\} \quad (6)$$

In the framework of the quark parton model the structure functions F_i are given by

$$F_2^{em}(x) = \sum e_i^2\,\{x\,q_i(x) + x\bar{q}_i(x)\} \quad (7)$$
$$2\,x\,F_i(x) = F_2(x) \quad (8)$$

$q_i(x)$ = probability to find a quark of type i and charge e_i with fraction x of the proton momentum. The Callan–Gross relation (8) follows from the coupling of the photon to a spin 1/2 parton. Taking into account gluon-radiation, QCD predicts a logarithmic Q^2 dependence of $F_2(x)$:

$$F_2(x) \to F_2(x,Q^2) \quad (9)$$

as well as the violation of the Callan-Gross relation. For $\alpha_S \ll 1$, $\alpha_S \ln\frac{1}{x} \ll 1$ the Q^2 dependence is given by the Altarelli–Parisi equation, which for the gluon evolution reads [11]

$$\frac{dG(x,Q^2)}{d\ln Q^2} = \frac{\alpha_S(Q^2)}{2\pi}\,\{P_{gq}\otimes q + P_{gg}\otimes G\} \quad (10)$$

$G(x)$ = gluon density; P_{Gq}, P_{GG} gluon splitting function (fig.3). Using this relation, the long lever arm in Q^2 accessible at HERA allows to check one of the basic predictions of QCD, namely the running of the coupling constant α_S, which is due to the gluon self–coupling, i.e.the nonabelian nature of gauge theory.

1.1.2 Semihard Physics

The study of deep inelastic scattering at $Q^2 \gtrsim 10\,GeV^2$ and low $x(10^{-2} \leq x \leq 10^{-4})$, which for the first time is possible at HERA, is of special importance:

- The measured structure functions are a basic input for all theoretical predictions of the physics at the new hadron colliders.

- The parton densities and hence the elementary cross sections are expected to be large and on the other side $\alpha_S/2\pi \ll 1$. Hence perturbative predictions in a totally new regime can be made.

- In this region the transition from free partons (current quarks) of deep inelastic scattering to heavy constituent quarks of sepctroscopy takes place, the latter are characterized by valence quarks surrounded by a parton cloud. A quantitative description in this new kinematical region [12] may provide a new approach to understand hadron masses.

A qualitative understanding of the phenomena expected at $x \leq 10^{-2}$ follows from the observation that effects of gluons should dominate in this interval (fig.3b), while quarks are coupled to gluons via processes given in fig.3c. In this kinematical region, due to the singularity of the gluon splitting function, the Altarelli–Parisi equations read [10]

$$\frac{\partial x\, G(x,Q^2)}{\partial \ln Q^2} \approx \frac{\alpha_S}{2\pi} \int_x^1 Pgg(\frac{y}{x}) y\, G(y,Q^2)\frac{dy}{y} \qquad (11)$$

$$\frac{\partial F_2(x,Q^2)}{\partial \ln Q^2} = \frac{\alpha}{2\pi} Pgg \otimes F_2(x,Q^2) \qquad (12)$$

At small x ($\ln 1/x \gg 1, Q^2 > 10\,GeV^2$) (11) has the solution

$$x\, G(x,Q^2) \underset{x \to 0}{\to} exp\left\{ 2\sqrt{\ln \frac{1}{x} \ln \frac{\ln Q^2/\Lambda^2}{\ln Q_0^2/\Lambda^2}} \right\}, \qquad (13)$$

which diverges for $x \to 0$. This behaviour is unphysical, since for gluons unitarity and for quarks the Pauli principle would be violated. Hence at low x besides the emission processes of fig.3, which are proportional to $\sim \alpha_S F$, also recombination processes $\sim \alpha_S^2 F^2$ are important. Gribov, Levin and Lipatov [12] have therefore proposed to replace the Altarelli–Parisi equation in this kinematical region by the following phenomenological ansatz:

$$\frac{\partial x\, G(x,Q^2)}{\partial \ln Q^2} = \frac{\alpha_S}{2\pi} P_{gg} \otimes x\, G(x,Q^2) - \frac{c\alpha_S^2}{R^2 Q^2} \int_x^1 \left(y\, G(y,Q^2)\right)^2 \frac{dy}{y}, \qquad (14)$$

where the second term describes the recombination. R can be interpreted as the correlation radius between partons: for large x the radius is $R \approx 1\,fm$, while at low x the correlation radius is $R \ll 1\,fm$ and recombination effects are large. Of course it is not clear a priori, if the density of partons is homogeneous or if the partons cluster ("hot spots") [13].

The predicted behaviour is summarized in fig.4. Perturbative techniques can be applied for the "parton gas" (Gribov-Levin-Lipatov) and in the regime of free partons (Altarelli-Parisi). It has to be investigated experimentally, which is the region covered by HERA precisely, but it is clear from these semiquantitative arguments that new phenomena are expected. Mueller [13] has proposed an interesting experiment, which allows to detect local high density fluctuations selecting special deep inelastic event topologies. To this end events of the type

$$ep \to e + jet + X \qquad (15)$$

are selected with the following kinematical constraints (fig.5):

$$Q^2 \geq 10\,GeV^2 \qquad (16)$$

$$k_{1T} < ... < k_{mT} \tag{17}$$

$$k_{1T} \approx Q \tag{18}$$

$$x_B \ll 1, \; x_B \ll x_1. \tag{19}$$

From (19) follows that many intermediate radiation steps between particle 1 and n are necessary, while (18) ensures that these partons are produced in the same (transverse) volume element. Hence from (18), (19) follows that the local parton density is high. One predicts [13] that

$$\frac{k_{1T}\, x_1\, dF_2(x_1, k_{1T}^2)}{dx_1\, dk_{1T}} \sim \alpha_s(q^2) \left[x_1 G(x_1, k_{1T}^2) + \frac{4}{9}\left(x_1\, q(x_1, k_{1T}^2) + x_1\, \bar{q}(x_1, k_{1T}^2)\right) \right] \cdot \frac{exp\left(\frac{12\alpha_s}{\pi} \ln 2 \ln \frac{x_1}{x_B}\right)}{\sqrt{\ln \frac{x_1}{x_B}}} \tag{20}$$

For $x_B/x_1 \to 0$ this function rises steeply and the saturation effect should show up. It has been shown [14] that this process can be studied at HERA already at modest luminosities, hence in the near future experimental results are expected.

1.2 Photoproduction

HERA is also a source of high energy photons, which are nearly on–shell:

$$10^{-8}\, GeV < Q^2 < 10^{-2}\, GeV^2 \;, \tag{21}$$

In this case the the total ep cross section σ_{ep} depends on the total cross section for real photons:

$$\sigma_{ep} = \int_{y_{MIN}}^{y_{MAX}} dy\, f_{\gamma/e}(y) \sigma_{\gamma p}(y \cdot s) \tag{22}$$

$$E_\gamma = y \cdot E_1, \tag{23}$$

In order to derive (22), the absorption cross section for longitudinal and transverse polarized photons are introduced [10], which for large photon energies Q_o and small Q^2 depend on the structure functions F_i, $i = 1, 2$ according to:

$$F_2(x, Q^2) = \frac{Q^2}{4\pi^2 \alpha}(\sigma_L + \sigma_T) \tag{24}$$

$$F_1(x, Q^2) = \frac{Q^2}{4\pi^2 \alpha}\frac{1}{2x}\sigma_T \;. \tag{25}$$

Introducing σ_L, σ_T into (6) one gets for the cross section

$$\frac{d^2\sigma}{dQ^2\, dy} = \frac{\alpha}{2\pi}\frac{1}{Q^2}\left\{\frac{1+(1-y)^2}{y}\sigma_T + 2\frac{1-y}{y}\sigma_L\right\} . \tag{26}$$

At small Q^2 σ_L can be neglected [10] and
$$\sigma_T(Q^2, y) \to \sigma_{\gamma p}(y) . \quad (27)$$

Integrating (26)
$$\frac{d\sigma(ep)}{dy} = \frac{\alpha}{2\pi} \frac{1 + (1-y)^2}{y} \sigma_T(y) \ln \frac{Q^2_{MAX}}{Q^2_{MIN}} \quad (28)$$

follows. The comparison of (28) and (22) shows that the photon flux in the Weizsäcker–Williams approximation is given by
$$f_{\gamma/e}(y) = \frac{\alpha}{2\pi} \frac{1 + (1-y)^2}{y} \ln \frac{Q^2_{MAX}}{Q^2_{MIN}} . \quad (29)$$

Note the typical $1/y$ dependence of the bremsstrahlung spectrum. In the experiments the forward scattered electron is detected at small scattering angles ($\theta_3 < 5\,mrad$) in the energy range
$$7\,GeV \leq E_3 \leq 22\,GeV, \quad (30)$$

the scaled photon energy is therefore
$$0.3 \leq y \leq 0.7. \quad (31)$$

The typical luminosity of tagged photon events is
$$L(\gamma p) = 0.03 L(ep) \quad (32)$$

and the γp cms energy interval accessible is
$$140\,GeV < \sqrt{s_{\gamma p}} < 250\,GeV \quad (33)$$

Many different reactions of real photons contribute at these high energies:

- Two classes of diffractive events are observable, namely "elastic" scattering (fig.6a)
$$\gamma p \to V p \quad (V = \gamma, \rho, \omega, \varphi, \Psi, ...) \quad (34)$$
and general single and double diffractive events (fig.6b)
$$\gamma p \to V X, X p, X_1 X_2. \quad (35)$$

Diffractive events can be described by Pomeron exchange. Experimentally they are characterized by a steep peak at forward scattering angles and a differential mass distribution
$$\frac{d\sigma}{dM} \sim \frac{1}{M} \quad (36)$$

Due to the M^{-1}-dependence of the differential cross section the multiplicity of diffractive events is expected to be small [15].

- The vector mesons, coupling to the photon, can also produce a state of high multiplicity and low transverse momentum of the produced particles. These events have the characteristics of typical hadronic events [16]. Presently they are not of topical interest but they have to be considered as a serious background source of hard processes discussed next.

- For the first time also hard reactions of real photons can be detected, which are a rich source of QCD reactions. Similar studies have already been performed in proton–proton interactions [17]. Two classes of hard photon processes can be distinguished (fig.7a-d)[18]. In the simplest reaction the photon couples directly to a parton of the proton, i.e. the photon reacts as a pointlike particle. These reactions include photon-gluon–fusion and QCD–Compton processes (fig.7a,b), the latter being a copious source of heavy quarks. The processes involving a parton from the photon usually are referred to as resolved photon processes (fig.7c-e). In the kinematical region accessible at HERA these processes dominate. The HERA experiments are in a unique situation, since for the first time a dominant contribution from the gluon component of the photon structure function is expected, which is detectable because of the large center of mass energies available.

1.3 Design Criteria for a HERA Detector

The physic processes discussed in chapters 1.1, 1.2 define stringent criteria, which a HERA detector should fulfill:

- Electrons and muons have to be detected with high efficiency and low misidentification probability, since they provide the optimal signal for new (exotic) physics events. Moreover, they allow to tag heavy flavour production via the semileptonic decay of the quarks.

- Electrons allow to select in addition deep inelastic neutral current events, hence the possibilities to detect and identify them unambiguously in the environment of quark and gluon jets should be optimized.

- Charge current and exotic events with a neutrino in the final state are characterized by a large missing transverse momentum. Hence the full coverage of the solid angle is necessary to measure all strong and electromagnetic interacting particles to derive all missing kinematical variables.

These conditions can be met by a calorimeter type detector (ch.2) covering nearly 4π of the solid angle, only a small hole for the beam tubes of the storage ring is not instrumented. In addition, good tracking in a magnetic field is necessary, since

- the charge of the leptons is an important ingredient, if one wants to disentangle different final states;
- the energy carried by a high energetic muon can only be measured with a tracking device;
- the flavour of leading particles in a jet allows to tag the scattered parton [17];
- the reconstruction of the interaction vertex is of utmost importance to reject background from beam–gas and beam–wall events, which are produced at HERA with a much higher rate than the physics events of interest.

This list of conditions already fixes to a large extent the layout of a HERA detector. The interaction point should be surrounded by a tracking device set up in a magnetic field of large volume. It is produced by a superconducting solenoid because of its small power consumption. In order to measure the sign of the charge of the produced particles up to momenta of $\sim 100\,GeV$, the momentum resolution should be better than $\sigma/p \lesssim 0.003\,p$. The tracking must be surrounded by a compact calorimeter covering as much as possible of the solid angle. It has to be followed by an instrumented iron shield, which serves the twofold purpose to identify muons and to close the magnetic circuit.

Further constraints follow from the discussion of the kinematics of the processes investigated at HERA. In deep inelastic scattering the most relevant kinematical variables are the four–momentum–transfer Q^2, the Bjorken scaling variable x and the fractional energy loss y (ch.1.1.1). These Lorentz invariants are given by the energies of the incoming electron E_1, the scattered electron E_3, the incoming proton E_2, the hadronic recoil jet E_4 and the corresponding polar scattering angles θ_3, θ_4 measured with respect to the proton direction (z–axis) in the lab–frame (fig.1a).

Electron measurement

$$Q^2 = 4\, E_1 E_3 \, \cos^2 \frac{\theta_3}{2} \tag{37}$$

$$x = \frac{Q^2}{4\, E_1 E_2 y} = \frac{E_1 E_3 \cos^2 \theta_3/2}{E_2 (E_1 - E_3 \sin^2 \theta_3/2)} \tag{38}$$

$$y = y_e = 1 - \frac{E_3}{E_1} \sin^2 \frac{\theta_3}{2}. \tag{39}$$

Hadron measurement:

$$Q^2 = \frac{E_4^2 \sin^2 \theta_4}{1 - y_h} = \frac{(\sum \vec{p}_{ti})^2}{1 - y_h} \tag{40}$$

$$y = y_h = \sum \frac{E_i - p_{zi}}{2\, E_1}. \tag{41}$$

In formula (40), (41) the sum is taken over all hadrons.
It is instructive to study the influence of uncertainties of the energy– and the angle–measurements on the Lorentz invariants.
As an example the case of electron detection is discussed:

$$\frac{\Delta Q^2}{Q^2} = \frac{\Delta E_3}{E_3} \otimes tg\frac{\theta_3}{2} \Delta \theta_3 \qquad (42)$$

$$\frac{\Delta x}{x} = \frac{1}{y}\frac{\Delta E_3}{E_3} \otimes tg\frac{\theta_3}{2}\left(x\frac{E_2}{E_3} - 1\right) \qquad (43)$$

From (42) follows that ΔQ^2 is dominated by the uncertainty of the electron energy measurement except at very forward scattering angles ($\theta_3 \approx 180°$). The second term in (43) has only an influence on Δx at large x and small Q^2. At low y the x–resolution is poor due to the y^{-1}–amplification factor, in this kinematical region hadronic measurements are more reliable.

Before one draws conclusions from these formulas and defines the region in the (x, Q^2) plane where reliable measurements of the structure function are possible, one has to consider migration effects and the influence of systematic errors. This can be studied by calculating the acceptance function

$$A(Q^2, x) = \frac{N_{true}}{N_{obs}}, \qquad (44)$$

where $N_{true(obs)}$ is the number of true (observed) events in a (x, Q^2)–bin. $A(Q^2, x)$ depends on the experimental resolution and on the cross section, hence the Q^{-4} dependence eq.(6) is important. To avoid a too large sensitivity of A on systematic uncertainties it is limited to values $0.9 < A < 1.1$. This condition strongly confines the allowed *systematic* uncertainties of the energy measurements and in addition requires good energy resolution [19], e.g. at $y = 0.1, x < 0.5$ a systematic energy shift of 1% results according to eq.(42), (43) in a 10% change of the cross section. These considerations and similar arguments for the hadron measurement limit the permitted *systematic* uncertainty of the energy measurements for the electromagnetic (hadronic) calorimeter to ±1% (±2%). In fig.8 the kinematical region in the (x, Q^2)–plane accessible to HERA experiments is given [19]. For comparison the region covered by previous experiments is also included.

2. The Art of Calorimetry

2.1 Basic Processes

Calorimetric devices are at the heart of the new detectors recently brought into operation at the HERA storage ring. In this chapter the physics underlying high energy calorimetry is discussed. The energy deposited by an energetic particle ($\sim 1\ TeV$) is too small to increase the temperature ($\Delta T \approx 10^{-14} K$) of the detector by a measurable amount. Therefore, the intermediate step between energy deposition and thermalization is exploited for particle calorimetry.

Three different processes contributing to the energy deposition can be distinguished. The most important process is ionization/excitation of matter by charged particles. The mean energy loss per unit length is given by the Bethe–Bloch formula [20]:

$$\frac{dE}{dx} = 4\pi\, N_A\, r_e^2\, m_e c^2\, z^2 \frac{Z}{A} \frac{1}{\beta^2} \left\{ \ln \frac{2 m_e c^2 \gamma^2 \beta^2}{I} - \beta^2 \right\} \tag{45}$$

$$4\pi N_A\, r_e^2\, m_e c^2 = 0.307\ \frac{MeV\, cm^2}{g} \tag{46}$$

z = charge of particle; $Z(A)$ = atomic number (weight) of material; I = ionization potential characteristic for the material [21]. In fig.9 dE/dx is plotted as function of the momentum of a particle [22]. A characteristic β^{-2} dependence at low momentum is followed by a logarithmic rise which saturates at high energies due to polarization of matter [23]. The energy loss is minimal for particles with a momentum of $p \approx 2.5\, Mc$ (M = rest mass; minimum ionizing particles). Typically a minimum ionizing particle loses $\sim 1.5\, MeV cm^2/g$ in matter. Muons deposit their energy in matter dominantly by ionization/excitation. As an example their energy loss in iron is ~ 1.1 GeV/m. Since for all other charged particles additional processes contribute to the energy loss their range is limited to ≤ 2 m. This fact is exploited in all modern detectors to identify high energy muons by their ability to penetrate large amounts of matter. The fluctuations of the energy loss due to ionization and excitation is governed by the Landau distribution [24].

The next important energy loss process is bremsstrahlung $e + Z \to e + Z + \gamma$. It dominates at high energies (fig.10b) and is of special importance for electrons since [25]

$$\frac{dE}{dx} = -\frac{E}{X_o} \sim \frac{1}{m_e^2} \tag{47}$$

X_o is the *radiation length*, which is characteristic for a material (table 2). At the *critical energy* ϵ_c the energy losses due to bremsstrahlung and ionization/excitation are equal

$$\left. \frac{dE}{dx}(\epsilon_c) \right|_{ion} = \left. \frac{dE}{dx}(\epsilon_c) \right|_{brems} \tag{48}$$

The photons, produced in the bremsstrahlung process, interact with matter. At high energies pair production $\gamma + Z \to Z + e^+\, e^-$ dominates, while at low energies the Compton– and the photoeffect take over (fig.10a). It turns out that the mean free path length for bremsstrahlung and pair production is nearly equal and is given by the radiation length X_o. The combination of bremsstrahlung and pair production by photons is the source of an electromagnetic shower in matter. The typical longitudinal extension of an electromagnetic shower amounts to

$$L \approx 10\ X_o + X_o\ \ln \frac{E}{\epsilon_c}, \tag{49}$$

while its width is given by the Moliere length R_M [26]

$$R_M = X_o \frac{21.2 MeV}{\epsilon_c}. \tag{50}$$

R_M can be traced back to the multiple scattering of electrons with an energy $E = \epsilon_c$, i.e. by those electrons which dominantly lose their energy by ionization/excitation. Values of X_o, R_M, ϵ_c for materials used in the construction of calorimeters are collected in table 2. Note that $\sim 90\%$ of the energy in an electromagnetic shower is deposited in a cylinder with the radius R_M around the direction of the primary electron. Typically 98% of the energy of an electromagnetic shower is deposited in a volume of

$$V_{em} \approx 20 X_o\ R_M^2. \tag{51}$$

Hadrons deposit their energy by nuclear reactions and particle production (fig.11). The extension of the shower is fixed by the mean free path λ for hadronic reactions (table 2). The typical longitudinal extension of a shower is

$$L = 7\ \lambda + \lambda\ \ln \frac{E}{1\ GeV}, \tag{52}$$

while its transverse size is $\sim \lambda$ (figs.12,13). Hence a hadron deposits its energy in a volume

$$V_{had} = 50\ \lambda^3. \tag{53}$$

According to eqs.(49), (50), (52) and table 2 the transverse and longitudinal extension of a hadron induced shower is larger than for an electromagnetic shower $V_{em} \ll V_{had}$. This observation is exploited in experiments to separate hadrons from electrons/photons [28, 29].

Table 2: Parameters characterizing elelctromagnetic and hadronic showers

	X_0	ϵ_c	R_M	λ_{int}
Scintillator	42.4 cm	85.4 MeV	10.4 cm	79.4 cm
liquid Ar	14 cm	21.5 MeV	13.7 cm	83.7 cm
Fe	1.76 cm	20.5 MeV	1.8 cm	16.6 cm
Pb	0.56 cm	7.2 MeV	1.6 cm	17.1 cm
U	0.32 cm	6.5 MeV	1 cm	10.5 cm

2.2 Calorimeter Versus Spectrometer

As discussed in chapter 2.3 the detectable signal in a calorimeter is proportional to the deposited energy

$$S \sim E \qquad (54)$$

and the resolution improves at high energies

$$\frac{\sigma}{E} \sim \frac{1}{\sqrt{E}}. \qquad (55)$$

Note that for a magnetic spectrometer a different behaviour is observed [30]

$$\frac{\sigma_p}{p} \sim \frac{p}{BL^2} \qquad (56)$$

B = field strength, L = length of the field, p = particle momentum.
From these results follows that calorimeters offer at high energies many advantages compared to other detectors:

- Their energy resolution improves with energy according to (55), while magnetic detectors show the opposite behaviour (56)

- With calorimeters energies can be measured in the full solid angle with constant resolution, since no preferred direction exists, while with a magnetic spectrometer only the momentum component transverse to the field direction can be determined.

- For constant relative resolution the length of a calorimeter has to increase logarithmically (eq.(49),(52)) with the detected energy, while for a magnetic spectrometer according to (56) $L \sim \sqrt{p}$.

- The values of X_o, λ given in table 2 demonstrate that calorimeters are compact detectors.

- Since $V_{had} \gg V_{em}$, a separation of hadrons and electrons/photons in a calorimeter is possible, if a fine granularity of the calorimeter read–out cells is chosen. Typically in longitudinal direction the electromagnetic (hadronic) energy is sampled in steps of 3-6 X_o (2-4 λ), while the transverse dimensions of the calorimeter read–out cells are chosen to be of the order of R_M.

- Calorimetric detectors are naturally adapted by their design to the fact that partons in the final state of high energy reactions show up as jets and, therefore, suggest a measurement of jet momenta as basic quantities.

In order to build a compact calorimeter high Z materials are the natural choice because of the smallness of X_o, λ (table 2). It turns out that these materials are not optimal to read out a signal. Therefore, the breakthrough of calorimetry as a detector technique followed from the invention of sampling calorimeters by Schopper and collaborators [31, 32]. In these detectors slices of high Z absorber– (Fe, Pb, U) and low Z read-out-materials (scintillator, liquid Argon) alternate. A few technical realizations of such sampling calorimeters are shown in fig.14.

2.3 Electromagnetic Calorimeters

The signal S of electromagnetic sampling calorimeters is given by

$$S = kE = k \frac{E_v}{E_v + E_{nv}} E = k\, a(e) \cdot E \tag{57}$$

$E_v\ (E_{nv})$ = visible (unvisible) energy; $a(e)$ = fraction of visible energy for an electron induced shower. The calibration constant k is derived from separate measurements of the calorimeter response with identified particles of known energy. Since only a fraction of the energy is detected, a further source of fluctuation turns up. Its influence on the achievable energy resolution follows from the following argument [33]. The total path length in the detector is given by

$$T = \frac{E}{\epsilon_c} X_o. \tag{58}$$

Hence the number N of visible particles is (d = thickness of a sandwich cell)

$$N = \frac{T}{d} = \frac{E}{\epsilon_c} \frac{X_o}{d}. \tag{59}$$

Since N fluctuates statistically, from Poisson statistics follows

$$\frac{\sigma_s}{E} \sim \frac{1}{\sqrt{N}} \sim \sqrt{\frac{d}{E}}. \tag{60}$$

In fig.15 this relation is checked experimentally [34]. Note that a constant term due to photoelectron statistics and leakage out of the finite size detector has to be added to eq.(60). Generally, the sampling fluctuations dominate for electromagnetic calorimeters.

2.4 Hadron Calorimetry

Far more elementary processes influence a hadron shower, compared to an electromagnetic one (fig.11). In order to evaluate the possibilities, which allow to optimize the response of a hadron calorimeter, the different processes contributing to the signal are discussed. The fraction of the visible energy of component i is

$$a(i) = \frac{E_v(i)}{E}. \tag{61}$$

It is advantageous to normalize $a(i)$ to the ratio $a(mip)$ of a minimal ionizing particle:

$$\frac{e}{mip} =: \frac{a(e)}{a(mip)} \tag{62}$$

$$\frac{h_i}{mip} = \frac{a(h_i)}{a(mip)} \tag{63}$$

h_i is the intrinsic hadronic – in distinction to the electromagnetic – component of a hadron shower The detected electron signal is

$$S(e) = \alpha\, E \frac{e}{mip}, \tag{64}$$

for a hadron the signal is

$$S(h) = \alpha\, E \left\{ f_{em} \frac{e}{mip} + (1 - f_{em}) \frac{h_i}{mip} \right\} \tag{65}$$

Monte Carlo calculations show that the fraction f_{em} of the total energy in a hadronic shower available as an electromagnetic shower is [35]

$$f_{em} \approx 0.1\, \ln \frac{E}{1\,GeV}. \tag{66}$$

The intrinsic hadronic component can be broken up into the following contributions

$$\frac{h_i}{mip} = f_{ion} \frac{ion}{mip} + f_n \frac{n}{mip} + f_\gamma \frac{\gamma}{mip} + f_B \frac{B}{mip} \tag{67}$$

f_i = fraction of the primary hadron energy converted to component i; i/mip = fraction of component i detected. The binding energy is lost

$$\frac{B}{mip} = 0. \tag{68}$$

The fractions of the intrinsic hadronic component are normalized

$$f_{ion} + f_n + f_\gamma + f_B = 1. \tag{69}$$

Electrons and hadrons produce the same signal according to (64), (65), if

$$\frac{e}{mip} = \frac{h_i}{mip}. \tag{70}$$

Any deviation from (70) is expected to deteriorate the resolution, furthermore according to (65), (66) a logarithmic deviation from linearity is expected in this case. Pioneering Monte Carlo studies by Gabriel [36], the extensive detector development program started by the HERA collaborations [37, 38] and the detailed Monte Carlo studies of Brückmann and Wigman [39, 40] resulted in a major breakthrough in the detection technique of high energy hadrons by developing means to fulfill (70).

2.4.1 Equalization of Electron and Hadron Signals

In this chapter the most important methods to equalize the electron and hadron response (70) and hence to achieve good energy resolution are discussed. Since for most combinations of absorber and active material the response to the electromagnetic component is larger than for the intrinsic hadronic contribution, eq.(70) can be fulfilled, if the visible electromagnetic contribution is reduced. As observed the first time by the ARGUS collaboration [34], an increase in the absorber thickness reduces e/mip (fig.15, insert). This decrease is due to the fact that the energy is deposited mainly by low energy electrons produced in the absorber plate of the calorimeter. These electrons, due to their low energy, can only leave the absorber, if they are produced in a thin layer near the absorber surface, while electrons from the bulk are stopped in the absorber [40]. The ZEUS collaboration has shown that this method can be used to fulfill (70) for a Pb–Sc calorimeter [41].

A second approach leaves e/mip constant and instead increases the intrinsic hadronic contribution h_i/mip. According to (67) there exist two ways to achieve the goal. Since the number of protons (neutrons) of an absorber decreases (increases) with Z, the choice of Z has an influence on f_{ion} and f_n. Note that f_n for U increases further due to fission. Moreover, the ratio ion/mip is small for large gap thickness d_{act} of the active medium and increases with decreasing d_{act}, since the ionization density of a track and hence the recombination losses [42] are high at the end of a track (fig.9). In case of large d_{act} the ions are stopped in the active gap, therefore compared to the mip signal, which is not influenced by recombination, a reduction is expected. A typical result of a Monte Carlo calculation [40] for this effect, supporting these qualitative arguments, is shown in fig.16.

A third method was pioneered by the CDHS collaboration [43] and optimized by the H1 collaboration [38, 44, 45]. In this case the electromagnetic component of a hadronic shower – characterized by a high local energy deposition – is identified event per event on software basis and a weight is applied. The weight is chosen such that the overall resolution for hadrons is optimized. This method can only be applied, if the calorimeter is segmented in longitudinal and transverse direction fine enough. The application of this method is demonstrated in fig.17a-e. As shown by fig.17a, the total energy increases with the amount of energy deposited in one cell. If one applies the weighting procedure, which optimizes σ/E, this dependence disappears (fig.17b). As follows from fig.17c,d also the signals for pions and electrons are equalized and the response becomes gaussian like. The resolution achieved after weighting amounts to $\sigma/E = \frac{50\%}{\sqrt{E}}$ (fig.17e).

The fine segmentation of the detector can be used to measure separately the sampling fluctuations (60) and the fluctuations due to the binding energy. To achieve this goal, consecutive active gaps 1, 2, 3, 4, ... are read out separately, in this way two calorimeters a, b with interleaved cells are available. The signals are given by

$$S_a = S_1 + S_3 + S_5 + ... \qquad (71)$$

$$S_b + S_2 + S_4 + S_6 + ... \qquad (72)$$

Note that the binding energy losses for S_a, S_b are strongly correlated, therefore, the fluctuation of the binding energy only contributes to the sum

$$S_\Sigma = S_a + S_b , \qquad \sigma_\Sigma^2 = \sigma_a^2 + \sigma_b^2 + \sigma_B^2, \qquad (73)$$

but not to the difference

$$S_\Delta = S_a - S_b , \qquad \sigma_\Delta^2 = \sigma_a^2 + \sigma_b^2 \sim \sigma_{sample}^2. \qquad (74)$$

As shown by fig.17e the sampling fluctuations in case of the H1 calorimeter are negligible compared to the binding energy fluctuations.

2.4.2 Compensation Calorimetry

It has been shown by Monte Carlo simulations [46] that the energy of neutrons in a hadron shower is correlated with the losses due to the binding energy (fig.18). Hence the contribution of this fluctuation source can be strongly reduced, if one uses scintillator as active material, which allows to detect neutrons. Indeed, the ZEUS collaboration has shown that one can achieve for an optimized detector an intrinsic resolution $\sigma_{int}/E = \frac{20\%}{\sqrt{E}}$ for a U-Sc and $\sigma_{int}/E = \frac{14\%}{\sqrt{E}}$ for a Pb-Sc calorimeter. The latter intrinsic resolution is better due to the fact that for spallation neutrons a stronger correlation between neutron- and binding-energy exists than for evaporation neutrons (fig.18). Because of fission more evaporation neutrons contribute in case of a U absorber.

The excellent performance achieved with the ZEUS calorimeter [47] is demonstrated by the results shown in fig.19. The response for hadrons has a gaussian shape, a fact of great importance, if one wants to unfold experimental distributions. Within 3% the electron and hadron signals for $2\,GeV < E < 100\,GeV$ are equal. The response is linear within 1% for $1\,GeV < E < 100\,GeV$ and a resolution of $\sigma/E = \frac{35\%}{\sqrt{E}}$ is achieved, which is dominated by sampling fluctuations [47]. These numbers demonstrate the great progress in calorimetry achieved in the last 5 years. The detector development work in connection with the preparation of the HERA experiments had a great impact on these improvements.

3. HERA Detectors

Electrons of $30\,GeV$ and protons of $820\,GeV$ are stored in the HERA rings. Hence the produced particles of a reaction are boosted in the proton direction leading to final states of high energy particles and high density in the proton direction. This basic asymmetry of the reaction reflects in the asymmetric lay-out of the two detectors.

3.1 H1–Detector

A schematic view of the H1 detector [38] cut along the beam $(z-)$axis is shown in fig.20a. A transverse cut is displayed in fig.20b. A detailed description of the full detector and its operation characteristics is given in ref.[48].
The electron and proton bunches overlap in an extended vertex region of $\leq \pm 25$ cm. Outside the vacuum tube this region is surrounded by a system of cylindrical driftchambers, which allows to track charged particles in the central region $25^0 \leq \theta \leq 153^0$. It consists of two concentric jet–chambers[49], which are complemented inside and outside by a flat driftchamber to measure the z–coordinate of the tracks. The forward tracking ($7^0 < \theta < 25^0$) consists of 3 supermodules of radial and planar driftchambers respectively, which provide accurate θ and $r\varphi$–information. The backward region ($155^0 < \theta < 175^0$) is covered by 4 plane multiwire proportional chambers, which provides an accurate space point. The latter chamber will be replaced in the future by a 8 plane driftchamber, which allows to determine a track segment, thus providing a more redundant information than presently available. A further improvement in this angular region is expected from a silicon tracker under design. The driftchambers in the forward and central region are backed up by multiwire proportional chambers, which are used primarily for trigger purposes. A superconducting coil with a radius of 3 m produces a field of $1.2\,T$. It provides the necessary bending of the charged particles, which in combination with the track information from the driftchambers allows to determine the particle momentum. The characteristics of the tracker are collected in table 3.

The radial chambers of the forward tracker fulfill a second purpose. A passive array of polyprophylene layers provides a sufficient number of dielectric surfaces for useful transition-radiation x-ray emission. The layer thickness and the mesh to space ratio were optimized to achieve a proper energy spectrum and x-ray yield. The radial chambers are designed such that efficient x-ray detection is possible. Test beam results show that for 90% electron efficiency 10% π–contamination is expected in the data.

A fine segmented liquid argon sampling calorimeter surrounds the tracking region starting at a radius of 1.2 m. The polar angle region $4^0 <$

Table 3: Parameters of the H1 tracking system, ( ) design values

Central tracking	
angular coverage	$25^0 < \theta < 155^0$
$\Delta p/p^2$	$< 0.008\ (0.003)$
jet chamber	
$\Delta \frac{dE}{dx}$	10 %
$\sigma_{r\varphi}$	170 μ
z–chamber	
σ_z	2 mm
Forward tracking	
angular coverage	$6^0 < \theta < 25^0$
$\sigma_{r\varphi}$	170 μm
σ_{xy}	210 μm
$\Delta p/p^2$	(0.006p)
Instrumented iron	
angular coverage	$4^0 < \theta < 171^0$
σ_θ	15 mm
σ_ϕ	10 mm
$\Delta p/p$	35 %
Forward μ toroid	
angular coverage	$3^0 < \theta < 17^0$
$\Delta p/p$	0.25 ... 0.32

$\theta < 153^0$ is covered. The calorimeter consists of an electromagnetic section with lead absorbers covered by G 10–plates for read–out and for high voltage supply. Its thickness varies between 20 X_o and 30 X_o. The hadronic section has steel absorbers interleaved with a double gap of 2 x 2.3 mm liquid argon and a double–sided pad read–out. The total thickness of the hadron calorimeter varies between 3.5 λ and 7 λ. The calorimeter is built in a modular way. In beam direction it consists of 8 wheels (fig.20a), each composed of 8 electromagnetic and 8 hadronic modules (fig.20b). In the first 4 rings in the forward region and the 8th ring in the backward the absorber plates are perpendicular to the beam, for the remaining wheels they are parallel. The calorimeter modules are housed in a cryostat filled with $4 \cdot 10^6$ Mol of liquid argon. The calorimeter is followed by a superconducting coil. This configuration of the magnet coil at the rear side of the calorimeter has the advantage to minimize the dead material in front of it and allows to detect low energy photons.

The backward region of the detector is covered by a warm lead–scintillator sandwich calorimeter to detect electromagnetic showers. The characteristic parameters of the calorimeters are collected in table 4.

The modules of the calorimeter were calibrated at the CERN SPS with electrons and pions with energies in the range $5-220\,GeV$. As the response ratio of electrons and pions for the chosen combination of active (liquid argon) and passive material (Pb/Fe) is not 1, a weighting procedure was applied (ch.2.2) to equalize the response [44].

The calibration is monitored during the run. The following methods are applied [50]:

- The liquid argon purity is continously measured with several liquid argon ionization chambers using $^{209}Bi-\beta$-probes, which are distributed over the calorimeter.

- Electronic calibration with capacitors with a precisely known capacity in liquid argon, which are fed with a known charge. Over periods of a few months a stability of $< 0.1\%$ is achieved.

- The energy signal is compared to the momentum of single particles (e, μ, π) measured with a tracker.

The present experience shows that the envisaged systematic uncertainty of $\leq 1\%$ for the electromagnetic and $\leq 2\%$ for the hadronic energy measurement with some additional work can be achieved.

The energy leaking out of the calorimeters is measured with a streamer tube system with analogue pad–read–out (tail catcher). It is sandwiched between iron plates, which serve in addition as a return yoke of the solenoid field. A strip–read–out of the streamer tubes provides ≤ 10 space points for muons punching through the iron. Note that pions are showering up and do not penetrate the iron yoke. These signals allow to identify muons and to perform a second independent measurement of the muon momentum. The latter can be used for momenta $p \leq 20\,GeV$ to reduce the background due to π/K-decay in the muon sample [51]. Forward going muons are detected with a muon–spectrometer consisting of a toroid magnet and a total of 12 planar driftchambers in front and behind the magnet respectively.

Two calorimeters systems near the incoming and outgoing beams close the detector. A (plug) sandwich calorimeter has the task to measure the p_t-balance in the region of the outgoing protons. It consists of 8 Cu plates (7.5 cm thick) with Si–read–out in the gaps. In the direction of the outgoing electrons and incoming protons an electron- and a photon-calorimeter are positioned respectively, which allow to detect the monitor reaction $ep \rightarrow ep\gamma$, see ch.3.3 for more details.

In table 5 the possibilities provided for particle detection by the H1 detector are collected. The many constraints available for each particle

Table 4: Parameter of the H1 calorimeter

	Main calorimeter
Stability	≤ 0.2 % variation/year
Noise	$10 - 30\ MeV$ / channel
Dead channels	≤ 0.3 %
Angular coverage	$4^0 < \theta < 155^0$
	Electromagnetic modules
Material	active: liquid argon; passive: Pb
σ/E	$11\ \%/\sqrt{E} \oplus 1\ \%$
Segmentation	transverse: $4 \times 4\ cm^2 (\sim 2 \times 2\ R_M^2)$
	longitudinal: $3 - 4$ segments
Calibration	2.5 % systematic uncertainty at present
	Hadronic modules
Material	active: liquid argon; passive: Fe
σ/E	$50\ \%/\sqrt{E} \oplus 2\ \%$
Segmentation	transverse: $8 \times 8\ cm^2 (\sim \ .....\lambda$
	longitudinal: $4 - 5$ segments
Calibration	8 % systematic uncertainty at present
	Backward calorimeter (BEMC)
Material	active: scintillator; passive: Pb
σ/E	$10\ \%/\sqrt{E} \oplus 4\ \%$
σ_x	$1.5\ cm$
Segmentation	transverse: $16 \times 16\ cm^2 (4.4 \times 4.4\ R_M^2)$
	longitudinal: 2 segments
Calibration	2 % systematic uncertainty at present
	Tail catcher $(6^0 < \theta < 172^0)$
Material	active: streamer tubes; passive: Fe
σ/E	$100\ \%/\sqrt{E}$
Segmentation	transverse: $30 \times 30\ ...\ 40 \times 50\ cm^2$
	longitudinal: 2 segments

species separately allow for a high sensitivity combined with a small misidentification probability.

3.2 ZEUS Detector

In fig.21a schematic view of the ZEUS detector is shown in a vertical cut along the beam tube [37]. The interaction vertex is surrounded by a tracking system. It consists of a high resolution vertex chamber and a cylindrical jet chamber. In the forward and backward region planar driftchambers close the volume. In the forward direction the electron identification capabilities are improved by transition radiation chambers. The whole tracking system is enclosed by a thin walled superconducting solenoid magnet, which provides a field of 1.8 T.

An uranium–scintillator sandwich calorimeter covers 99.7% of the solid angle as the main component of the detector. It is subdivided in a forward (FCAL), barrel (BCAL) and backward (RCAL) calorimeter covering the polar angle interval $2.6^0 < \theta < 176.1^0$ in a homogeneous way. The calorimeter moduls are made out of a sandwich of 3.3 mm depleted uranium plus a 2.6 mm thick scintillator. Wave length shifters absorb the light leaving the scintillator plates and transport it to the photomultipliers. Longitudinally the moduls are divided into one electromagnetic and two longitudinal hadronic sections with a total length varying between 4λ in the rear to 7λ in the forward direction. The transverse segmentation in the electromagnetic (hadronic) part is $5 \times 20\,cm^2$ ($10 \times 20\,cm^2$). To improve the electron identification capabilities the transverse segmentation in the forward and backward part of the calorimeter will be improved by inserting a plane of silicon diodes with $3 \times 3\,cm^2$ segmentation after the first $3\,X_o$ of the calorimeter.

The calibration of the photomultipliers is monitored with the signal due to the natural radioactivity of U. The ZEUS collaboration has taken great care to equalize the signal for electrons and pions, they achieve signal ratio of $e/\pi = 1.00 \pm 0.03$ for energies $E > 3\,GeV$ (ch.2.3). The energy resolution measured in the test beam is $\sigma/E = \frac{18\%}{\sqrt{E}}$ $\left(\frac{35\%}{\sqrt{E}}\right)$ for electrons (pions) [52, 53]. A big advantage of the scintillator calorimeter with photomultiplier read–out is the excellent time resolution possible ($< 1\,ns$ for $E > 3\,GeV$), which provides a powerful tool to reduce background from beam-gas and beam–wall interactions (fig.22) [54].

The calorimeter is surrounded by an iron yoke, which serves as an muon filter and tail catcher. It consists of sandwiches of iron and proportional tube chambers. The iron is magnetized with a copper coil to allow for a rough momentum measurement. In the forward region a toroid spectrometer with limiting streamer tubes as tracking device serves as a μ–spectrometer.

A unique feature of the ZEUS detector is a leading proton spectrometer, which allows to detect very forward scattered protons with a transverse momentum of $p_T < 1\,GeV$. They either are produced in diffractive events or are fragments of diquark systems [17]. This spectrometer uses the proton ring as spectrometer magnets, the scattered protons are detected in 8 Si-strip detectors mounted on the beam tube between 26 m to 96 m from the interaction point.

Finally, a comparison of the strong (weak) points of the two detectors is of interest. Note that general purpose detectors as the H1- and the ZEUS-detector are always a compromise between the aim to have optimal conditions for a specific process and the detection of as many reaction channels as possible. H1 has the better electron detection due to the better energy resolution and identification possibilities because of its higher granularity. The advantage of ZEUS is the better hadron calorimetry due to the better energy resolution and the equalization of electron and hadron response (eq.(70)) by hardware compensation (ch.2.3). The use of scintillators and photomultipliers for calorimetry has the advantage to provide a precise fast signal, which is a powerful tool to suppress background (fig.22). The good tracking information of the H1 detector already available at the fast trigger level serves the same purpose.

3.3 Luminosity Measurements

The reaction rate N_r observed depends on the luminosity L of the machine and the cross section σ of the respective reaction

$$N_r = L \cdot \sigma. \tag{75}$$

The luminosity L is a property of the machine, it depends on the transverse beam dimensions and on the number of circulating electrons and protons. It is measured using a monitor reaction with a known cross section. At HERA the bremsstrahlung process

$$ep \rightarrow ep\gamma \tag{76}$$

at small angles is used, which can be calculated in QED, since the momentum transfer is small and hence contributions from proton structure can be neglected. The process is selected by measuring with electromagnetic calorimeters the electron and photon energy E_3 and E_γ respectively in coincidence. The following relation has to be fulfilled

$$E_3 + E_\gamma = E_1. \tag{77}$$

The H1 collaboration uses two calorimeters out of $TlBr/TlCl$ crystals, which detect electrons produced in the electromagnetic shower via their Čerenkov radiation [55, 60]. The calorimeters are placed 34 m (electron) and 105 m (photon) from the interaction point (fig.23). The electron calorimeter consists of a matrix of 5×5 crystals with dimensions $22 \times 22 \times 200\,mm^3$

each. In fig.24 the validity of eq.(77) is proven for bremsstrahlung events. The density of points in the scatter plot reflects the acceptance of the two calorimeters, typically 70% of all bremsstrahlung events in the energy interval $8\,GeV < E_3 < 19\,GeV$ are accepted. ZEUS uses a similar set up, the calorimeters are of the Pb-scintillator sandwich type [56]. As electron–gas interaction has the same signature as reaction (76), its contribution has to be determined on a statistical basis. An electron–bunch, which does not collide with a corresponding proton–bunch ("pilot-bunch") allows to determine experimentally the beam–gas background to reaction (76).
In the year 1992 an integrated luminosity of $30 - 50\,nb^{-1}$ was collected depending on the specific trigger configuration chosen.

Note that the electron calorimeter of the luminosity system is also used in anticoincidence with the photon detector to tag the photon–energy in photoproduction experiments (ch.4).

3.4 Trigger and Data Selection

Interactions of electrons and protons with the rest gas of the storage ring vacuum and of halo particles of the machine near the detector produce a high background, which can deposit much more energy in the detector than most ep events. Note that the cross section for p–gas interactions is 0 $(200\,mb)$, while the typical cross sections of interest are smaller than $150\,\mu b$ (γp total cross section), hence the background rates are orders of magnitude higher than the signal of interest. The triggering and data reduction is therefore of utmost importance. It is performed in several steps summarized in fig.25 for the H1 experiment. After 25 bunch crossings ($2.5\,\mu s$) the trigger stops the pipeline. The most important trigger elements used in the analysis of ch.4-5 are [48]:

- A global veto derived from two planes of time–of–flight counters, which are set up behind the backward calorimeter (fig.20.a) to reject background from upstream proton interactions. ZEUS uses instead the fast time–of–flight information of its forward (FCAL) and backward (RCAL) calorimeter. Particles produced at the vertex are clearly separated from the background signal (fig.23).

- Electron triggers are based on clusters of energy in the electromagnetic part of the calorimeter.

- Transverse and missing energy signal are derived from the calorimeter signal by carefully combining read-out channels.

- A muon–trigger is available from the instrumented iron and the μ–spectrometer respectively.

Table 5: Information available from the different detector components used to identify particles in the H1 detector

e	$\frac{dE}{dx}$ from driftchamber; signal of transition radiation detector ($5^0 < \theta < 25^0$); transverse and longitudinal shower profile in electromagnetic calorimeter; no (small) signal in hadron calorimeter
μ	$\frac{dE}{dx}$; signal of a minimum ionizing particle in the electromagnetic and hadronic calorimeter; signal of instrumented iron; momentum measurement in driftchamber and instrumented iron, μ-spectrometer
charged hadrons	$\frac{dE}{dx}$; transverse and longitudinal shower profile in electromagnetic and hadronic calorimeter; shower tail in tail catcher
γ	no signal in driftchamber; transverse and longitudinal shower profile in electromagnetic calorimeter; no (small) signal in hadron calorimeter

- H1 in addition triggers on the pad signal of the two multiwire proportional chambers, which cover the central region $25^0 < \theta < 155^0$ of the detector. The trigger selects events with a significant peak in the vertex histogram, which is built for each event from the chamber information.

In the future with higher rates a more sophisticated trigger scheme will be set up (fig.25), which uses more detailed information from the front end electronics to reduce the data flow to the filter farm. As an example H1 presently develops a level 2 ($L\,2$)-trigger (fig.25) based on neural networks. The filter farm L4 reduces the data based on a fast filter algorithm and a partial reconstruction of the event, e.g. presently a full reconstruction of the tracks in the central tracks and a vertex fit are performed and proper cuts on the information achieved are applied. Moreover, cosmic muons are rejected and energy cuts on the thresholds of the calorimeter triggers are applied.

4. Photoproduction

4.1 Total Photoproduction Cross Section

Besides $p\bar{p}$ interactions, studied at the large hadron colliders, photoproduction is the only hadronic interaction, where cross sections in the few

Table 6: Measured total photoproduction cross section

	Reference	$<W_{\gamma p}>$	$\int L\, dt$	$\sigma_{\gamma p}$
H1	[62]	$195\, GeV$	$(1.5 \pm 0.1)\, nb^{-1}$	$(159 \pm 7 \pm 20)\, \mu b$
H1	[60]	$197\, GeV$	$(21.9 \pm 1.5)\, nb^{-1}$	$(156 \pm 2 \pm 18)\, \mu b$
ZEUS	[61]	$209\, GeV$	$227\, \mu b^{-1}$	$(154 \pm 16 \pm 32)\, \mu b$

$100\, GeV$ cms energy range can now be studied. Hence one expects more detailed information on the source of the rising cross sections observed for cms energies above $20\, GeV$ [57]. No consistent theoretical interpretation of this phenomenon in the framework of the standard model exists, rather one has to rely on phenomenological Regge fits for a coherent description of hadronic cross sections [58].

H1 and ZEUS have recently published first results on the total photoproduction cross section at $\sqrt{s_{\gamma p}} \approx 200\, GeV$, a factor of 10 higher than achieved in previous fixed target experiments [60, 61, 62]. The basic signature of a photoproduction event (fig.26) in both experiments is an electron detected in the electron tagger of the luminosity monitor (fig.23) in anticoincidence to the time–of–flight information available, if it signals a background event. Moreover, the photon calorimeter of the luminosity system should show no photon signal. Finally, in both experiments a signal in the main detector is required: while H1 selects events with at least 1 track in the central tracker, ZEUS demands an energy deposition of $> 1.1\, GeV$ in the backward calorimeter RCAL (fig.21).

Further weak cuts have to be applied to suppress the contribution from beam–gas interactions in the vertex region of the detector. Since this method is exploited in more detail by the H1 collaboration, their procedure is discussed here in some detail. Proton gas interactions due to the Lorentz boost cluster at large values of $\sum p_z / \sum p$ and small values of y_h (defined by eq.(41)). This is demonstrated by fig.27a, where the results of a measurement with the pilot bunch is plotted. Photoproduction results (fig.27b) pass this cut. Still a background from electron–gas interactions remains, which is measured with the help of the electron pilot bunch (fig.28a). The electron spectrum of photoproduction events is shown in fig.28b, it nicely agrees with the result of a Monte Carlo simulation included in the figure.

The major source of uncertainty is the acceptance, which is sensitive to the relative contribution of different photoproduction reactions discussed in chapter 1.2. The total acceptance considers the geometrical acceptance, the trigger efficiency and the event selection efficiency. In the H1 experiment the total acceptance amounts to $<A> = 0.27 \pm 0.023$. The simulated p_t– and θ–dependence of the produced particles agree perfectly with the measured

one and support the assumptions the simulation is based on.
Using the measured luminosity one gets the cross sections summarized in table 6. In fig.29 the experimental result is compared to model predictions. Note the good agreement with the Regge pole ansatz, which at the same time also describes other hadronic reactions [58]. The dotted curves represent the predictions of a model using the ansatz

$$\sigma_{\gamma p} = \sigma_{soft} + \sigma_{jet} . \tag{78}$$

σ_{soft} describes the VDM contribution, while σ_{jet} results from contribution of direct and resolved photons [18]. They depend critically on the photon structure function and on a cut–off parameter p_{TMin} which represents the lower integration limit for the hard process. The data exclude a low cut–off, thus extreme minijet models [63] are excluded by these new results.

4.2 Hard γp–Scattering Reactions

The study of $\gamma\gamma$–interactions at PETRA and other e^+e^-–machines has established the parton structure of the photon [64]. Besides a pointlike component the photon can be considered as a system of quarks and gluons. It turns out that $\gamma\gamma$–reactions at the present generation of e^+e^-–storage rings allow only for small statistics and moderate values of cms–energies, hence the (x, Q^2) values, which can be probed for the photon are severely limited. The high cms–energy of $\sim 200\,GeV$ of the γp–system, which can be achieved at HERA, changes the situation dramatically and allows for the first time to investigate thoroughly the photon structure function at low x and high Q^2, which is of special interest for its gluon component hardly known up to now. Two classes of events contribute to hard γp interactions in leading order (fig.7). In the simplest case the photon couples directly to the partons of the proton. These processes include the QCD Compton scattering $\gamma q \rightarrow q g$ and the photon gluon fusion process $\gamma g \rightarrow Q \bar{Q}$. The latter being a copious source of heavy quarks and gives direct access to the gluon structure function of the proton, which otherwise can only be probed indirectly. The processes involving partons of the photon are termed resolved photon processes, they include in leading order the scattering of quarks/gluons of the photon on quarks/gluons of the proton (fig.7b). The cross section of resolved photon processes exceeds the direct one for the presently achievable energies by an order of magnitude (fig.30). Both types of reaction are characterized by two jets at large angle plus the spectator jet of the proton collimated into the beam–tube, which therefore are hardly detectable. For the resolved photon processes in addition the spectator jet of the photon exists (fig.7b) in the backward region. A candidate for such an event as observed by the H1 collaboration is shown in fig.31.
H1 uses a similar trigger as discussed in chapter 4.1 for the measurement of

the total photoproduction cross section [60, 65]. The background of beam–gas interactions faking a hard process, characterized by a large transverse energy deposition in the calorimeter, is negligible. It is estimated from the studies using the electron– and photon–pilot bunch. The ZEUS trigger for the study of hard photon–proton reactions is based on the energy deposition in the calorimeter [67]. The measured energy has to exceed a given threshold, which depends on the polar angle. In addition, ZEUS exploits its precise time–of–flight information to suppress background.

The transition from the soft to hard γp–interactions has been studied by the H1 collaboration [66, 68] using the measured single charged particle spectrum. They were analyzed in the central region $30^0 < \theta < 150^0$ for $p_t > 0.3\,GeV/c$. As a measure of the "hardness" of an interaction the transverse energy

$$E_T = \sum E_i \sin \theta_i \qquad (79)$$

recorded with the calorimeter is taken. The comparison of the p_t–spectra for soft ($E_T < 5\,GeV$) and hard γp–interactions ($E_T > 10\,GeV$) shows large qualitative difference (fig.32a,b). A tail at large p_t values shows up for hard processes. The measured spectra are nicely reproduced by a Monte Carlo simulation using the PYTHIA [69] code. It includes soft VDM like contributions and a hard component due to resolved photons. Note, however, that the spectra depend on the photon–structure function chosen and on the p_t cut–off for the hard process (ch.4.1).

The evolution of the hard process is visible in fig.32c,d, where the distance in ϕ (pseudorapidity η) between a detected particle and that of highest p_t in a jet ("jet leader") is plotted. The pseudorapidity is defined by

$$\eta = -\ln tg\,\theta/2 \qquad (80)$$

and is a good approximation of the rapidity

$$y = \frac{1}{2} \ln \frac{E + p_z}{E - p_z} \qquad (81)$$

of a particle, if $p >> m$ and $\theta >> m/E$ [57]. With increasing E_T a clustering around the direction of the jet–leader is observed. In ϕ the characteristic back–to–back configuration of a two–body reaction is observed for $E_T > 20\,GeV$. A similar behaviour is known from hard proton–proton interactions [17].

ZEUS selects their hard photoproduction events by a cut on the energy deposited in the calorimeter. The event selection is described in detail in ref.[67]. The measured integrated cross section for transverse energies $E_T > 10\,GeV$ is shown in fig.33a, it is in good agreement with the predictions of a Monte Carlo simulation. Note, however, that the prediction depends on the photon–structure function chosen and on the p_T cut–off applied to the hard process. The angular correlation between two jets in the transverse plane is displayed in fig.33c. Again a back–to–back configuration as expected for a two–body process, is observed. In fig.34a,b an example for a 2–jet and for a 3–jet event is shown in a lego–plot. In fig.34 a jet of the photon remnant shows up in the direction of the electron beam. On a statistical basis the remnant is observed at large energy ($4\,GeV < E_T < 8\,GeV$) in the backward calorimeter RCAL in coincidence with a large pseudorapidity of two other jets ($\eta > 0.5$) [54, 67].

H1 has performed a jet–analysis of the full data set collected 1992 [70]. The measured differential jet cross section is compared in fig.35a with the prediction of a Monte Carlo simulation. Note, however, that also in this case the predictions are sensitive to the photon structure function chosen and depend on the p_T cut–off in the hard matrix element. The pseudorapidity distribution for jets with a transverse energy $E_T > 7\,GeV$ is shown in fig.35b. A steep gluon distribution of the photon structure function seems to be excluded, but the main message of the results is that the gluon distribution in the photon can be measured in this way once higher statistics is available. Finally, it turns out that the dependence of the photon–structure function on the Bjorken variable x_γ can be extracted from the data, if both jets at large scattering angle are detected. If q, p are the four–momenta of the photon and proton respectively and j_1, j_2 those of the final state partons, from four–momentum conservation follows

$$x_\gamma \cdot q + x \cdot p = j_1 + j_2 \,. \tag{82}$$

Multiplying with p and neglecting terms of the order p^2 one gets in the laboratory frame

$$2\,x_\gamma\,E_\gamma = E_1 - E_{1T} \sin h\,\eta_1 + E_2 - E_{T2} \sin h\,\eta_2 \,. \tag{83}$$

Since [57]
$$E_1 = E_{1T} \cdot \cos h\,\eta \tag{84}$$

the final expression

$$x_\gamma = \frac{\sum E_{iT}\,e^{-\eta_i}}{2\,E_\gamma} \tag{85}$$

follows. Since E_γ, η_i, E_{iT} are measured, x_γ can be reconstructed. In fig.36 the number of 2–jet events as a function of x_γ is plotted and compared to the expectation as derived from different photon structure functions [68]. Note that no corrections for acceptance etc. were applied to the preliminary data, rather the raw data are compared to the result of the simulation, which includes the full detector Monte Carlo. In the x_γ–range covered by these preliminary results the parametrization of Glück et al.[71] gives the best description of the data, while models with a large gluon component are not favored.

5. Deep Inelastic Scattering

Deep inelastic scattering is at the heart of HERA physics. As discussed in ch.1.1.1 the following topics can be studied:

- structure functions at high Q^2 and $10^{-4} \leq x \leq 1$ can be measured, hence the transition between the regime where perturbative QCD is applicable to the region were nonperturbative phenomena dominate can be investigated.

- hadron final states, the colour flux in the event and the production of gluons can be studied.

- electroweak parameters can be measured once enough luminosity is available.

A typical deep inelastic scattering event as recorded with the H1 detector is shown in fig.37. As a trigger H1 demands an energy cluster $E \geq 4 GeV$ in the backward calorimeter BEMC (fig.20a). The beam-gas background is characterized by an off-time signal of the time-of-flight system positioned behind BEMC [72, 73, 74, 75, 77, 76]. ZEUS triggers on a minimal energy deposition in the calorimeter [78, 79, 80, 81] and applies a soft cut on the time-of-flight information of the calorimeter (fig.22).

In the off-line analysis H1 requires a space-point in front of the backward calorimeter which is correlated with the energetic cluster. The transverse size of the latter should agree with the expectation for an electromagnetic shower. In addition a vertex, reconstructed from the tracks in the central tracker, should agree with the nominal one within ±50cm. Studies exploiting the pilot bunches show that after these cuts the beam-gas background is negligible.

Still a background from photoproduction reactions survives these cuts, where a high energy hadron fakes an electron. This background is suppressed by a cut on the variable

$$\delta = \Sigma(E - p_z) \qquad (86)$$

The sum runs over all energies deposited in the calorimeter. Particles leaving the detector through the beam tube in forward (proton) direction do not contribute to δ, while for scattered electrons the two terms sum up, i.e. for deep inelastic scattering $\delta = 2E_1$. For photoproduction events, where the scattered electron leaves the detector through the rear beam tube δ is small. Therefore a cut on this variable strongly suppresses the background (fig.38) [54]. It turns out that the photoproduction background after these cuts is small and negligible for $y \leq 0.4$.

5.1 Jets in Deep Inelastic Scattering

The analysis of the final state in deep inelastic scattering allows to study basic topics:

- The different partonic processes leading to a given final state can be described in the framework of QCD. Initial and final state radiation can be detected and be compared to QCD predictions.

- Since the coupling of quarks and gluons depends on α_s, the running of the coupling constant as the basic property of a nonabelian gauge theory can be tested.

- The energy flow in the event can be connected to the colour flow in the event.

- In the analysis one has to consider the fragmentation of the parton system into the observed final state hadrons. Presently one has to rely on phenomenological models for this step.

The parton models used have been successfully applied to final states observed in $e^+ e^-$ at LEP and in deep inelastic scattering at lower energies. HERA opens a new regime compared to these studies. In contrast to the foregoing DIS experiments the scales characteristic for the parton shower namely Q^2 and the invariant mass of the hadronic system

$$W^2 = (p_2 + q)^2 = s \cdot y_h - Q^2 + m^2 = Q^2 \cdot \frac{1-x}{x} + M^2 \tag{87}$$

differ for HERA experiments by orders of magnitude. Since the probability for gluon radiation increases with increasing values of the scale due to the phase space available, the model predictions depend strongly on the scale chosen. Typical values at HERA are $Q^2 = 15 GeV^2$ and $W^2 = 10^4 GeV^2$ while in DIS fixed target experiments $Q^2 \approx W^2$, hence the new data allow to discriminate between different scales proposed in the literature. Compared to jet studies at LEP the underlying colour flow is expected to differ at HERA since no parton with quantum numbers corresponding to the spectator jet are produced in $e^+ e^-$ reactions. Unfortunately this is a source of uncertainty since no detailed studies of the spectator jet have so far been performed at lower energies [17].

The QCD effects in jets are simulated in three ways [82]:

- The quark parton model is combined with leading log parton showers (PS, HERWIG).

- $O(\alpha_s)$ matrix elements with matched parton showers (ME+PS) have been studied.

- The colour dipole model including $O(\alpha_s)$ matrix elements (CDM) is used.

The measured energy flow in pseudorapidity η and in the azimuthal angle Φ is shown in fig.39 [72] . Clearly the extreme assumptions on the energy scale (Q^2, W^2) can be excluded, while the other models are in good agreement with the data. Note however that for large values of η the data systematically lie above the predictions. This may indicate that the spectator fragmentation is not properly parametrized by the present models. As follows from eq.(87) the largest differences between W^2 and Q^2 and hence the greatest sensitivity to the scale chosen is expected at low x. This conclusion is supported by recent ZEUS results [80] shown in fig.40.

To study details of the produced jets on the event basis it is natural to choose the cms system of the final hadronic state with the direction of the virtual photon as the z-axis. In this frame the Feynman variable

$$x_F = \frac{p_z^*}{p_{zmax}^*} \qquad (88)$$

is defined. The transverse momentum with respect to the z-axis p_t^{*2} as a function of the Feynman variable x_F ("sea-gull" plot) is particularly sensitive to QCD effects. It is shown in fig.41. The ME+PS and the CDM model describe the measured distribution [72, 77].

Jet rates up to now have only been measured in $e^+ e^-$ and $p\bar{p}$ collisions since only for these reactions high enough energies are achieved to allow for well developed jets. In $O(\alpha_s)$ order one expects in ep reactions the $1 + 1$ and $2 + 1$ configuration (fig.42), where the spectator jet in most cases is lost in the beam tube. A typical example for a $2 + 1$ jet is shown in fig.37, where two well separated jets are observed, with a possible signal of the spectator jet near the beam tube in forward direction. Besides the current jet a second jet is observed in fig.37 due to gluon radiation in the initial and final state respectively or due to quark-gluon fusion (fig.42).

In the jet algorithm applied (JADE algorithm [83]) , which is an almost Lorentz invariant cluster algorithm, a cut (y_{cut}) dependent jet multiplicity is determined by calculating the scaled invariant mass

$$y_{ij} = \frac{m_{ij}^2}{W^2} \qquad (89)$$

$$m_{ij}^2 = 2 \cdot E_i \cdot E_j (1 - \cos(\theta_{ij})) \qquad (90)$$

Calorimeter cells are combined, starting with the minimal y_{ij} of all cells. The clustering progresses until $y_{ij} \geq y_{cut}$. The remaining number of clusters is counted as the jet multiplicity. The clustering uses all calorimeter cells of the liquid argon and the backward calorimeter above a given threshold [77, 75].

The results of this analysis are shown in fig.43 and compared to model predictions. The ME+PS model describes the data best, both the PS(QW) and the CDM model predictions are too high. Combined with the conclusions drawn from the energy flow measurements the PS and the CDM models can be excluded.

The 3–jet rate increases with Q^2 due to the increase of the phase space available according to (87). Moreover this rate depends on α_s. In fig.44 the measured Q^2 dependence of the 3–jet rate [77, 84] is compared with model predictions (ME+PS) assuming $\alpha_s = const$ and $\alpha_s(Q^2)$. Clearly the running coupling constant can be measured if higher statistic is available in the near future.

5.2 Measurement of the Structure Function $F_2(x, q^2)$

According to eq.(7) structure functions are the basis for the determination of the parton distributions $q(x, q^2)$. The deep inelastic scattering cross section in Born approximation (6) is given by

$$\frac{d^2\sigma}{dx dQ^2} = \frac{2\pi\alpha^2}{Q^4 x} \cdot 2(1-y) + \frac{y^2}{1+R} F_2(x, Q^2) , \qquad (91)$$

where instead of the Callan-Gross relation (8) the QCD prediction

$$2x F_1(x, Q^2) = F_2(x, Q^2) \cdot \frac{1}{1 + R(x, Q^2)} \qquad (92)$$

is used. $R(x, Q^2)$ presently has not been measured, therefore the value predicted by QCD is taken in the analysis using the MRSD–parametrisation of the structure functions [85]. The maximum change of the cross section due to this correction amounts to 8% at the highest y–value.

The cross section was derived by two methods [74]. In analysis 1 the electron information has been used to evaluate x, Q^2 according to eqs.(37),(38). The cross sections and the efficiencies were determined in $\sqrt{E_3}, \theta_3$ bins being matched to the resolution and the acceptance of the detector. This detector oriented binning was transformed into the x, Q^2 binning. The kinematic region covered by this analysis is $E_3 \geq 10.4\,GeV$, $160° \geq \theta_3 \geq 172.5°$. In analysis 2 bins in x, Q^2 were used to calculate the acceptance and the efficiency. The Q^2 value was derived from the electron information according to eq.(37), while y was calculated from the calorimeter information using eq.(41). The kinematic region covered by this analysis differs from analysis 1: $0.02 \leq y_h \leq 0.3$. The final sample comprises about 1000 events.

Table 7: Systematic errors of F_2

	analysis 1	analysis 2
shift of E_3 scale (2%)	20%...4%	$\leq 6\%$
BEMC 2% change of resolution	6%	-
hadron energy scale	-	10% ... 25%
$\Delta\theta_3 = 5$ mrad	$\leq 8\%$	$\leq 8\%$
event selection	10%	10%
beam-gas background	4%	4%
γ p contamination	$\leq 10\%$	
radiative corrections	$\leq 8\%$	$\leq 3\%$
bin wise correction	$\leq 5\%$	$\leq 5\%$

The two analyses differ in the photoproduction background to be subtracted, they use different components of the detector and the corrections applied are unlike. Hence the good agreement between the results of analysis 1, 2 is an important cross-check of the analysis procedure.

Radiative corrections have been applied using different programs and different parametrisations of the parton structure functions. The correction is maximal for analysis 1 at low x (20% for the MRSD-parametrisation), for analysis 2 the maximal radiative correction amounts to 8%. The result of the two analyses are compared in fig.45. They are in good agreement. Within the (presently still large) errors they agree in the largest x-bin achieved in the H1 experiment with recent results from fixed target experiments [86, 87]. The systematic error sources are collected in table 7. Note that in addition a global 8% uncertainty due to the luminosity measurement and the event selection has to be added.

The most important feature of the data is the steep rise of $F_2(x, Q^2)$ at small x, which is a promising hint to new QCD phenomena as saturation (ch.1.1.2). The new results are compared (fig.46) to various parametrisations of parton densities which fit the low energy data from fixed target experiments. Due to lack of data before HERA these parametrisations usually make assumptions on the behaviour of the parton distributions at $x \leq 10^{-2}$. Obviously the new HERA data constrain the range of acceptable parton distributions strongly and will be helpful guidelines for the development of new theoretical tools .

Physics with the ARGUS Detector

The idea to build a new detector to study $e^+ e^-$–interactions in the energy region of the Y–resonances was born a few month after the observation of the two lowest $b\bar{b}$ >–bound states [91]. Few months later physicists from different countries were convinced by the optimism of the founding fathers and joined the ARGUS collaboration (**A** **R**ussian **G**erman **U**SA **S**wedish collaboration).[2] The proposal was accepted 1979, the detector started 1982 with the data taking. A strong Canadian group joined the collaboration and contributed a high resolution vertex chamber, later a Yugoslavian group took part in the common effort. It turned out that ARGUS could contribute to many fields of high energy physics, some of them of fundamental interest:

- In $\gamma\gamma$–physics the $\Gamma_{\gamma\gamma}$ width of many 0^{--} and 2^+–meson states were measured. Special effort was put into the study of $\gamma\gamma$–reactions with 2 vector mesons in the final state, since many models predicted a large production cross section for $4q$–states in these reactions [92]. ARGUS could exclude the dominance of this reaction channel.

- Since the $\Upsilon(1S)$–meson dominantly decays into 3 gluons, the study of this decay offers the unique chance for gluon fragmentation studies. ARGUS has performed a systematic study of the spectra of all octet and decuplet mesonic and baryonic states [59]. These results have a great impact on the formulation of fragmentation models used in all present day QCD studies.

- Since the $\Upsilon(1S)$–meson is a narrow state, it offers a unique chance to search for exotic particles. Limits for the production of free quarks, light Higgs mesons, axions etc. were determined, which constrain severely the free parameters of models.

- B–physics is at the heart of the ARGUS research program, the results achieved will be discussed in the following chapter.

- Finally ARGUS performed a comprehensive study of τ–decays. The big advantage of the ARGUS detector is the high statistics available and the good particle identification capability of the experiment. A short selection of the results emphasizing the study of the Lorentz-structure of the interaction underlying the τ–decay will be given in the last chapter of the lecture.

---

[2]Some people claim that in reality ARGUS is the abbreviation for „Alle Richtigen Genies Unter Sich".

6. B–Physics with the ARGUS Detector

6.1 Motivation

The $e^+ e^-$ cross section as measured in the energy region of the Y-resonances [93] is shown in fig.47. The width of the $Y(nS)$-resonances (n = 1,...,3) is compatible with the energy resolution of the storage–ring, while the $\Upsilon(4S)$–meson has a larger width. This observation points to different decay mechanisms and indeed it has been shown that the $\Upsilon(4S)$–meson decays into a $B\bar{B}$–pair, while the other resonances dominantly decay electromagnetically or into a 3 gluon final state. Moreover, the $\Upsilon(4S)$–resonance is produced near the $B\bar{B}$–threshold

$$m(\Upsilon(4S)) - 2\, m(B) \approx 23\ MeV, \qquad (93)$$

hence no further particles exist in the final state. These facts provide important advantages for the experiment:

- The energy of the B–meson is known with high precision ($\sigma = 4\ MeV$) from the beam energy
$$E_B = E_{beam}\ . \qquad (94)$$

- The B–meson is produced nearly at rest
$$E_B \approx M_B, \qquad (95)$$
hence no correlation between the decay products of the two B–mesons exists, while the hadrons produced in the fragmentation of $q\bar{q}$ states of the continuum background are strongly correlated in their direction. This fact is heavily exploited in the analysis (ch.6.3.1).

- The signal to background ratio is $\sim 1:3$. No other source of B–mesons in this respect can compete with $\Upsilon(4S)$–decays.

- Since the number of $\Upsilon(4S)$–mesons produced can be calculated from the integrated luminosity, the number of B–mesons produced is known. Therefore, branching ratios can be derived from the data without further input.

B–physics in the recent decade had an important impact on the development of the standard model, because it allows to determine 4 basic parameters of the standard model. In addition, interesting QCD studies are possible on the borderline of hard and soft physics.

In the standard model the Higgs mechanism is responsible for mass generation and for flavor mixing [94]. Therefore, one hopes to cast some light on the problem of mass generation by the measurements of the Cabibbo–Kobayashi–Maskawa (CKM) matrix–elements [95]. In fig.48 the mass spectrum of quarks is plotted, in addition the observed weak decays are included. Note that the strength of a transition is largest within a generation, it decreases with the number of generations skipped by a transition. The strength is given by a unitary 3×3 matrix (CKM–matrix):

$$V_{CKM} = \begin{pmatrix} V_{ud} & V_{us} & \mathbf{V_{ub}} \\ V_{cd} & V_{cs} & \mathbf{V_{cb}} \\ \mathbf{V_{td}} & V_{ts} & V_{tb} \end{pmatrix} \qquad (96)$$

The boldface matrix–elements can be measured in B–decays. Since the CKM–matrix is unitary three angles and a phase δ fix all matrix–elements in the 3–generation version of the standard model. In addition, an interesting triangular relation holds

$$V_{ud} V_{td}^* + V_{us} V_{ts}^* + V_{ub} V_{tb}^* = 0 \qquad (97)$$

The norm conditions, the orthogonality of the rows and measurements allow to simplify this relation:

$$V_{tb} \approx V_{ud} \approx 1 \qquad (98)$$

$$V_{ts}^* \approx -V_{cb} \qquad (99)$$

Hence one gets

$$V_{ub} + V_{td}^* = V_{us} V_{cb} \qquad (100)$$

The test of this triangular relation will be discussed in ch.6.4.

Note that the standard model predicts CP–violation in B–decays, if the phase is

$$\delta \neq 0, \pi \qquad (101)$$

The present generation of experiments because of lack of statistics is not sensitive enough to check this important prediction. B–factories [96] or dedicated experiments [97] are presently under discussion.

The study of strong interaction effects is another field of application. A new approach which allows to approximate systematically QCD applied to heavy quark bound states (Heavy Quark Effective Theory, HQET) has recently been proposed [98]. It allows to test QCD in the transition region from hard to soft physics. Moreover, it is of great importance as an input information, if one wants to extract the V_{CKM} matrix elements from the data, since the influence of strong interaction dynamics on the transition rate has to be known. The basic idea of HQET is well–known from atomic physics, where the influence of the nucleus can be separated in 1st order. In analogy the ansatz is made

$$\Psi_{Meson} = \Psi_Q \cdot \Psi_q, \qquad (102)$$

where Ψ_Q is a plane wave description of a free heavy quark. Corrections in Λ/m_Q can be systematically calculated [98]. Since the meson wave function factorizes, the same is true for the transition rate. Therefore, the heavy meson decay can be described in the framework of the HQET by the free quark decay

$$b \to c \; l^- \; \bar{\nu}_l \;, \qquad (103)$$

which is calculable, times an overlap integral which reproduces the overlap probability of the light quark clouds in the initial and final states. It depends on the velocities of the initial/final quarks. This overlap integral is called the Isgur-Wise function [99] $\xi(v_B \cdot v_D)$. It is universal but unknown except for the case, when the final/initial heavy quarks are at rest, i.e. for maximum recoil of the $(l\nu_l)$-system. In this limit $\xi(v_B \cdot v_D = 1) = 1$, since the lightquark cloud does not feel that the b-quark decays into a c-quark and hence the overlap is optimal. E.g. V_{cb} can be derived from the measured lepton spectrum [100] in the limit $y \to 1$:

$$\mid V_{cb} \mid^2 \xi(y) = \frac{1}{\sqrt{1-y^2}} \frac{d\Gamma}{dy} F(y) \qquad (104)$$

$$y = v_B \; v_D \approx \frac{E_D}{m_D} \qquad (105)$$

$F(y)$ is a known function, dM/dy is measured.
Other approaches [101] describe the strong interaction dynamics by a formfactor, the color matching in the final state is parametrized by two free parameters corresponding to the two contributions of fig.49. An impressive amount of branching ratios for two-body decays has been successfully predicted by this ansatz and verified in the experiment [101].

6.2 Exclusive B–Decays

As pointed out in ch.6.1, eq.(94) is heavily exploited in many studies of B-decays performed by the ARGUS collaboration. As a first example we discuss the exclusive B-decays. The mass of reconstructed B-mesons is constrained by

$$m_B^2 = E_{beam}^2 - \vec{p}^{\;2}. \qquad (106)$$

In this case the mass resolution is given by the expression

$$\sigma_m = \sqrt{\beta^2 \gamma^2 \sigma_p^2 + \sigma_{beam}^2} \approx 4 \; MeV, \qquad (107)$$

since according to (93),(95) $\beta = 0.06, \gamma \approx 1$.
If only the momenta of the reconstructed decay particles of the B-meson would have been used to determine its invariant mass, the mass resolution would be of the order

$$\sigma_m = \sigma_p \approx 50 \; MeV \;. \qquad (108)$$

It is clear that the signal to background ratio for exclusive decays is improved appreciably by the decrease of the mass uncertainty.

A typical invariant mass distribution for reconstructed B–meson using eq.(106) is shown in fig.50. A clear signal above a smooth background is observed [102]. It can be shown that the nonresonant background is due to continuum events, hence the shape of this background is known from independent measurements performed at $e^+ e^-$ cms–energies slightly below the B–meson production threshold. ARGUS derived from its measurement the following B–meson masses:

$$m(B^0) = (5278.2 \pm 2) MeV \qquad (109)$$

$$m(B^+) = (5279.2 \pm 2) MeV . \qquad (110)$$

Within the error limits the masses are equal. Note that CLEO [103] arrives at the same conclusion.

Since the decay $\Upsilon(4S) \rightarrow B\bar{B}$ is phase space dominated, we assume for the following analysis

$$Br(\Upsilon(4S) \rightarrow B^+B^-) = Br(\Upsilon(4S) \rightarrow B^0\bar{B}^0) = 0.5 . \qquad (111)$$

The analysis of the angular distribution of the produced B–mesons with respect to the beam direction as quantization axis provides a nice consistency check of the analysis. Since the $e^+ e^-$ pair annihilates into a ± 1 helicity state and the B–meson is expected to have spin 0, the two B–mesons are produced in a $l = 1$ angular momentum state. Hence their angular distribution with respect to the beam–axis is given by

$$\frac{d\sigma}{d\cos\theta_B} \sim | Y_{11}(\cos\theta_B) |^2 \sim \sin^2\theta_B \qquad (112)$$

in good agreement with the data.

As discussed in ch.6.1 formfactor models [101] allow to predict branching ratios for two–body decays with two free parameters a_1, a_2. In fig.51 the measured branching ratios are compared with the model predictions [101]. The agreement between theory and experiment is satisfying. Of course one can use the data to determine a_1, a_2. The results of such an analysis are given in ref.[102]. As mean values ARGUS finds

$$a_1 = 1.18 \pm 0.08 \qquad (113)$$

$$a_2 = -0.22 \pm 0.03 . \qquad (114)$$

The decay

$$B \rightarrow D_S^{(*)} D^{(*)} \qquad (115)$$

can be analyzed in the same way as discussed for the example given in fig.50. We discuss it here in some detail, since it provides further important information on weak hadronic decays of heavy quarks. It has been shown [101] that the branching ratio for this decay, in case factorization of the two hadronic currents holds (fig.52), is given by

$$Br \sim q_1^2 \mid V_{cb} \mid^2 \tau_B \, f_{D_S}^2 \qquad (116)$$

f_{DS} is the weak decay coupling constant, it is a measure of the wave function of the $\mid \bar{c}s >$ system at the origin of the D_S^+-meson. In HQET, due to its spin symmetry, the formfactors of the D_S^+- and the D_S^{+*}-meson should be equal in the limit of large charm quark mass:

$$f_{D_S^*} = f_{D_S} \,. \qquad (117)$$

Using this ansatz ARGUS has derived a value [104] of

$$f_{D_S} = (267 \pm 28) MeV \frac{2.7\%}{Br(D_S^+ \to \phi \pi^+)} \,. \qquad (118)$$

The second factor in (118) appears, since the D_S^{+*}-meson has been detected via the decay chain $D_S^{+*} \to D_S^+ \gamma \to \phi \pi^+ \gamma$ and the branching ratio of the decay $D_S^+ \to \phi \pi^+$ has not directly been measured yet. Many theoretical predictions for f_{D_S} exist based on formfactor models, QCD sum rules and lattice gauge calculations. These predictions range between $150\,MeV - 500\,MeV$, clearly this new result is a challenge for all theorists, especially those who try to solve QCD on the lattice.

Before leaving this topic it should be mentioned that extensive studies of inclusive B-decays have been performed [106, 107]. The results of these studies are collected in table 8. Clearly transitions $b \to c$ are the dominant decay mode. ARGUS, in addition, counted the fraction of charmed mesons in semileptonic B-decays:

$$Br(B \to D^0 l^- X) = (72 \pm 6 \pm 6)\%$$
$$Br(B \to D^+ lX) = (29 \pm 5 \pm)4\% \,. \qquad (119)$$

These numbers again demonstrate the dominance of $b \to c$ transitions. The sum of the experimental branching ratios is in agreement with the theoretical expectation of $Br(b \to c l^- \bar{\nu}) = 98\%$.

6.3 Semileptonic B–Decays

6.3.1 Semileptonic $b \to c$ Decays

Semileptonic B-decays from the beginning have been used as a clear signal of B-decays. In first order one expects that the branching ratio for this decay channel can reliably be predicted, since the spectator model should be a good approximation. But it turns out that in reality the problem is far from being easy. In fig.53a the measured lepton spectrum from data collected at the resonance peak of the $\Upsilon(4S)$–meson are shown [105]. Three problems can be read off:

Table 8: Measured inclusive branching ratio

$B \to D^c + X$	$(60 \pm 4 \pm 4)\,\%$
$B \to D^+ + X$	$(29 \pm 4 \pm 4)\,\%$
$B \to \Lambda_c^+ + X$	$(6.7 \pm 0.7 \pm 1.2)\,\%$
$B \to \Psi + X$	$(1.7 \pm 0.2 \pm 0.2)\,\%$
	$(97.4 \pm 6 \pm 6)\,\%$ expect $98\,\%$
$B \to D_s^+ X$	$(9 \pm 2 \pm 4)\,\%$

- A strong background from continuum events exists. It can be measured independently by running the experiment at an $e^+\,e^-$ cms-energy below the B-meson threshold.

- Moreover, a background from cascade decays

$$B \to D + X, D \to l\,Y \qquad (120)$$

exists. They can be simulated using measured lepton branching ratios [57].

- For lepton momenta $p_l \leq 1.4\,GeV$ the b-decays contribute only a small fraction of all decays, hence in this region a reliable measurement is hardly possible. One has to rely on model predictions to extrapolate down to $p_l = 0$.

The lepton spectrum after all corrections is shown in fig.53b. Good agreement between muon and electron spectra is observed. Using the model of Altarelli et al. [108] for extrapolation ARGUS arrives at a branching ratio of

$$Br(B \to l\,\nu_l\,x) = (10.2 \pm 0.4 \pm 0.2)\,\% \,. \qquad (121)$$

A recent detailed analysis of Altarelli et al. [109] shows that this result is not compatible with model predictions. Varying the quark mass and the QCD coupling constant in a reasonable interval they conclude that one expects values of $Br(B \to l\,\nu\,x) = 12\%...14.5\%$. Note that the recent LEP average amounts to $Br(B \to l\,\nu\,x)_{LEP} = (11.2 \pm 0.6)\%$. It is not clear, if this discrepancy between data and model predictions is a real problem or unknown corrections can unravel it.

One of the uncertainties of all experiments results from the fact that in the analysis one has to extrapolate over a large momentum interval using theoretical input for the shape of the electron spectrum. ARGUS has developed a new method to reduce the influence of the extrapolation. They have tagged the decay

$$\Upsilon(4S) \to B\bar{B} \qquad (122)$$

by selecting events with a high momentum electron, e.g. $\bar{B} \to l^- \bar{\nu}_l X$ ($p_l = 1.4\ GeV...2.3\ GeV$). The lepton spectrum of the second B-meson ($B \to l^+ \nu_l x$) is analyzed. Still a small background from cascade decays, Ψ-decays, γ-conversion etc. remains, which contribute mainly at low momenta [110]. This background is either measured or derived from Monte Carlo simulations. The measured spectrum extends down to $p_l = 0.6\ GeV$. The contribution to the branching ratio from the extrapolation is small. From this analysis again a small branching ratio

$$Br(B \to l\ \nu_l\ x) = (9.6 \pm 0.5 \pm 0.4)\% \qquad (123)$$

is derived. Hence it is unprobable that the discrepancy between data and theoretical prediction mentioned above is due to the extrapolation procedure used.

6.3.2 Semileptonic $b \to u$ Transitions

The determination of the coupling strength between first- and third-generation quarks is of fundamental importance for our understanding of weak interactions. It is known from the prolific charm production in B-decays [106] that the $b \to c$ dominates over the $b \to u$ transition. As discussed in ch.6.3.1 already for $b \to c$ decays the continuum background has to be carefully measured. In order to increase the sensitivity for $b \to u$ transitions proper cuts have to be applied, which strongly suppress the continuum contribution but have a large efficiency for $b \to u$ decays. These cuts exploit the more spherical shape of b–decays as compared to continuum events (ch.6.1) and the existence of energetic leptons including neutrinos in the event.

The event shape can be characterized by the scalar sum of the momentum perpendicular to the lepton direction for all particles, which have an angle between 60^0 to 120^0 with respect to the lepton direction ($\sum p_T$). For continuum events which have a jet–like shape $\sum p_T$ should be smaller in average than for B–decays (fig.54). A cut $\sum p_T > 2\ GeV$ reduces the continuum background by a factor 11, while 40% of all $Y(4S)$-decays pass the cut.

The missing momentum of the neutrinos produced in semileptonic decays is a further mean to suppress the continuum contribution. Note in this context that this cut can only be applied, if the detector is hermetic. ARGUS, though designed 15 years ago, fulfills this condition [111]. Since the cms- and the lab system coincide the missing momentum is given by the expression

$$\vec{p}_{Miss} = \sum \vec{p}\ . \qquad (124)$$

The measured distribution agrees with the predicted one (fig.55a). In the endpoint region of the lepton spectrum the direction of the $e(\mu)$-momentum and $\vec{p}_{Miss}$ are correlated, i.e.

$$\cos \beta = \hat{\vec{p}}_{Miss} \cdot \hat{\vec{p}}_l \qquad (125)$$

peaks at $\cos\beta = -1$ for semileptonic b-decays (fig.55b). The comparison with Monte Carlo prediction demonstrates the good quantitative agreement. In fig.55d for a subsample of reconstructed $B^0 \to D^* \, l^+ \, \nu_l$ decays (ch.6.4) the correlation between $\vec{p}_{Miss}$ and $\vec{p}_\nu$ is demonstrated. A cut $p_{Miss} > 1 \, GeV$ and $\cos\beta < -0.5$ reduces the background by a factor 3.6, while the efficiency for $b \to u \, l^- \, \bar{\nu}_l$ transition is 81%.

The final result of this [112] and a similar analysis [113] of the ARGUS collaboration is shown in fig.56. A clear enhancement beyond the momentum interval populated by $b \to c$ transition is visible. In total 77.1 ± 13.4 $b \to u$ candidates were observed in the momentum interval $2.3 \, GeV < p_l < 2.6 \, GeV$. They are compared to the $b \to c \, l \, \nu_l$ rate in the interval $2.0 \, GeV < p_l < 2.3 \, GeV$. Using models the measured ratio can be interpreted in terms of the ratio $\mid V_{bu}/V_{bc} \mid$. Values between $\mid V_{ub}/V_{bc} \mid = 0.11 \pm 0.012$ [108] and $\mid V_{ub}/V_{bc} \mid = 0.20 \pm 0.023$ [114] follow from the analysis for different models. From these numbers it is clear that the theoretical uncertainty already now dominates.

Further evidence for $b \to u$ transitions comes from 2 fully reconstructed events (fig.57). All decay particles besides the ν from semileptonic decays have been detected. The possible alternative explanations for this decay are carefully discussed in[113], their probability is always smaller than 10^{-3}. Note that for the event of fig.57 in addition a $B^0 \to \bar{B}^0$ transition is observed (ch.6.6).

6.4 Exclusive Semileptonic Decays

Since more specific model predictions exist for exclusive semileptonic decays [115] and besides the lepton spectrum also angular distributions can be measured, more sensitive tests of models follow from the study of exclusive semileptonic B-decays. ARGUS has developped a new method [116] to identify these exclusive semileptonic decays. It relies heavily on the fact that the $\Upsilon(4S)$-meson decays into a B-pair, hence eqs.(93),(94) hold. For the decay

$$B^0 \to D^{*+} l^- \bar{\nu}_l \qquad (126)$$

one can reconstruct the recoil mass

$$M_{rec}^2 = (E_B - E_{D \cdot l})^2 - (\vec{P}_b - \vec{P}_{D*} - \vec{p}_l)^2 \qquad (127)$$

which due to (93) and (94) can be approximated by

$$M_{rec}^2 = (E_{beam} - E_{D \cdot l})^2 - (\vec{p}_{D*} + \vec{p}_l)^2 \qquad (128)$$

For reaction (126) one expects

$$M_{rec}^2 \approx m_\nu^2 \qquad (129)$$

Table 9: Measured branching ratios for exclusive semileptonic B-decays

Decay	Br	ref
$B^- \to D^{*0} l^- \bar{\nu}$	$(4.9 \pm 0.7 \pm 0.7)\,\%$	[119]
$\bar{B}^0 \to D^{*+} l^- \bar{\nu}$	$(5.2 \pm 0.8 \pm 0.8)\,\%$	[119]
$B^- \to D^0 l^- \bar{\nu}$	$(1.8 \pm 0.6 \pm 0.4)\,\%$	[119]
$\bar{B}^0 \to D^+ l^- \bar{\nu}$	$(1.9 \pm 0.6 \pm 0.5)\,\%$	[119]
$\bar{B}^0 \to D^{**+} l^- \bar{\nu}$	$(2.3 \pm 0.6 \pm 0.4)\,\%$	[120]

A typical recoil mass distribution is shown in fig.58 for the decay (126). As expected a peak at zero recoil mass sticks out. The background in the peak region is mainly due to the background in the D^* invariant mass distribution. A collection of branching ratios obtained with this techniques and a similar method using the missing energy is collected in table 9.

As pointed out in ch.6.1, the HQET can be used to derive V_{bc} exploiting eq.(104). For the $B^- \to D^{*0} l^- \bar{\nu}$ decay such an analysis has been performed. The decay has been identified using the recoil method. The velocity variable y in the Isgur-Wise function is given by (105). The data are plotted in fig.59, the different curves entered into the figure are parametrisations of the Isgur-Wise function [98]. Note that $\xi(1) = 1$ and therefore V_{bc} can directly be read off from the figure. ARGUS [117] derives in this way

$$V_{bc} = 0.044 \pm 0.009 \qquad (130)$$

This value depends only weakly on the parametrisation of the Isgur-Wise function used (fig.59).
The branching ratios allow to determine the lifetime ratio of charged and neutral mesons:

$$\frac{\tau(B^+)}{\tau(B^0)} = 0.95 \pm 0.15 \pm 0.19 \qquad (131)$$

in good agreement with the theoretical expectations based on the spectator model. Note that in contrast to D-decays, where the lifetimes of D^+ and D^0 differ by a factor of 2 [118], the result for B-mesons is compatible with 1.

It is known that the most reliable way to determine the CKM matrix element V_{bc} is the measurement of the decay width of the decay $B \to Dl\nu$. Using the ARGUS measurement of the branching ratio for this channel and the lifetime of ref.[57], one finds

$$V_{bc} = 0.044 \pm 0.007 \qquad (132)$$

based on the model of ref.[101]. Within the error limits values of V_{bc} using other models agree with (132).

ARGUS has gone one step further and analyzed in more detail exclusive

Table 10: Possible decay channels contributing to the recoil spectrum of fig.60

I	$\bar{B}^0 \to \boxed{D^{*+} \ell^-}\, \bar{\nu}$
II	$\bar{B}^0 \to D^{**+}\boxed{\ell^-\,\bar{\nu}}$; $\;\hookrightarrow D^{*+}\pi^0\quad\quad B^0 \to D^{**0}\boxed{\ell^-\,\bar{\nu}}$; $\;\hookrightarrow D^{*+}\pi^-$
III	$B_1 \to \boxed{\ell^-\,X_1}$; $B_2 \to \boxed{D^{*+}\,X_2}$
IV	$e^+e^- \to c\bar{c} \to \boxed{D^{*+}\ell^-}\,X$

semileptonic decays with a D^{*+}-meson in the final state. Many processes contribute to the $D^{*+}\,l^-$ recoil mass distribution (table 10), which is shown in fig.60 as predicted by a Monte Carlo simulation. The measured distribution is shown in fig.61 for two decay channels of the D^0-meson reconstructed in the analysis. Clearly a contribution from the channel II in table 10 $\bar{B}^0 \to D^{**+}\,l^-\,\nu_l$ is observed, it is also visible in the (D^*,π) invariant mass distribution [120]. The unexpected large branching ratio is included into table 9. It is interesting to note that the sum of all exclusive branching ratios observed in this analysis is

$$Br(\bar{B} \to (D, D^*, D^{**})\, l^-\, \bar{\nu}_l) = ((9.4\ldots 9.8) \pm 1.0 \pm 0.9)\,\% \quad (133)$$

in good agreement with the result of the inclusive analysis (eqs.(121),(123)). This analysis has been used to perform a detailed angular analysis. The decay

$$\bar{B}^0 \to D^{*+}\,l^-\,\bar{\nu}_l \quad (134)$$

is characterized by 3 helicity amplitudes H_0, H_+. The angular distribution is given by

$$\frac{d^4\Gamma}{dq^2\,d\cos\theta\,d\cos\theta^*\,d\lambda} \sim \frac{q^2}{M_B^2}\,|\,\sum d^1_{0\lambda}(\theta^*)\,d^1_{\lambda 1}(\theta)\,e^{-i(\lambda+1)\chi}H_\lambda(q^2)\,|^2 \quad (135)$$

The angles are defined in fig.62. Parity violation means that $|\,H_-\,|^2 > |\,H_+\,|^2$, since the c quark is predominantly produced with negative helicity. Two angular distributions can be measured:

$$\frac{d\Gamma}{d\cos\theta} \sim \alpha\sin^2\theta - \frac{4}{3}A_{FB}(3+\alpha)\cos\theta + 2 \quad (136)$$

and
$$\frac{d\Gamma}{d\cos\theta^*} \sim 1 + \alpha \cos^2\theta^* \qquad (137)$$

with
$$\alpha = \frac{2\Gamma_0}{\Gamma_+ + \Gamma_-} - 1 \qquad (138)$$

$$A_{FB} = \frac{\int_{-1}^{0}\frac{d\Gamma}{d\cos\theta}d\cos\theta - \int_{0}^{+1}\frac{d\Gamma}{d\cos\theta}d\cos\theta}{\int_{-1}^{+1}\frac{d\Gamma}{d\cos\theta}d\cos\theta} = \frac{3}{4}\frac{\Gamma_- - \Gamma_+}{\Gamma} \qquad (139)$$

A_{FB}, the forward–backward asymmetry, is a measure of the strength of parity violation. The measured distributions are compared in fig.63 with model predictions. Note that no acceptance corrections have been applied. From the fit one gets
$$A_{FB} = 0.20 \pm 0.08 \pm 0.06 \qquad (140)$$
and
$$\alpha = 1.1 \pm 0.4 \pm 0.2 . \qquad (141)$$

These data are in good agreement with recent model predictions $\alpha = 1.1$ [121]. Note that the first measurement of α by the ARGUS collaboration [122] at that time was an effective filter for models on the market and was an essential milestone on the way to the HQET. Also the measured value of A_{FB} is in good agreement with the prediction $A_{FB} = 0.2$ [121].

6.5 $B^0\bar{B}^0$–Mixing

The first observation of $B^0\bar{B}^0$–mixing by the ARGUS collaboration [123] is clearly the highlight of the physics program of this successful experiment. This effect has its analogue in K^0–decays and the same formalism can be used to describe the effect [124].
The Hamiltonian matrix of the problem is given by

$$H = \begin{pmatrix} M - \frac{1}{2}i\Gamma & M_{12} - \frac{1}{2}i\Gamma_{12} \\ M_{12}^* - \frac{1}{2}i\Gamma_{12}^* & M - \frac{1}{2}i\Gamma \end{pmatrix} \qquad (142)$$

M mass and Γ decay width of the flavor eigenstates. M_{12} corresponds to virtual $B^0\bar{B}^0$–transition, $\Gamma_{12} \approx 0$ describes real transitions due to common decay modes $B^0, \bar{B}^0 \to u\bar{d}d\bar{u}, \ldots$. M_{12}, Γ_{12} can be calculated from box-graphs (fig.64). Diagonalisation of (142) leads to CP eigenstates.

$$|B_{1,2}\rangle = \frac{1}{\sqrt{2}}(|B^0\rangle \pm |\bar{B}^0\rangle) \qquad (143)$$

$$M_{12} = M \pm \frac{\Delta M}{2} \tag{144}$$

$$\Gamma_{12} = \Gamma \pm \frac{\Delta \Gamma}{2} . \tag{145}$$

Due to the many open decay channels

$$\Delta \Gamma \approx \Gamma_1 - \Gamma_2 = 0 . \tag{146}$$

If a B^0-meson is produced at $t = 0$, the time development is given by

$$W_{B^0}(t) = \frac{1}{4} \left\{ e^{-\Gamma_1 t} + e^{-\Gamma_2 t} + 2 e^{-\Gamma t} \cos \Delta M t \right\} \tag{147}$$

$$W_{\bar{B}^0}(t) = \frac{1}{4} \left\{ e^{-\Gamma_1 t} + e^{-\Gamma_2 t} - 2 e^{-\Gamma t} \cos \Delta M t \right\} . \tag{148}$$

Integration of (147) gives

$$N(B^0) \sim \int_0^\infty W_{B^0}(t) = \frac{1}{4} \left\{ \frac{1}{\Gamma_1} + \frac{1}{\Gamma_2} + \frac{2\Gamma}{\Gamma^2 + (\Delta M)^2} \right\} \tag{149}$$

and a corresponding number for $N(\bar{B}^0)$ with a minus sign in the third factor. The mixing parameter is defined by

$$r = \frac{N(\bar{B}^0)}{N(B^0)} = \frac{x^2}{2 + x^2} \tag{150}$$

$$x = \frac{\Delta M}{\Gamma} . \tag{151}$$

The analysis of the box diagram leads to

$$\Delta M = \frac{G_F^2 M_W^2}{8\pi^2} <\bar{B}^0 \mid j_\mu j^\mu \mid B^0> \sum V_{ib}^* V_{id} V_{jb}^* V_{jd} A_{ij} . \tag{152}$$

Due to unitarity of the CKM-matrix one gets

$$\sum V_{ib}^* V_{id} = 0 \tag{153}$$

A_{ij} are loop integrals. Using the unitarity relation and explicit expression for the loop integrals

$$\sum V_{ib}^* V_{id} V_{jb}^* V_{jd} A_{ij} \approx \mid V_{tb}^* V_{td} \mid^2 m_t^2 F\left(\frac{m_t^2}{M_W^2}\right) \frac{1}{M_W^2} \tag{154}$$

F is a slowly decreasing function with $F(0) = 1$.

$$<B^0 \mid j_\mu j^\mu \mid B^0> = B \frac{4}{3} f_B^2 m_b \tag{155}$$

$B \approx 1$ is a bag constant, f_B is the B–decay constant describing the B wave–function at the origin.
Using this result one gets for the mixing parameter

$$r = \frac{G_F^2}{6\pi^2} B f_B^2 m_b \tau_b \mid V_{tb}^* V_{td} \mid^2 m_t^2 F\left(\frac{m_t^2}{m_W^2}\right) \eta \qquad (156)$$

η is a QCD correction:

$$\eta \approx 0.78 \ldots 0.85 \; . \qquad (157)$$

The interest in the observation of $B^0 \bar{B}^0$–mixing follows directly from eq.(156). It allows the measurement of V_{td}, once f_B, B are known from theory (other experiments) and m_t is measured. Already in the original publication the conclusion was drawn that only $m_t > 55\,GeV$ or weak interaction theory was in trouble.

ARGUS uses three methods to establish $B^0 \bar{B}^0$–mixing. They observe a few fully reconstructed events, fig.57, where one of the B^0–mesons has changed into its antiparticle. They used the recoil method discussed in ch.6.4. Finally, $l^- l^-$–events from $\Upsilon(4S)$–decays were observed, their rate can directly be compared to the $l^+ l^-$–rate of unmixed events.

A note of warning has to be applied. In $\Upsilon(4S)$–decays one has a coherent $B^0 \bar{B}^0$–system in a state with well defined quantum numbers. Hence before the decay of the first particle always one of them has to be in the B^0– the other in the $\bar{B}^0$–state. If the first B–meson decays the other starts to oscillate according to (147),(148). Hence one gets the same relation as for 1 oscillating B^0–meson also for the $B^0 \bar{B}^0$–system from $\Upsilon(4S)$–decay. The rate is

$$R = \frac{N(B^0 B^0) + N(\bar{B}^0 \bar{B}^0)}{N(B^0 \bar{B}^0)} = \frac{x^2}{2 + x^2} \; . \qquad (158)$$

Often the variable

$$\chi = \frac{N_{BB} + N_{\bar{B}\bar{B}}}{N_{b\bar{b}}} \qquad (159)$$

is used to parametrise oscillations. If B_d^0– and B_s^0–mesons are produced and oscillate as at LEP and at the $p\bar{p}$ colliders one gets the incoherent sum

$$\chi_{eff} = p_d \chi_d + p_s \chi_s \qquad (160)$$

$p_d(p_s)$ is the fraction of $B_d^0(B_S^0)$–mesons produced in the fragmentation of b–quarks. ARGUS gets [125]

$$r_d = 0.21 \pm 0.07 \; , \qquad (161)$$

where the error is dominantly systematic. In fig.65 the 1992 data are entered into a $\chi_d - \chi_s$ plot. It shows that the present data are compatible with the 3–generation standard model. Furthermore, one derives

$$\frac{\mid V_{td} \mid}{\mid V_{ts} \mid} < 0.21 \qquad (162)$$

and
$$0.21 \leq \frac{V_{td}}{V_{cb}} \leq 0.65 \ . \tag{163}$$

Using eq.(156), the measured value of r gives a connection between V_{td} and m_t shown in fig.66. Taking the interval of m_t derived from the LEP data, one gets the CKM–matrix element of eq.(162).

One can summarize the results of the experiments by checking the unitarity triangle (100). This is shown [126] for three values of the top mass in fig.67. Clearly the measurements of V_{ub} and x_d are consistent and they agree with the conclusions drawn from CP–violation.

6.6 Summary

The ARGUS experiment is one of the main contributors to present day knowledge of B–physics. The main results of this first generation B–experiments are:

- Exclusive two–body decays are in good agreement with QCD inspired model predictions.

- Inclusive decays into charmed mesons and baryons have been observed to dominate.

- The semileptonic branching ratio
$$Br(B \to l\,\nu_l\,X) = (10.2 \pm 0.4 \pm 0.2)\,\% \tag{164}$$
has been measured for $p_l \geq 0.6\,GeV$, hence the result is not any more dependent on the extrapolation assumption. Spectator models including QCD corrections seem hardly to be compatible with this small branching ratio.

- Exclusive semileptonic B–decays have been studied in detail. The lifetime ratio of charged to neutral B–mesons is compatible with 1. The CKM–matrix element
$$V_{bc} = 0.044 \pm 0.007 \tag{165}$$
is derived from these measurements. The semileptonic decay into excited D^{**+}–mesons is unexpected large
$$Br(B^0 \to D^{**+}\,l^-\,\bar{\nu}_l) = (2.3 \pm 0.6 \pm 0.4)\ . \tag{166}$$

Parity violation in B–decays has been observed.

- ARGUS has observed the inclusive $b \to u\, l^- \bar{\nu}_l$ decay and gets for the CKM–matrix elements
$$\frac{V_{bu}}{V_{bc}} = 0.1 \ldots 0.2 \qquad (167)$$
depending on the model used to interprete the measured rates.

- $B_d^0 \bar{B}_0^d$–mixing was observed for the first time. Its large strength is only compatible with a large top–quark–mass. For the CKM–matrix elements one gets
$$0.21 \leq |V_{td}/V_{bc}| \leq 0.65 \,. \qquad (168)$$

7. τ-Physics with the ARGUS Detector

The analysis of τ-decays is a major part of the ARGUS physics program. Experimenting in the energy region accessible at the DORIS storage ring ($\sqrt{s} \approx 10 GeV$) offers a few advantages compared to previous experiments:

- τ-leptons are copiously produced, hence studies are possible which previously were excluded by statistics.

- the energy is large enough such that the multiplicities of τ-events and $q\bar{q}$-events - the major background source - differ appreciably and hence can be used to separate the two processes.

- the ARGUS detector allows to detect photons down to very low energies since the electromagnetic shower counters are positioned inside the magnet coil, hence the shower development is not distorted by material in front.

In this lecture the major results of the ARGUS experiment in the field of τ-physics are summarized. They include the following topics:

- Mass, spin and helicity measurements of the τ-lepton and the ν_τ.

- the Lorentz-structure of the weak interaction.

- QCD tests

- measurements of the branching ratios for specific decay channels.

- analysis of the spin parity of hadronic final states including channels which allow to search for second class currents and to test the CVC hypothesis.

7.1 Properties of the τ-Lepton and the ν_τ

The decay of the τ-lepton is described by the Feynman graph shown in fig.68. Assuming lepton universality, the coupling strength of the $\tau\nu_\tau$-pair to the W-boson is the same as for the other sequential leptons:

$$g_\tau = g_\mu = g_e \qquad (169)$$

The decay width is given by [127]

$$\Gamma(\tau^- \to l^- \bar\nu_l \nu_\tau) = \frac{1}{\tau_\tau} = \frac{2g_\tau^2 g_l^2}{192\pi^3 m_W^4} m_\tau^5 f(\frac{m_l^2}{m_\tau^2}) r_\tau \qquad (170)$$

where f is a known function [127] $f(\frac{m_\mu}{m_\tau}) = 0.973$ and $f(\frac{m_e^2}{m_\tau}) = 1$ and $r_\tau \approx 1$ [128] takes into account electroweak corrections. From π-decay the ratio

$$\frac{g_\mu}{g_e} = 1.0031 \pm 0.0023 \qquad (171)$$

is known. The relation corresponding to eq.(170) for μ-decay is given by

$$\Gamma(\mu \to e^- \bar\nu_e \nu_\tau) = \frac{1}{\tau_\mu} = \frac{2g_\mu^2 g_e^2}{192\pi^3 m_W^4} m_\mu^5 r_\mu \qquad (172)$$

Combining eqs.(170) and (172) one gets

$$\tau_\tau = \frac{g_\mu^2}{g_\tau^2}(\frac{m_\mu}{m_\tau})^5 \tau_\mu Br(\tau^- \to e^- \nu_e \nu_\tau) \qquad (173)$$

Taking the 1991 world average [57]

$$m_\tau = (1784.1^{+2.7}_{-3.6}) MeV \qquad (174)$$

$$\tau_\tau = (302.5 \pm 5.9) fs \qquad (175)$$

$$B_e = (17.73 \pm 0.23)\% \qquad (176)$$

one gets

$$\frac{g_\tau}{g_\mu} = 0.967 \pm 0.017 \qquad (177)$$

i.e. a 2σ discrepancy graphically demonstrated in fig.69.
Since only one precision measurement of the τ−mass existed, performed many years ago by the DELCO collaboration [129], ARGUS has developed a new pseudomass technique, which allows an independent measurement of this important parameter [130].
τ-decays at $\sqrt{s} \approx 10 GeV$, where one τ–lepton decays into 1, the other into 3 charged particles were selected. A typical event is shown in fig.70, where the hits of tracks in the transverse plane are shown. A clear 1–3 topology is

observed which allows to separate τ-decays from background which is mainly due to the reaction $e^+e^- \to q\bar{q}$. Dominantly the decay into the $(3\pi)^-$-system proceeds via the a_1^--resonance as intermediate state (ch.7.4.3). Due to the Lorentz boost from the τ-rest-system to the laboratory system the angle between the a_1-meson and the τ-lepton in the latter frame is small. Hence we assume

$$\cos(\vec{p}_{a_1}, \vec{p}_\tau) = 1 \tag{178}$$

The following relations hold

$$m_\tau^2 = E_\tau^2 - \vec{p}_\tau^2 \tag{179}$$

$$E_\tau = E_{beam} \tag{180}$$

$$|\vec{p}_\tau| = |\vec{p}_{3\pi}| + |\vec{p}_\nu| \tag{181}$$

$$|\vec{p}_\nu| = \sqrt{(E_\tau - E_{3\pi})^2 - m_\nu^2} \tag{182}$$

If one neglects the ν_τ-mass, one gets for the pseudomass the following expression

$$m_\tau^{*2} = 2(E_{3\pi} - |\vec{p}_{3\pi}|)(E_\tau - E_{3\pi}) + m_{3\pi}^2 \tag{183}$$

The mass resolution achievable decreases with the beam energy

$$\delta m_\tau \sim \sqrt{E_{beam}} \tag{184}$$

Events with a photon candidate on the 3-prong side were rejected. 10943 events out of 325000 produced τ-pairs passed the selection criteria. The background was determined by a Monte Carlo calculation to be 23%. $\gamma\gamma$- and Bhabha-events turn out to be negligible. 360 background events due to $q\bar{q}$- and 3 gluon-decays of the Υ-meson are estimated by a Monte Carlo calculation to pass the selection criteria. The main background of 2160 events is due to τ-decays, e.g. the decays $\tau^- \to \pi^-\pi^-\pi^+\pi^0\nu_\tau$, which are not rejected by the cuts applied. The measured pseudomass distribution together with the background is shown in fig.71. A steep threshold is observed at $m_\tau^* = m_\tau$ for the $\tau^- \to \pi^-\pi^-\pi^+\nu_\tau$ data, while the background shows a smooth behavior in the threshold region. A maximum likelihood fit is applied to the data. The shape of the signal is taken from a Monte Carlo simulation using the TAUOLA event generator [131]. The background is included, the relative normalisation of signal and background is derived from the Monte Carlo simulation, hence the only free parameters of the fit are the absolute normalisation and the τ-mass. We find $m_\tau = (1784.1 \pm 2.4)$ which has to be compared with the previous value of eq.(174). As shown by fig.71 the fit describes the data perfectly. The tail of the distribution for $m_\tau^* > m_\tau$ is due to initial state radiation; in this case eq.(180) overestimates the τ-energy and according to eq.(183) m_τ^* is too large.

A detailed study of the systematic errors has been performed. The main

error source is the uncertainty of the absolute momentum scale of the AR-
GUS detector derived from a study of the invariant mass distribution of
the $K_S^0 \to \pi^-\pi^+$- and $D^0 \to K^-\pi^+$-decays. It leads to a systematic uncertainty of $\delta m_\tau = 1.2\,MeV$. Other sources of systematic errors studied
are the uncertainty of the beam energy $\delta E = 3\,MeV$ ($\delta m_\tau = 0.5\,MeV$),
the choice of the fit interval ($\delta m_\tau = 0.5\,MeV$) and the neglect of the ν_τ-
mass (δm_τ) = $0.3\,MeV$. In total we estimate the systematic error to be
$\delta m_\tau = \pm 1.4\,MeV$. The final result is

$$m_\tau = (1776.3 \pm 2.4 \pm 1.4)\,MeV \tag{185}$$

After this analysis was finished a new value of the τ-mass was determined
in a high precision experiment in the threshold region [132], which arrives at
a value of

$$m_\tau = (1776.9 \,{}^{+0.4}_{-0.5} \pm 0.2)\,MeV \tag{186}$$

Plugging this new value into eq.(173) and using the 1992 values of the τ-
lifetime and branching ratios [133] one gets for the ratio of coupling constants

$$\frac{g_\mu}{g_\tau} = 0.990 \pm 0.007 \tag{187}$$

The good agreement between the theoretical prediction of eq.(173) – based
on lepton universality – and the data (fig.69) demonstrates that the τ-lepton
is a sequential lepton [134].

ARGUS has used this new value of the τ-mass to improve its 1988 limit of the
ν_τ-mass limit [135]. It is derived from a study of the $\tau^- \to \pi^-\pi^-\pi^-\pi^+\pi^+\nu_\tau$
decay. Again the selection criteria exploit the characteristic 1-5 topology of
τ-decays. The 5π-invariant mass distribution peaks at large masses. Due to
energy conservation

$$m(\tau) = m(\nu_\tau) + m_{max}(5\pi) \tag{188}$$

the maximal achievable 5π-mass is limited by the mass of the ν_τ. From Monte
Carlo calculation we expect a background of 0.2 events from $q\bar{q}$-events. To
avoid uncontrollable uncertainties due to the $q\bar{q}$-generator used to determine
the background, we do not consider the event with the highest invariant
5π-mass in the analysis. The conservative limit thus achieved is [130]

$$m(\nu_\tau) < 31\,MeV \text{ at } 95\%CL\,. \tag{189}$$

For high enough statistics the limit can be reduced to 15 MeV as demonstrated by the fact that the inclusion of the event rejected leads already now
to a limit of 20 MeV at 95% confidence level.

7.2 Lorentz–Structure of the Interaction

The standard model predicts that the Langrangian of weak interactions is of the current–current type. It is assumed to be the same for

all lepton and quark generations. Specifically for the charged current the Lorentz-structure is supposed to be of the V–A type. This assumption has been verified with high precision for the leptons and quarks of the first two generations, experimental results for the third generation just start to become available. The ARGUS collaboration has exploited its large statistic to study carefully this problem. Four results achieved will be discussed in detail in this lecture.

7.2.1 Michel Parameter

The shape of the measured lepton spectrum from τ-decays

$$\tau^- \to l^- \nu_l \nu_\tau \qquad (190)$$

is given by the expression [127]

$$\frac{d\Gamma}{dx} \sim x^2 \{ \frac{1 + h(x)}{1 + 4\eta \frac{m_l}{m_\tau}} I(x) - P_\tau \xi \cos\theta A(x) \} \qquad (191)$$

$$I(x) = 12(1-x) + \frac{4}{3}\rho(8x-6) + 24\frac{m_l}{m_\tau}\frac{1-x}{x}\eta \qquad (192)$$

$$A(x) = 4(1-x) + \frac{4}{3}\delta(8x-6) + \frac{g(x)}{x^2} \qquad (193)$$

where

$$x = \frac{p_l}{p_{l max}} \qquad (194)$$

h(x) and g(x) describe radiative corrections [136]

$$\cos\theta = \vec{s}_\tau \vec{P}_l \qquad (195)$$

$\vec{P}_\tau$ τ-polarisation, $\vec{S}_\tau$ τ-spin, $\vec{p}_l$ lepton momentum [127]. It can be shown that for V–A interactions

$$\rho = \frac{3}{4}, \ \eta = 0, \ \xi = 1, \ \delta = \frac{3}{4} \qquad (196)$$

Note that for other combinations of V, A one gets

$$\rho(V+A) = 0, \ \rho(V,A) = \frac{3}{4} \qquad (197)$$

According to eq.(191) the low energy part of the lepton spectrum ($x \leq 0.2$) is sensitive to η, in addition due to the $\frac{m_l}{m_\tau}$ term η is only observable for muon decays. Unfortunately the detector does not allow this study, because a minimal penetration length of at least 1 m of iron is necessary to separate experimentally muons from hadrons (ch.2.1), which limits the momentum interval accessible.

Events were detected using the 1-1 prong topology of pure leptonic τ-decays. In addition we require one of the tracks to fulfill the criteria of an electron with $p_e > 0.65\,GeV$ and the second that of a muon with $p_\mu > 1.5\,GeV$. Since for an event, where both τ-leptons decay leptonically, 4ν are emitted, we expect a large missing mass. We therefore cut at a missing mass of $M^2_{miss} > 4\,GeV^2$. The result of the analysis is shown in fig.72, where the measurements are compared to the predictions for a V–A and a V + A current. We find [137]

$$\rho_e = 0.747 \pm 0.045 \pm 0.028 \tag{198}$$

$$\rho_\mu = 0.734 \pm 0.055 \pm 0.027 \tag{199}$$

The average is $\rho = 0.742 \pm 0.035 \pm 0.020$. This result is in good agreement with the expectation for a V–A current. Note, however, that proper chosen combinations of scalar and tensor interactions allow to reproduce this value.

7.2.2 Spin-Correlations of the Produced τ–Lepton

Due to helicity conservation in electromagnetic interactions the helicities of the produced τ-leptons are correlated (fig.73). If the weak interactions are of the V–A type, this fact induces a helicity correlation of the leptons in the final state which finally transforms into a momentum correlation of the decay leptons as indicated in fig.73. This correlation is described by the threefold differential cross section [127]

$$\frac{d^3\sigma}{dp_e\,dp_\mu\,d\cos\theta_{acol}} = I(P_e, P_\mu, \theta_{acol}) + \xi_e\,\xi_\mu\,A(p_e, p_\mu, \theta_{acol})\,, \tag{200}$$

where θ_{acol} is the acolinearity angle of electron and muon. I, A are known functions. Comparing data with the results of a Monte Carlo calculation which varies $\xi = \sqrt{\xi_e\,\xi_\mu}$ and takes for δ, η the values predicted for a V–A current one gets [138]

$$\xi = 0.90 \pm 0.13 \pm 0.13 \tag{201}$$

The systematic errors are due to the fit stability (± 0.07), acceptance (± 0.10) and background (± 0.06). The value achieved is in good agreement with the expectation of the standard model. Note however that this result is clearly only a first step, since it is logically unsatisfying to vary only one out of three independent parameters and fix the others to the value predicted by the theory one wants to test.

7.2.3 Spin Alignment of the ρ–Meson in τ–Decays

The decay

$$\tau^- \to \rho^- + \nu_\tau \tag{202}$$

is induced by a pure vector interaction (ch.7.4), hence due to helicity conservation the ρ–meson and the τ–lepton should have the same helicity. V–A interactions predicts that the ν_τ–lepton is lefthanded (ch.7.2.4). Since $m_\tau \neq 0$, the τ–lepton is not in a pure lefthanded state but has also a righthanded component. For a V–A interaction one gets

$$\frac{L}{R} = \frac{1+\beta_\tau}{1-\beta_\tau} = \frac{m_\tau^2}{m_\rho^2} \tag{203}$$

where in the ρ–rest–frame

$$\beta_\tau = \frac{m_\tau^2 - m_\rho^2}{m_\tau^2 - m_\rho^2} \tag{204}$$

In the τ–rest–frame only two values of the ρ^-–helicity $H_\rho = 0$ (lefthanded τ) and $H_\rho = -1$ (righthanded τ) are allowed. This polarization of the ρ–meson translates into the following angular distribution of the π–meson in the ρ–rest–frame

$$\frac{dN}{d\cos\theta_H} \sim \frac{L}{R}|Y_1^0|^2 + |Y_1^{-1}|^2 = 1 + b_\tau \cos^2\theta_H \tag{205}$$

$$b_\tau = \frac{m_\tau^2 - m_\rho^2}{m_\rho^2} = 4.4 \tag{206}$$

θ_H is the helicity angle of the π^-–meson measured with respect to the ρ–direction in the τ rest–system. Since the τ–direction is not measurable as long as the ν_τ is not detected, we use the ρ–direction in the lab–frame as reference (angle θ) and transform it into the ρ–rest–frame. The resulting angular distribution

$$\frac{dN}{d\cos\theta} \sim 1 + b_{lab} \cos^2\theta \tag{207}$$

is of course diluted in comparison to eq.(205). From a Monte Carlo simulation follows

$$b_{lab} = 0.57 \pm 0.01 \tag{208}$$

for a V–A interaction. The measured distribution is shown in fig.74. Fitting eq.(207) to these data one gets

$$b_{lab} = 0.57 \pm 0.12 \tag{209}$$

in good agreement with the standard model prediction. The probability for helicity $H_\rho = 0$ is

$$0.77 < \text{prob}(H_\rho = 0) = \frac{b_\tau + 1}{b_\tau + 2} = \frac{L}{L+R} < 0.92 \tag{210}$$

with 95% confidence level. Hence no helicity state of the ρ–meson is exclusively populated. From this observation follows that a spin $J(\nu_\tau) = \frac{3}{2}$ is excluded, since in this case only the helicity $|H_\rho| = 1$ would be populated in contradiction to the observation.

In conclusion this measurement is in good agreement with a vector like coupling of the $\tau - \nu_\tau$ vertex. In addition spin $\frac{3}{2}$ for the ν_τ is excluded.

7.2.4 First Measurement of the ν_τ Helicity

The measurement described in ch.7.2.3 does not allow to measure the polarisation of the ρ-meson and derive the helicity of the participating leptons, since only the absolute value of $|H_\rho|$ can be derived as long as the squares of the amplitudes are added (eq.(205)) to achieve the cross section. For polarisation measurements two amplitudes have to contribute to the same subprocess which first are added according to the rules of quantum mechanics and then are squared. The interference term carries the essential information for a polarisation measurement.

In case of the decay chain

$$\tau^- \to a_1^- \nu_\tau$$
$$\hookrightarrow \rho^0 \pi^-$$
$$\hookrightarrow \pi^+ \pi^- \qquad (211)$$

the π^+ in the final state can combine with either of the two π^--mesons to obtain a ρ^0-meson. The two amplitudes are added and then squared to get the cross section. In this case the population of the different helicity states of the a_1^--meson can be measured as pointed out by Kuehn and Wagner [139]. The measured $m_{2\pi}$ invariant mass distribution is shown in fig.75. A clear ρ^0-signal is observed in the $\pi^+\pi^-$ channel, the shoulder at small masses is due to nonresonant combinations as demonstrated by the fact that this part of the spectrum approximately coincides with the $\pi^-\pi^-$ invariant mass distribution. It has been shown [140] that the small differences between the $\pi^-\pi^-$- and the $\pi^-\pi^+$-spectrum is due to the necessary Bose–symmetrisation of the two amplitudes contributing.

Analyzing the possible spin combinations in the a_1^--rest–system and assuming that the ν_τ has negative helicity it follows that only the $J_z(a_1^-) = -1$ and the $J_z(a_1^-) = 0$ are allowed, while the $J_z(a_1^-) = +1$ component is not populated. This can experimentally be checked in the following way.

The matrix element for the process is given by the expression

$$|M|^2 = |L_\mu J^\mu|^2 \qquad (212)$$

$$L_\mu = \frac{G_F}{\sqrt{2}} \bar{u}_\nu (g_V + g_A \gamma_5) \gamma_\mu u_\tau \qquad (213)$$

$$J^\mu = <3\pi|A^\mu|0> \qquad (214)$$

$$p_\tau = p_\nu + Q \qquad (215)$$

If $G(q^2)$ describes the a_1 Breit–Wigner and $B(s_i)$ the ρ^0 Breit–Wigner distribution then

$$J^\mu = G(q^2)\left\{\left(q_1^\mu - q_+^\mu + \frac{Q(q_1+q_+)}{q^2}\right)B(s_2) + (1 \leftrightarrow 2)\right\} \quad (216)$$

$$Q = q_1 + q_2 + q_+ \quad (217)$$

where q_1, q_2 are the 4-vectors of the negative mesons, while q_+ that of the π^+, all measured in the a_1^--rest-frame. Using this ansatz, which considers the contribution of the two amplitudes and takes into account the Bose-symmetrisation, one arrives at the following expression:

$$\sigma = \sigma_{PC} + \sigma_{PV} \quad (218)$$

The term of interest for the analysis is

$$\sigma_{PV}^\pm = \pm 3\gamma_{AV}\vec{P}_\tau \vec{n}_{3\pi} Im\left(B(s_1)B^*(s_2)\right)\sqrt{\frac{s_1 s_2 s_3 - m_{3\pi}^2(q^2 - m_\pi^2)}{Q^2}} \quad (219)$$

$$\gamma_{AV} = \frac{2g_A g_V}{g_A^2 + g_V^2} \quad (220)$$

$$Im\left(B(s_1)B^*(s_2)\right) \sim (s_1 - s_2) \quad (221)$$

$$\vec{n}_{3\pi} = \frac{\vec{q}_1 \times \vec{q}_2}{|\vec{q}_1 \times \vec{q}_2|} \quad (222)$$

If one associates the index 1 such that

$$s_1 = (q_1 + q_+)^2 > s_2 = (q_2 + q_+)^2 \quad (223)$$

it follows that

$$\vec{n}_{3\pi}\, sign(s_1 - s_2) \quad (224)$$

is defined in a Lorentz invariant way. $\vec{p}_\tau$ is not measurable. One convinces oneself easily that the

$$<\vec{p}_\tau \vec{n}_{3\pi}>_\phi = <\vec{Q}_{3\pi}\vec{n}_{3\pi}\cos\psi>_\phi \quad (225)$$

where Ψ is the angle between the τ-direction and $\vec{Q}$ and ϕ is the azimuthal angle of $\vec{p}_\tau$ with $\vec{Q}$ as axis. In the experiment the expression

$$\frac{<\vec{Q}_{3\pi}\vec{n}\, sign(s_1 - s_2)>_x}{<\cos\Psi>_x} = \pm\gamma_{AV} A_{RL}(q^2) \quad (226)$$

is measured which on the other side can also be calculated. A_{RL} turns out to be model dependent, specifically it depends on the S/D-wave ratio in a_1-decay [142]. Typical results are shown in fig.77. ARGUS has recently [141] determined the S/D-wave ratio to be (ch.7.4.3)

$$\frac{D}{S} = -0.11 \pm 0.02 \quad (227)$$

Plugging this result into model calculations, one gets for the results shown in fig.77. Three observations can be made:

- ν_τ and $\bar{\nu}_\tau$ have opposite helicities as expected in the framework of the standard model.
- ν_τ is lefthanded.
- In accordance with the standard model expectation $\gamma_{AV} = 1$ one gets

$$\gamma_{AV} = 1.25 \pm 0.23 \pm^{0.15}_{0.08} \tag{228}$$

With this result ARGUS for the first time has detected parity violation in τ–decays and hence checked an essential input of the standard model. The results of this section are summarized in table 11 and compared to the standard model predictions. Good agreement is observed. Presently these measurements are still limited by statistic, further progress can be expected from the new high statistic data of CLEO and the LEP experiments.

Table 11: Tests of Lorentz–structure of the $\tau\,\nu_\tau$–current performed by the ARGUS collaboration

	ARGUS	Standard Model
Michel parameter	$\rho = 0.742 \pm 0.035 \pm 0.020$ $\xi = 0.90 \pm 0.13 \pm 0.13$	$\rho = 0.75$ $\xi = 1$
ρ Spin–alignment	$0.77 < \frac{L}{L+R} < 0.92$	$\frac{L}{L+R} = 0.85$
ν_τ–helicity	negative	negative
γ_{AV}	$1.25 \pm 0.23^{+0.15}_{-0.08}$	1
$\rho^-\rho^+$ correlations	observed $\gamma_{AV} \approx 1$	$\gamma^2_{AV} = 1$

7.3 Branching Ratios

In this chapter the branching ratios will be summarized as measured by the ARGUS collaboration. For comparison the world averages are included in table 12. Comparing the errors given it follows that the ARGUS results had a major impact on the present knowledge. The measured branching ratios for leptonic decays on one side allow to test lepton universality in the three generation standard model. On the other side they can be used to test QCD. It has been shown [143] that the ratio

$$R_\tau = \frac{\Gamma(\tau^- \rightarrow \nu_\tau hadron)}{\Gamma(\tau^- \rightarrow \nu_\tau e^- \bar{\nu}_e)} = \frac{1 - B_e - B_\mu}{B_e} \tag{229}$$

Table 12: Comparison of branching ratios as measured by ARGUS with world averages

	Br (%)	Dallas 92/ PDG 92
$\tau^- \to e^- \bar{\nu}_e \nu_\tau$	$17.3 \pm 0.4 \pm 0.5$	$17.6 + 0.16$
$\tau^- \to \mu^- \bar{\nu}_\mu \nu_\tau$	$17.2 \pm 0.4 \pm 0.5$	17.37 ± 0.18
$\tau^- \to \pi^-/K^- \nu_\tau$	$11.7 \pm 0.6 \pm 0.8$	12.48 ± 0.35
$\tau^- \to \rho^- \nu_\tau$	$22.6 \pm 0.4 \pm 0.9$	24 ± 0.6
$\tau^- \to a_1^- \nu_\tau$	$6.8 \pm 0.1 \pm 0.5$	7.92 ± 0.26
$\tau^- \to \pi^- \pi^- \pi^+ \pi^0 \nu_\tau$	$4.2 \pm 0.5 \pm 0.9$	5.3 ± 0.4
$\tau^- \to \omega \pi^- \nu_\tau$	$1.65 \pm 0.3 \pm 0.2$	1.6 ± 0.4
$\tau^- \to K^{*-} \nu_\tau$	$1.23 \pm 0.21 \pm 0.2$	1.43 ± 0.17
$\tau^- \to \pi^- \pi^- \pi^- \pi^+ \pi^+ \nu_\tau$	$(6.4 \pm 2.3 \pm 1)10^{-4}$	$(5.6 \pm 1.6)10^{-4}$

can be calculated in QCD [143]

$$R_\tau = 3\left\{1 + \frac{\alpha_s}{\pi} + 5.2(\frac{\alpha_s}{\pi})^2 + 26.37(\frac{\alpha_s}{\pi})^3 + ...\right\} . \qquad (230)$$

The corrections due to confinement effects have been calculated using QCD sum-rules and are estimated to be small

$$\Delta R_\tau = c\frac{m_q^2}{m_\tau^2} + \frac{<m_{q\bar{q}}>}{m_\tau^4} + ... \qquad (231)$$

From the ARGUS results one derives

$$R_\tau = 3.79 \pm 0.15 \qquad (232)$$

compared to the world average [57] of $R_\tau = 3.59 \pm 0.04$. Pich [143] has used the world average to derive from eq.(230) the strong coupling constant

$$\alpha_s = 0.32 \pm 0.04 . \qquad (233)$$

Altarelli in his summary at the Aachen QCD conference arrives at a more pessimistic estimation of the theoretical uncertainties

$$\alpha_s = 0.32 \pm 0.10 . \qquad (234)$$

The importance of this result follows from fig.78 which demonstrates the running of the strong coupling constant. Note that ARGUS data enter twice into this plot: besides the τ-results discussed also the results from Υ-decay [144] are entered into the figure.
Furthermore ARGUS has studied 29 forbidden τ-decays, where no ν_τ exists

in the final state [145]. The decay $\tau^- \to e^- e^- \mu^+$ may serve as an example. τ-decays are characterized by their typical 1-1 and 1-3 topology respectively. As an additional constraint the relation

$$m \text{ (final state e.g. } e^- e^- \mu^+) = m_\tau \tag{235}$$

is used in the analysis. The limits for the various decay channels lie in the interval $1.3 \cdot 10^{-5} (\tau^- \to e^- e^- \mu^+)$ to $1.3 \cdot 10^{-3} (\tau^- \to \bar{p}\eta)$ at 90% confidence level. Most of these branching ratios have been measured for the first time. They provide a stringent constraint for models beyond the standard model.

7.4 Analysis of Final States

The hadronic current coupling to the W–boson has the general form

$$J_h^\mu = \bar{u}\gamma^\mu (1 - \gamma_5)(d \cos\theta_c + s \sin\theta_c) \tag{236}$$

The vector current $\bar{u}\gamma^\mu d$, $\bar{d}\gamma^\mu u$ of weak interactions can be related to the isovector electromagnetic current $\bar{u}\gamma^\mu u - \bar{d}\gamma^\mu d$, since they form an isovector triplet [127]. This very fact enforces weak vector current conservation

$$q_\mu V^\mu = 0, \tag{237}$$

i.e. the current is transverse. Hence the timelike component of the vector current, which transforms like a scalar for spacelike rotations, disappears. Therefore the vector current can only couple to a $J^\pi = 1^-$ state. No constraint of this kind exists for the axial vector current A^μ, it couples to hadronic final states with $J^\pi = 0^-$ (time like component) and $J^\pi = 1^+$ (space like component). A further constraint follows from the behavior of the V^μ, A^μ with respect to G–parity transformations. For 1st class currents

$$G V^\mu G^{-1} = V^\mu \tag{238}$$

and

$$G A^\mu G^{-1} = -A^\mu, \tag{239}$$

therefore $J^\pi = 1^-$ states decay into an even number of π-mesons while for $J^\pi = 0^-, 1^+$ an uneven number of π-mesons is observed. For 2nd class currents the inverse holds. Note that in the standard model only 1st class currents exist [146]. Therefore, the observation of $2\pi, 4\pi$ in a $0^-, 1^+$ state would signal new physics.

In his pioneering and thorough theoretical investigation of heavy lepton decay, which predated the observation of τ-decays, Tsai showed that for a sequential lepton the hadronic decay width is given by [134]

$$\frac{d\Gamma}{dq^2} = \frac{G_F^2}{32\pi^2 m_\tau}(m_\tau^2 - q^2)^2$$
$$\left(\cos^2\theta_C \left\{(m_\tau^2 + 2q^2)(v_1(q^2) + a_1(q^2)) + m_\tau^2 a_0(q^2)\right\}\right.$$
$$\left.+ \sin^2\theta_C \left\{(m_\tau^2 + 2q^2)\left(v_1^s(q^2) + a_1^s(q^2)\right) + m_\tau^2\left(v_0^s(q^2) + a_0^s(q^2)\right)\right\}\right) \quad (240)$$

The a_i, v_i are spectral functions describing the influence of soft hadronics in the final state, θ_c is the Cabibbo angle. The rest are kinematical factors. The CVC hypothesis can be phrased in the following way [127]

$$v_1(q^2) = \frac{\sigma_{e^+e^- \to 2n\pi}^{I=1}}{\sigma_{point} 3\pi} \quad (241)$$

7.4.1 Decay $\tau^- \to \pi^-\pi^0\nu_\tau$

Kühn et. al.[140] have carefully analysed the available data for the reaction $e^+e^- \to \pi^+\pi^-$ and parametrized the corresponding formfactor (fig.79). ARGUS uses this parametrization as an input for its Monte Carlo program in order to describe the measured invariant mass distribution, taking acceptance and resolution effects into account [141]. A fit of the CVC prediction to the ARGUS data is shown in fig.80, the only free parameter is the overall normalization. The fit interval covers the mass region $0.28 GeV \leq m_{\pi^-\pi^0} \leq 1.4 GeV$. One gets a $\chi^2 = 22.5$ for 22 degrees of freedom using the Kühn parametrization. If one considers only the $\rho(778)$ contribution (fig.79, broken line) the fit is worse ($\chi^2 = 38.7$ per 22 dgf).
Also the absolute prediction of CVC theory has been checked by the ARGUS collaboration. Using the measured branching ratios they get

$$\frac{Br(\tau^- \to \pi^-\pi^0\nu_\tau)}{Br(\tau^- \to e^-\bar{\nu}_e\nu_\tau)} = 1.31 \pm 0.06 \quad (242)$$

while Kühn et al. [140] predict 1.32 ± 0.05. The good agreement between the data and theory shows that the CVC hypothesis can be extended into the region of large momentum transfers characteristic for τ-decays.

7.4.2 Decay $\tau^- \to \pi^-\pi^-\pi^+\pi^0\nu_\tau$

Again this decay channel is selected exploiting the characteristic 1–3 topology, but this time no cut is applied on the number of photons on the 3-prong side of the event (ch.7.2.4). ARGUS [147] for the first time investigated the resonance substructure in this channel (fig.81). The important role which vector mesons play in the final state of this reaction follows from the fact that in $> 81\%$ of all cases (95% confidence level) a $\rho-$ or ω-meson is produced as a substructure in the final state.

For the $\omega\pi$-system the angular distribution $\frac{dN}{d\cos\Psi}$ has been studied [148], where
$$\cos\Psi = \vec{n}_\omega \vec{p}_\pi \qquad (243)$$
$\vec{n}_\omega$ is the unit normal vector to the ω-decay plane, $\vec{p}_\pi$ is the unit vector in the direction of the bachelor π of the $\omega\pi$ final state, both measured in the ω-rest-frame. As shown by fig.82 this angular distribution depends on the spin of the final state. Only $J^\pi = 1^-$ is compatible with the data in agreement with the standard model predictions that only 1st class currents contribute to τ-decays. In fig.83 the data are compared with the results available for the reaction $e^+e^- \to \omega\pi^0$. Good agreement is observed, hence again the CVC hypothesis has been confirmed.

7.4.3 Decay $\tau^- \to \pi^-\pi^-\pi^+\nu_\tau$

As discussed in ch.7.2 the decay $\tau^- \to \pi^-\pi^-\pi^+\nu_\tau$ is of special interest since it provides the possibility to measure the ν_τ-helicity. But it has also an interest in itself since it allows to study in a clean way the decay of the a_1-meson. This is demonstrated by fig.84, which shows a background free resonance signal. The full line is a fit of an isobar model of Isgur et al. [149]. From its data ARGUS [141] derives the following parameters of the resonance

$$m(a_1) = (1211 \pm 7) MeV \quad , \Gamma(a_1) = (446 \pm 21) MeV \qquad (244)$$

using the model of Isgur et al. Note however that the result for the resonance parameters depends on the model used for the analysis. because the a_1-meson is a very broad resonance. It is worthwhile to mention that pure hadronic data are much harder to analyse since a strong background from the Deck-effect competes with the a_1-signal [57].

As pointed out in ch.7.2.4 the ratio of the D- and S-wave in the a_1-decay is important since it is an essential ingredient in the analysis of the ν_τ-helicity. To get this information we have analyzed the Dalitz plot distribution for four bins of the 3π invariant mass distribution (fig.85). The experimental distributions [141] are compared with the theoretical expectations [149] for pure S- and D-waves and for an amplitude ratio of $\frac{D}{S} = -0.11$. The data are in good agreement with

$$\frac{D}{S} = -0.11 \pm 0.02 \qquad (245)$$

Note, however, that this result is slightly model dependent, for the analysis ARGUS used again the model of Isgur [149]. A full partial wave analysis is presently performed.

7.5 Summary

- ARGUS has developped a new method to measure the τ-mass, the result is in good agreement with recent low energy data and appreciably smaller than the previous value. The new world average for the lepton mass is

$$m_\tau = (1777.1 \pm 0.5) \; MeV \tag{246}$$

Using the 1992 value of the τ-lifetime $(295.7 \pm 3.2) fs$ and the world average of the electron branching ratio of $B_e = (17.76 \pm 0.15)\%$ one arrives at a ratio of the τ- and μ-coupling constants of

$$\frac{g_\tau}{g_\mu} = 0.990 \pm 0.007 \tag{247}$$

- The helicity of the ν_τ is determined to be negative and for the ratio of τ coupling constants one gets

$$\gamma_{AV} = \frac{2 g_A g_V}{g_A^2 + g_V^2} = 1.25 \pm 0.23 ^{+0.15}_{-0.08} \tag{248}$$

- The Michel parameters are determined with high precision

$$\rho = 0.742 \pm 0.035 \pm 0.026 \tag{249}$$

$$\xi = 0.90 \pm 0.13 \pm 0.13 \tag{250}$$

They agree with the standard model predictions eq.(196).

- The CVC hypothesis has been tested for two decay channels of the τ-lepton: $\tau^- \to \pi^- \pi^0 \nu_\tau$ and $\tau^- \to \pi^- \pi^- \pi^+ \pi^0 \nu_\tau$. Good agreement between data and theory on the 5% level is observed.

- Finally τ-decays turn out to be an ideal laboratory for the study of $J^\pi = 1^-, 0^-$ and 1^+ hadronic final states. These studies just started.

In 10 years of running ARGUS has pioneered B-meson and τ-physics and established a plenitude of new and partially unexpected strong effects. It turns out that only a door to a new field has been opened and many interesting results are to be expected in the future, as one example let me mention the search for CP violation in B^0-decays [97] hence:

De scientiae naturae non omnes factae sunt sed plures restant adhuc inveniendae [150].

Ackowledgement: I thank Prof. J.L. Lucio and his collegues for the invitation to this wonderful place and the stimulating atmosphere at the school. These lectures are based on results of the ARGUS and the H1 collaborations, I am grateful to my collegues for the long and successful cooperation. Special thanks are due to the students and young scientist of Dortmund university who shared with me the fun to work in high energy physics. Last but not least I thank my wife Beate for her untiring help to finalize this written version of my talk.

References

[1] P. Lenard Ann. Phys. 51 (1894) 225

[2] J. Franck and G. Hertz VDpG 14 (1912) 457, 512

[3] R. Hofstadter Nuclear and Nucleon Structure, Benjamin 1963

[4] M. Breidenbach et. al. Phys. Rev. Lett. 23 (1969) 935

[5] R.D. Peccei ed. Proc. HERA workshop Vol. I,II, 1987

[6] W. Buchmüller and G. Ingelman ed. Physics at HERA Vol. 1-3, 1991

[7] H. Hertz Ann. Phys. 34 (1888) 551, 36 (1889) 769

[8] M. Böhm, A. Fernandez, M.A. Perez, A. Rosado Z. Phys. C53 (1992) 135

[9] B. Schremp in Physics at HERA 1991,Vol.2 p.1034

[10] R.G. Roberts The Structure of the Proton, Cambridge University Press 1990

[11] G. Altarelli and G. Parisi Nucl. Phys. 126 (1977) 297

[12] L.V. Gribov et. al. Phys. Rep. 100 (1983) 297

[13] A.H. Mueller Nucl. Phys. 18C (1991) 125

[14] J. Bartels et. al. Physics at HERA Vol. I, 1991 p.203

[15] A. Breakstone et al. Phys. Rev. D30 (1984) 528

[16] D. Wegener in Workshop on Future Relativistic Heavy Ion Experiments, Darmstadt 1980 p.133

[17] W. Geist et al. Phys. Rep. 197 (1990) 265

[18] M. Drees et al. BU-TH-92/5 preprint

[19] J.Feltesse Proc.HERA workshop 1987 Vol.1 p.33

[20] U. Fano Ann. Rev. Nucl. Sci. 13 (1963) 1

[21] R. Sternheimer and R.F. Peierls Phys. Rev. B3 (1971) 3081

[22] H. Aihara et al. Phys. Rev. Lett. 61 (1988) 1263

[23] E. Fermi Phys. Rev. 57 (1940) 485

[24] L.D. Landau J. Exp. Phys.(USSR) 6 (1944) 201

[25] E. Fermi Nuclear Physics University of Chicago Press 1949

[26] G. Moliere Z. Naturf. 3a (1948) 78

[27] C. Leroy et al. Nucl. Instr. Meth. A252 (1986) 4

[28] A. Babaev et al. Nucl. Instr. Meth. 160 (1979) 427

[29] B. Andrieu et al. Proc. 3rd Intern. Conf. Advanced Technology and Particle Physics 1991

[30] R.L. Gluckstern Nucl. Instr. Meth. 24 (1963) 381

[31] J.Engler et al. Phys. Lett. 29 (1969) 321, Nucl. Instr. Meth. 106 (1973) 189

[32] D. Wegener Phys. Bl. 45 (1989) 358

[33] U. Amaldi Phys. Scripta 23 (1981) 409

[34] W. Hofmann et al. Nucl. Instr. Meth. 195 (1982) 475

[35] C.W. Fabjan in Experimental Techniques in High Energy Physics, T. Ferbel ed.

[36] T. Gabriel et al. Nucl. Instr. Meth. 134 (1976) 271, 150 (1976) 145, 195 (1982) 961

[37] Technical Proposal for the ZEUS Detector, DESY 1986

[38] Technical Proposal for the H1 Detector, DESY 1986

[39] H. Brückmann et al. Nucl. Instr. Meth. A263 (1988) 136

[40] R. Wigman Nucl. Instr. Meth. A259 (1987) 389

[41] E. Bernardi et al. Nucl. Instr. Meth. A262 (1987) 229

[42] J.B. Birks Theory and Practice of Scintillation Counting, Pergamon Press 1964

[43] H. Abramowicz et al. Nucl. Instr. Meth. 180 (1981) 429

[44] W. Braunschweig et al. DESY 89-022

[45] B. Andrieu et al. DESY 93-047

[46] J. Brau Nucl. Instr. Meth. A312 (1992) 483

[47] U. Behrens et al. Nucl. Instr. Meth. A289 (1990) 115

[48] I. Abt et al. DESY 93-103, Nucl. Instr. Meth. A in print

[49] H. Drumm et al. Nucl. Instr. Meth. A176 (1980) 333

[50] B. Andrieu et al. DESY 93-078, Nucl. Instr. Meth. A in print

[51] F. Eisele and D. Wegener H1 Internal Report H1-04/85-15

[52] A. Bernstein et al. DESY 93-076

[53] G.D. Agostini et al. Nucl. Instr. Meth. A274 (1989) 134

[54] G. Wolf DESY 92-190

[55] S. Levonian in Physics at HERA 1991, Vol.1 p.499

[56] M. Costa et al. in Physics at HERA 1991, Vol.1 p.509

[57] Particle Data Group Phys. Rev. D45 (1992) I1

[58] A. Donnachie and P.V. Landshoff Phys. Lett. B296 (1992) 277

[59] D. Wegener in A. Ali and P. Söding High Energy Electron Positron Interactions, World Publishing, Singapore 1989

[60] S. Levonian DESY 93-077

[61] M. Derrick et al. Phys. Lett. B293 (1992) 465

[62] T. Ahmed et al. Phys. Lett. B299 (1993) 374

[63] M. Drees et al. Phys. Rev. Lett. 61 (1988) 275

[64] H. Kolanoski Springer Tracts on Modern Physics, Vol.105, 1984

[65] M. Erdmann DESY 93-077

[66] M. Colombo, Thesis Dortmund 1993

[67] M. Derrick et al. Phys. Lett. B297 (1992) 404

[68] A. De Roeck Rapp. Talk EPS Conf. High Energy Phys., Marseille 1993

[69] T. Sjöstrand in Physics at HERA 1991, Vol.3 p.1405

[70] I. Abt et al. Phys. Lett. B314 (1993) 436

[71] M. Glück et al. Phys. Rev. D46 (1992) 1973

[72] T. Ahmed et al. Phys. Lett. B298 (1993) 469

[73] T. Ahmed et al. Phys. Lett. B299 (1993) 385

[74] T. Ahmed et al. DESY 93-117

[75] A. De Roeck DESY 93-077

[76] C. Vallee DESY 93-077

[77] H. Küster DESY 93-077

[78] M. Derrick Phys. Lett. B306 (1993) 158

[79] M. Derrick et al. Phys. Lett. B303 (1993) 183

[80] M. Derrick et al. DESY 93-068

[81] M. Derrick et al. DESY 93-093

[82] W. Buchmüller and G. Ingelmann ed. Physics at HERA 1991 Vol.3

[83] W. Bartel et al. Z. Phys. C33 (1986) 23

[84] I. Abt et al. DESY 93-137

[85] A.D. Martin et al. Phys. Lett. B306 (1993) 145, B309 (1993) 492

[86] A. Benvenuti et al. Phys. Lett. B233 (1989) 485

[87] P. Amandrez et al. Phys. Lett. B295 (1993) 159

[88] J. Botts et al. Phys. Lett. B304 (1993) 159

[89] M. Glück et al. Z. Phys. C53 (1992) 127, Phys. Lett. B306 (1993) 391

[90] A. Donnachie and P.V. Landshoff DAM TP 93-23 preprint

[91] S.W. Herb et al. Phys. Rev. Lett. 39 (1977) 252

[92] H. Kolanoski et al. in A. Ali and P. Söding High Energy Electron Positron Interactions, World Publishing, Singapore 1989

[93] D. Andrews et al. Phys. Rev. Lett. 44 (1980) 1108

[94] C. Jarlskog in C. Jarlskog ed. CP Violation, World Publishing 1989, p.1

[95] M. Kobayashi and T. Maskawa Progr. Theor. Phys. 49 (1973) 652

[96] H. Albrecht et al. DESY 92–041 HELENA A Beauty Factory at Hamburg

[97] H. Albrecht et al. DESY PRC 93–4

[98] N. Isgur and M. Wise Phys. Lett. B232 (1989) 113, B237 (1990) 527 and B. Grinstein in Proc. Workshop on High Energy Phenomenology, Mexico City 1991, World Publishing

[99] N. Isgur in Proc. XXVI International Conference on High Energy Physics, Dallas 1992 p.33

[100] M. Neubert Phys. Lett. B262 (1991) 455

[101] M. Bauer et al. Z. Phys. C34 (1987) 103

[102] H. Albrecht et al. Z. Phys. C48 (1990) 543

[103] A. Barteletto et al. Phys. Rev. D45 (1992) 21

[104] H. Albrecht et al. Z. Phys. C54 (1990) 1

[105] H. Albrecht et al. Phys. Lett. B249 (1990) 359

[106] H. Albrecht et al. Z. Phys. C52 (1991) 353

[107] H. Albrecht et al. Z. Phys. C56 (1992) 1

[108] G. Altarelli et al. Nucl. Phys. B208 (1982) 365

[109] G. Altarelli et al. Phys. Lett. B261 (1991) 303

[110] H. Albrecht et al. DESY 93–104

[111] H. Albrecht et al. Nucl. Instr. Meth. A275 (1989) 1

[112] H. Albrecht et al. Phys. Lett. B234 (1990) 409

[113] H. Albrecht et al. Phys. Lett. B255 (1991) 297

[114] B. Grinstein et al. Phys. Rev. Lett. 55 (1986) 298, N. Isgur et al. Phys. Rev. D39 (1989) 799

[115] M. Wirbel Progr. in Part. and Nucl. Phys. 21 (1988) 33

[116] H. Albrecht et al. Phys. Lett. B197 (1987) 452

[117] H. Albrecht et al. Phys. Lett. B275 (1992) 195

[118] J. Appel Lectures at this School

[119] H. Albrecht et al. DESY 92-029

[120] H. Albrecht et al. Z. Phys. C57 (1993) 533

[121] J.G. Koerner et al. Phys. Lett. B226 (1989) 185

[122] H. Albrecht et al. Phys. Lett. B219 (1989) 121

[123] H. Albrecht et al. Phys. Lett. B192 (1987) 245

[124] H. Schroeder in S. Stone ed. B–Decays, World Publishing 1992

[125] H. Albrecht et al. Z. Phys. C55 (1992) 357

[126] A. Ali in S. Stone ed. B–Decays, World Publishing 1992

[127] L.B. Okun Leptons and Quarks North Holland, Amsterdam 1982

[128] W.J. Marciano and A. Sirlin Phys. Rev. Lett. 41 (1988) 185

[129] S. Baciano et al. Phys. Rev. Lett. 41 (1978) 13

[130] H. Albrecht et al. Phys. Lett. B292 (1992) 221

[131] S. Jadach and Z. Was Comp. Commun. 36 (1985) 483(E), S. Jadach, J.H. Kuehn, Z.Was Comp. Phys. Comm. 76 (1993) 361

[132] J.Z. Bai et al. Phys. Rev. Lett. 69 (1992) 3021

[133] P.S. Drell Proc. XXVI Intern. Conf. High Energy Physics, Dallas 1992 p.3

[134] Y.S. Tsai Phys. Rev. D4 (1971) 2821

[135] H. Albrecht et al. Phys. Lett. B202 (1988) 149

[136] A. Ali and Z.Z. Aydin Nuovo Cim. A43 (1978) 270

[137] H. Albrecht et al. Phys. Lett. B246 (1990) 278

[138] H. Albrecht et al. DESY 93-108

[139] H. Kuehn and F. Wagner Nucl. Phys. B236 (1984) 16

[140] J.H. Kuehn and A. Santamaria Z. Phys. C48 (1990) 445

[141] H. Albrecht et al. Z. Phys. C56 (1992) 339

[142] M. Feindt Z. Phys. C48 (1990) 681

[143] A. Pich Rapp. Talk 2nd Workshop on τ-Lepton Physics, Ohio 1992 CERN TH 6758/92

[144] H. Albrecht et al. Phys. Lett. B199 (1987) 291

[145] H. Albrecht et al. Z. Phys. C55 (1992) 179

[146] S. Weinberg Proc. XIX International Conference on High Energy Physics, Tokyo 1978, p.917

[147] H. Albrecht et al. Phys. Lett. B260 (1991) 259

[148] H. Albrecht et al. Phys. Lett. B185 (1987) 223

[149] N. Isgur et al. Phys. Rev. D39 (1989) 1357

[150] Albertus Magnus Liber Primus Posteriorum Analyticorum tract.1, cap.1

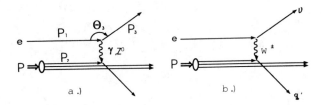

Figure 1: Diagram describing deep inelastic scattering by a) neutral current and b) charged current exchange

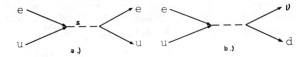

Figure 2: Diagram contributing to s–channel leptonquark production

Figure 3: Splitting function of quarks and gluons

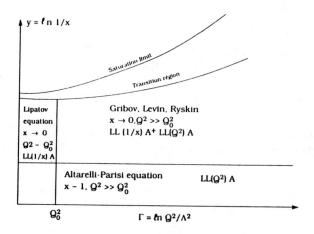

Figure 4: Regions in the $\ln 1/x - Q^2$ plane, where different approximations are necessary to describe deep inelastic scattering

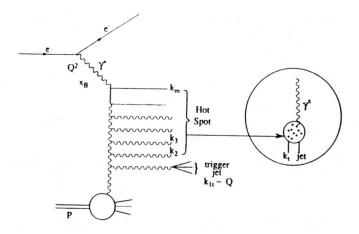

Figure 5: "Hot spot" production in deep inelastic scattering

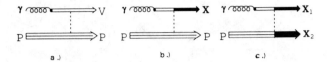

Figure 6: Processes contributing to low p_t photoproduction

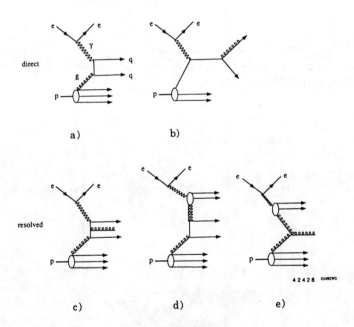

Figure 7: Diagrams contributing to direct and resolved hard photon reactions

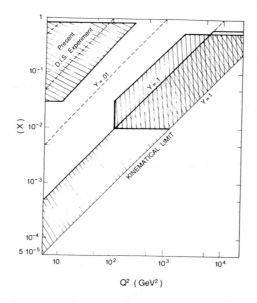

Figure 8: Regions in the $x - Q^2$ plane, where reliable measurements of structure functions at HERA are possible

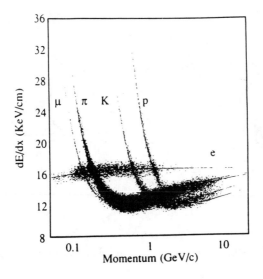

Figure 9: Specific energy loss dE/dx for different particle species as function of their momentum

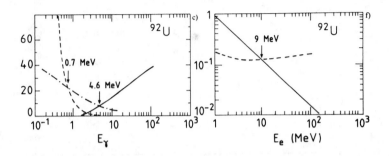

Figure 10: a) Cross section (barn/atom) for photon reactions in U, b) fractional energy loss of electrons (full line ionization and excitation, broken line radiation)

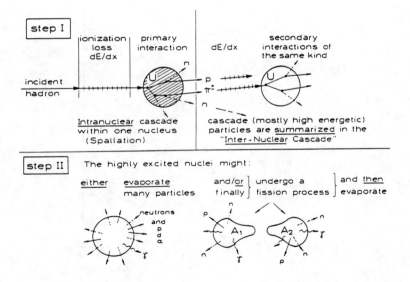

Figure 11: Processes contributing to the energy loss of hadrons in matter

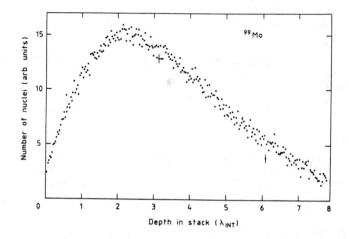

Figure 12: Longitudinal hadron shower development in U

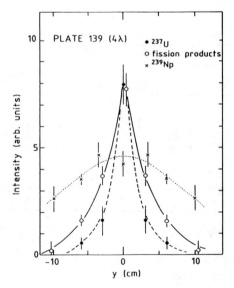

Figure 13: Transverse hadron shower development in U

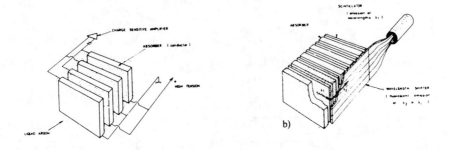

Figure 14: Technical realizations of the read-out for sandwich calorimeters

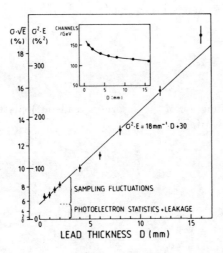

Figure 15: Energy resolution of an electromagnetic sandwich calorimeter as a function of the absorber thickness. Insert: visible energy as a function of the absorber thickness

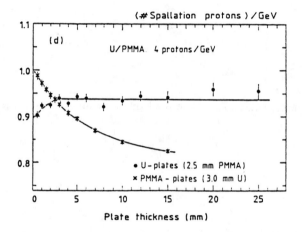

Figure 16: p/mip as a function of the absorber (read-out) plate thickness

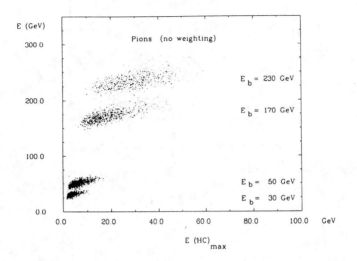

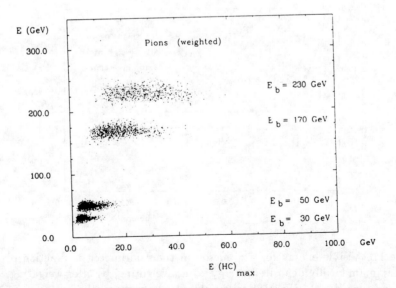

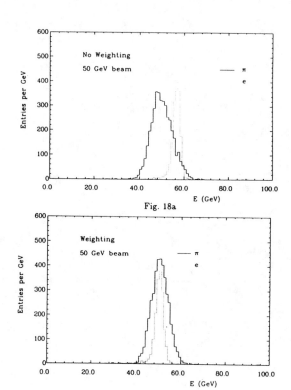
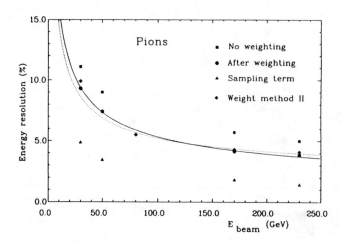

Figure 17: Visible energy for a π-meson in the calorimeter as function of the maximum locally deposited energy, a) no weighting, b) after weighting, c)–d) recorded signal before (after) weighting, e) energy resolution

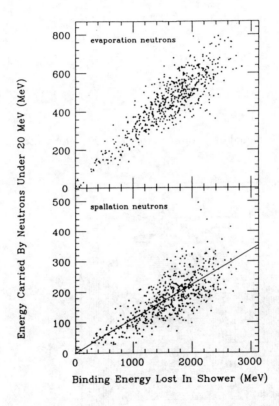

Figure 18: Correlation between energy carried by neutrons and the binding energy losses in a hadronic shower

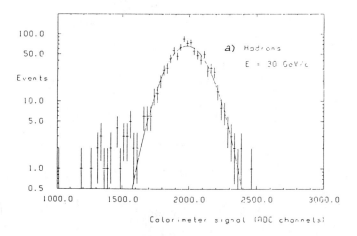

Figure 19: Signal of a compensated $U - Sc$ calorimeter

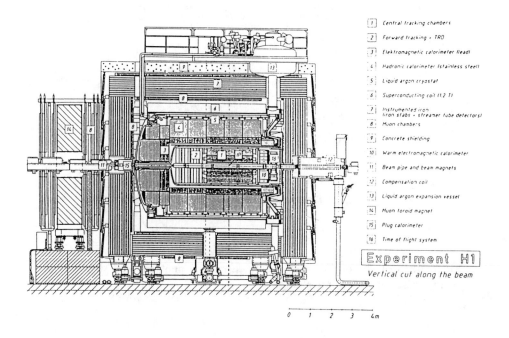

Figure 20: Schematic view of H1 detector

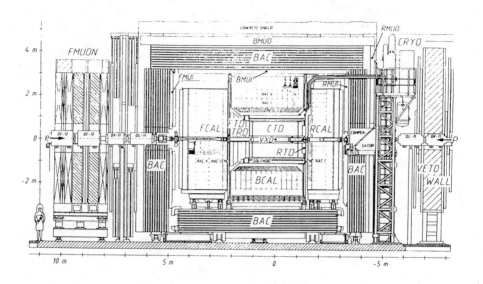

Figure 21: Schematic view of ZEUS detector

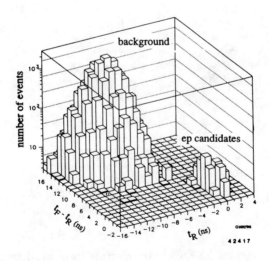

Figure 22: Time–of–flight distribution measured with the ZEUS detector between the forward rnd the rear calorimeter

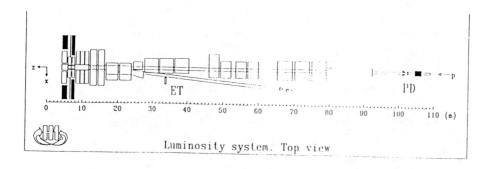

Figure 23: Luminosity monitor of the H1 detector

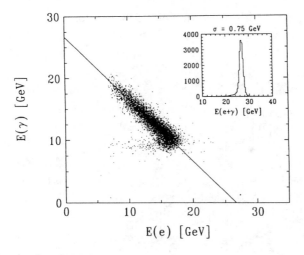

Figure 24: Correlation of energies detected in the electron (photon) detector of the luminosity monitor. The line corresponds to eq.(77)

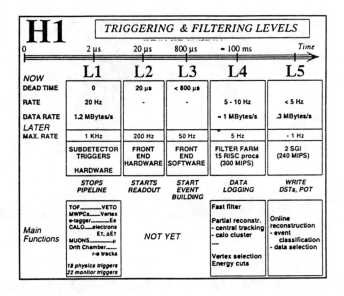

Figure 25: Trigger scheme of the H1 experiment

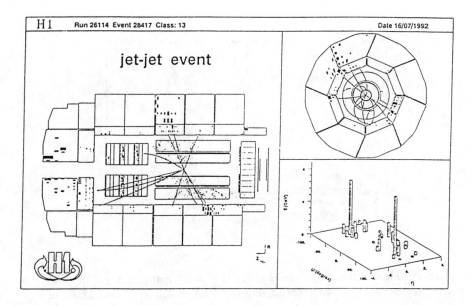

Figure 26: Typical hard photon event

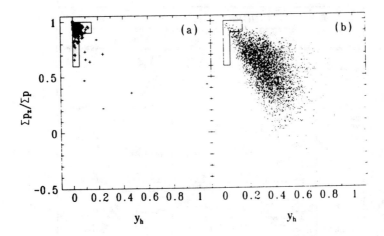

Figure 27: Separation of beam–gas events, studied with the pilot bunch (a) and (b) for photoproduction events. The region suppressed by a cut is indicated by the box

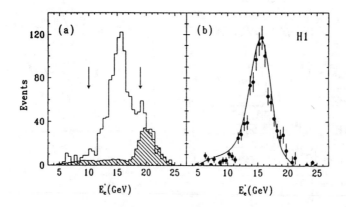

Figure 28: Measured energy-sepctrum in the electron tagger (fig.23). The shaded region in a) was measured using the pilot bunch, it is subtracted in b)

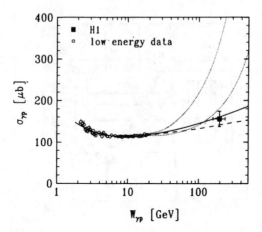

Figure 29: Measured photoproduction cross section compared to theoretical predictions: – Regge model, ... minijet models

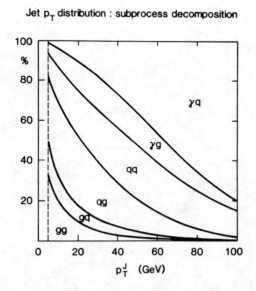

Figure 30: Relative contribution of different subprocesses to hard γp interactions as a function of p_{TJET}

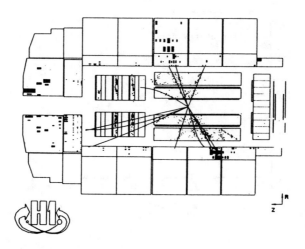

Figure 31: Candidate for a resolved photon event

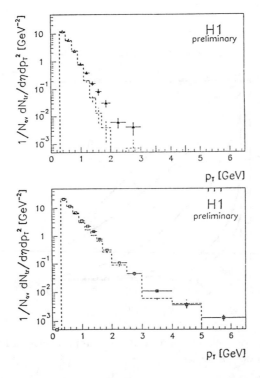

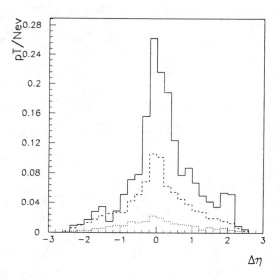

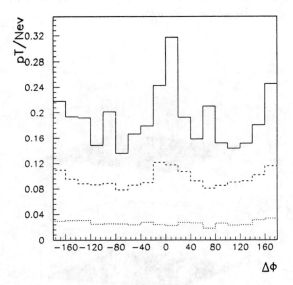

Figure 32: Measured p_T spectrum of hadrons in a) soft and b) hard γp interactions characterized by a small and large transverse energy respectively. c) and d) show the measured angular distribution of the produced particles with respect to the jet leader for different E_T of the event

318 Experiments at the HERA Storage Ring

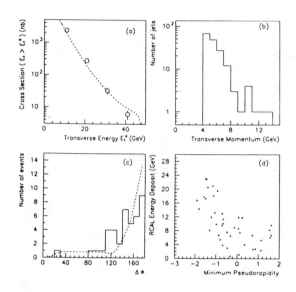

Figure 33: a) Measured cross section of hard γp interactions, b) measured azimuthal angular correlation between jets

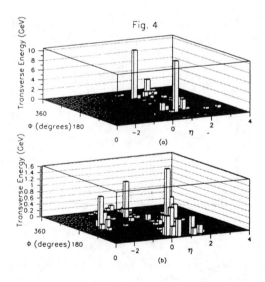

Figure 34: Examples of a 2- and a 3-jet event in hard γp interactions

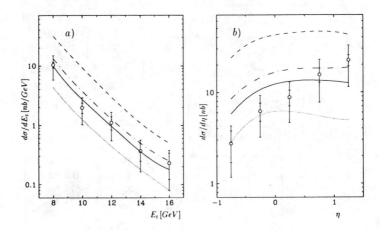

Figure 35: Measured a) E_T and b) η-distribution in hard photon reactions compared to theoretical predictions using different photon structure functions

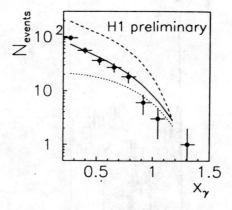

Figure 36: Momentum fraction of partons in the proton, the prediction of a leading order calculation is included

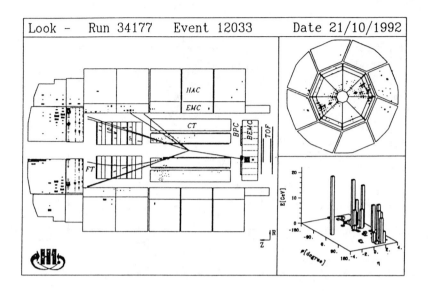

Figure 37: Typical deep inelastic electron scattering event

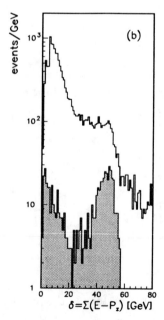

Figure 38: Distribution of events in δ (eq.(86)) at preselection level. The shaded region shows the same distribution after all cuts but the δ-cut is applied

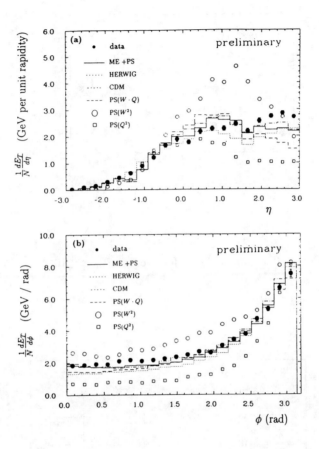

Figure 39: Measured energy flow in ϕ and η compared in deep inelastic ep–events to different model predictions

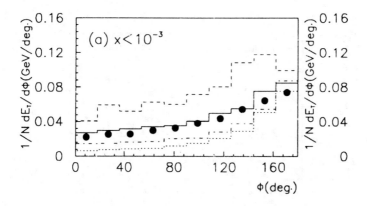

Figure 40: Same as fig.39 for events with $x < 10^{-3}$

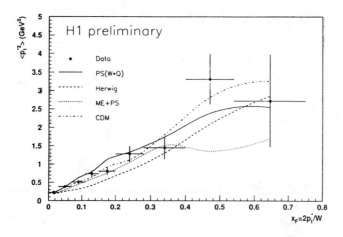

Figure 41: Sea gull plots as measured in deep inelastic scattering events

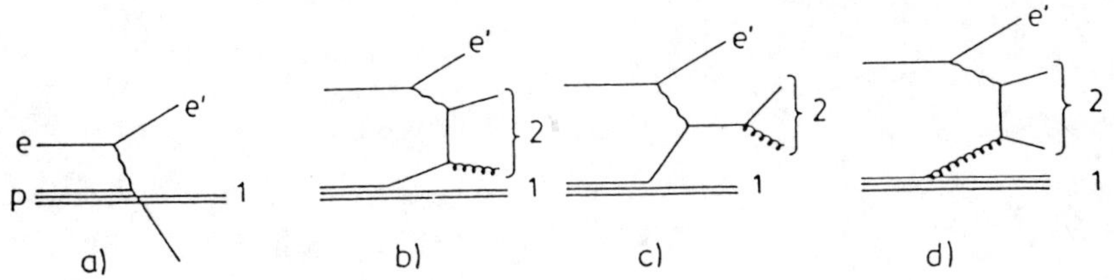

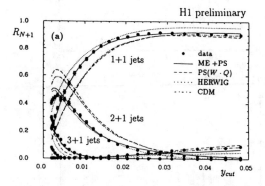

Figure 42: Expected jet configurations in deep inelastic scattering events

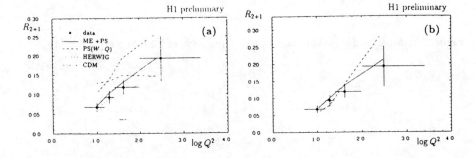

Figure 43: Measured jet rates in deep inelastic scattering events

Figure 44: Measured $2 + 1$ jet rate as a function of Q^2 compared to model predictions for a) running and b) constant α_s

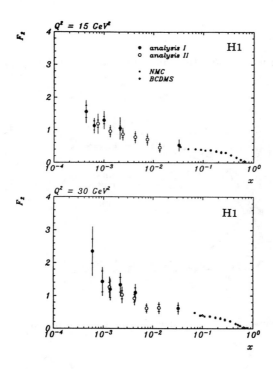

Figure 45: Measured structure function as a function of x for two Q^2 intervals

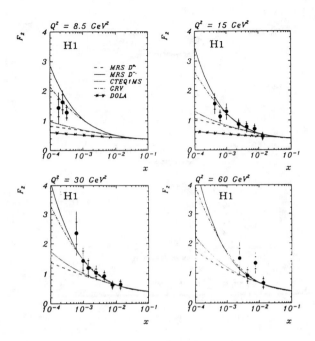

Figure 46: Comparison of measured structure function with model predictions

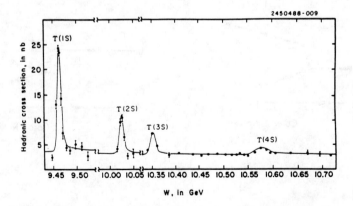

Figure 47: Measured cross section in the region of the Υ-meson

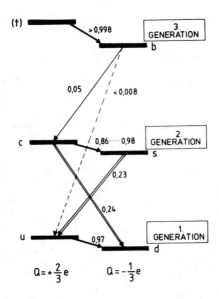

Figure 48: Mass spectrum of observed quarks. 1st order weak transitions are indicated

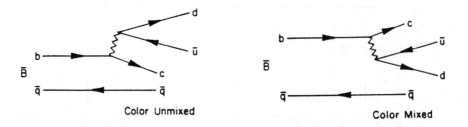

Figure 49: Contributions of allowed weak B-decays

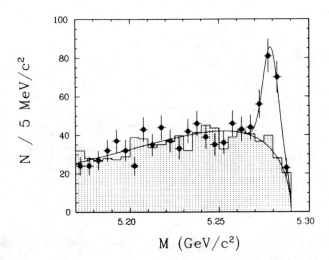

Figure 50: Measured invariant mass distribution for the decay $B \to D n\pi$

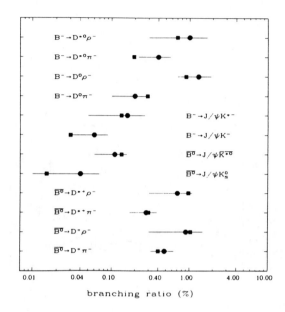

Figure 51: Comparison of measured branching ratios of two–body B–decays with model predictions

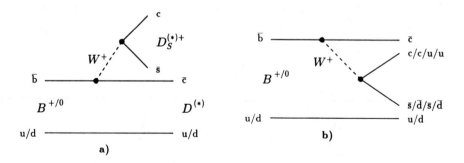

Figure 52: Diagrams contributing to the decay $B \to D_s D$

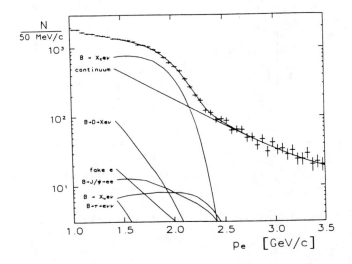

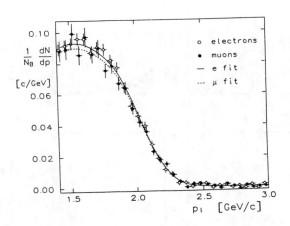

Figure 53: a) Measured electron spectrum and its decomposition into different contributions, b) Lepton spectrum for $B \to l\nu_l X$ decay

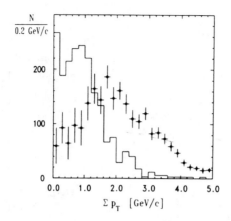

Figure 54: $\sum p_t$ disktribution for the continuum (histogram) and direct $\Upsilon(4S)$-decays

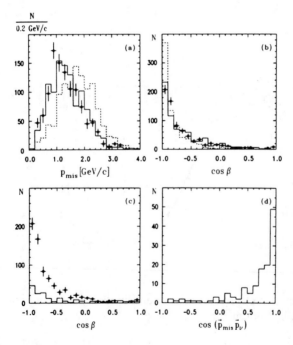

Figure 55: Distribution of a) $|\vec{p}_{MISS}|$, b) $\cos\beta$, d) $\cos(\vec{p}_{MISS},\vec{p}_\nu)$ for electrons in the $b \to c$ region. Monte Carlo predictions for $b \to c$ transitions are given as a full histogram, for $b \to u$ as broken line

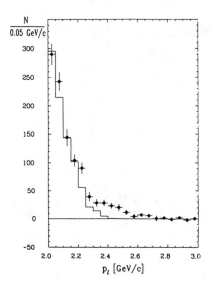

Figure 56: Measured lepton spectrum for direct $\Upsilon(4S)$-decays near the kinematical limit. The theoretical expectation for $b \to c$ transitions is given as a histogram

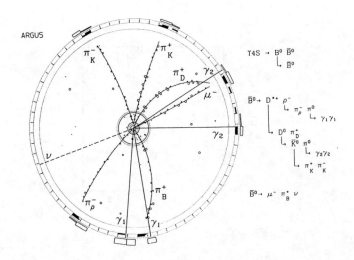

Figure 57: Fully reconstructed B-event. The most probable decay chain is indicated

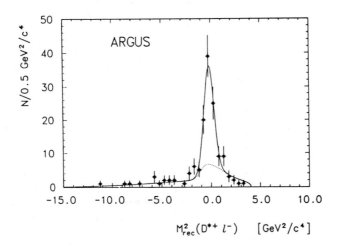

Figure 58: Measured recoil spectrum for the decay (125)

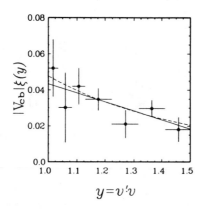

Figure 59: Comparison of eq.(104) with data

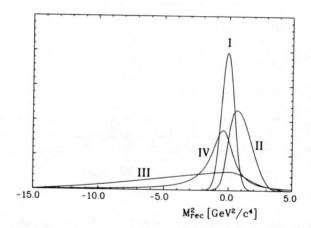

Figure 60: Expected recoil spectrum for the decays of table 10

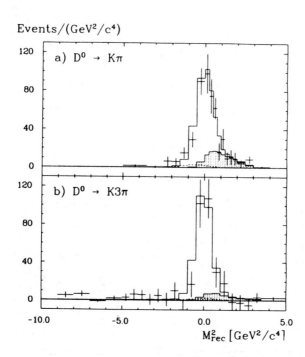

Figure 61: Measured recoil spectrum, the D^{**} contribution is indicated by the shaded region

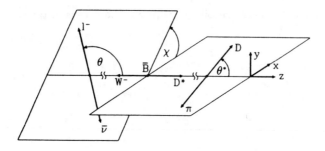

Figure 62: Definition of the polar angle θ, θ^* used in the analysis of $B^0 \to D^{**} l^- \bar{\nu}_l$ decays

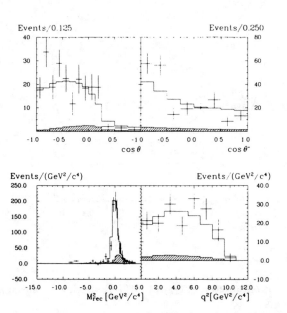

Figure 63: Measured angular distribution compared to the expectations of eq.(136), (137). No acceptance correction was applied

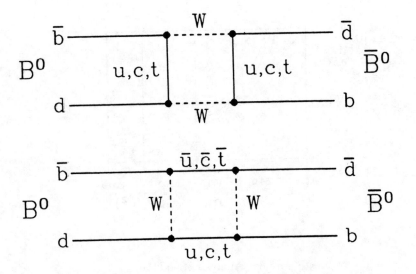

Figure 64: Box diagram describing $B^0\vec{B}^0$-mixing

336 Experiments at the HERA Storage Ring

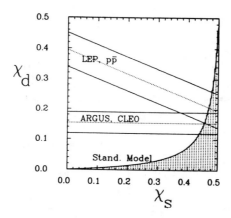

Figure 65: Mixing parameter $\chi_s(\chi_d)$ of $B_s^0(B_d^0)$-mixing as measured in different experiments

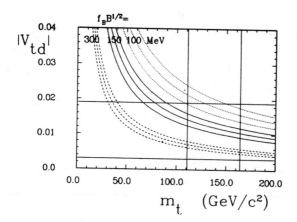

Figure 66: Analysis of eq.(156)

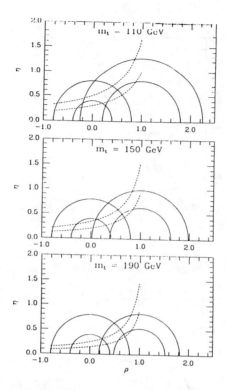

Figure 67: Test of the unitarity triangle (eq.(100)) for three values of the top mass

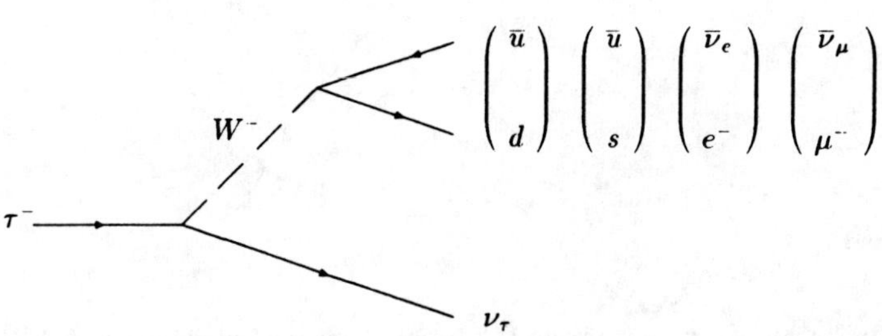

Figure 68: Feynman diagram for τ-decays

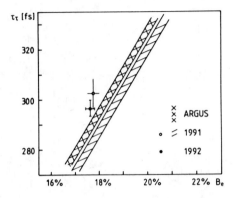

Figure 69: Comparison of eq.(173) with recent data

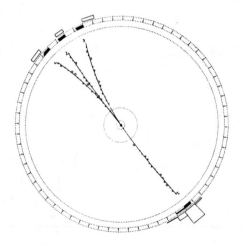

Figure 70: Decay of a τ-pair in the 1–3 topology as recorded with the ARGUS detector

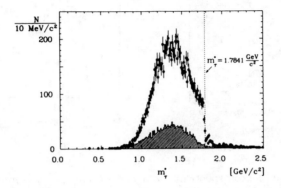

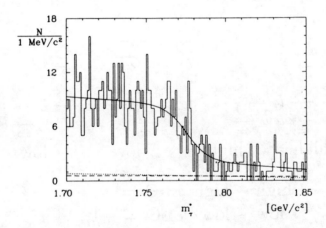

Figure 71: Pseudomass distribution (eq.(183)) for the decay $\tau^- \to \pi^-\pi^-\pi^+\nu_\tau$

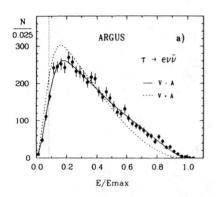
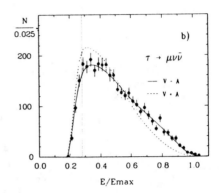

Figure 72: Measured lepton spectrum from $\tau^- \to l^- \bar{\nu}_l \nu_\tau$ decay compared to theoretical expectations for a V − A and V + A current

$$\text{high} - \text{high} \quad \cos\Theta = -1$$
$$\text{low} - \text{low} \quad \cos\Theta = +1$$

Figure 73: Helicity and energy correlations for a τ-pair decay into leptons (V − A interactions)

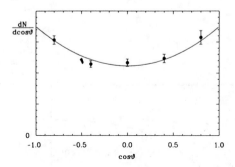

Figure 74: Angular distribution (eq.(207)) of a π^--meson from ρ^--decay

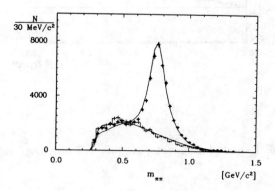

Figure 75: 2π invariant mass from the decay chain eq.(211)

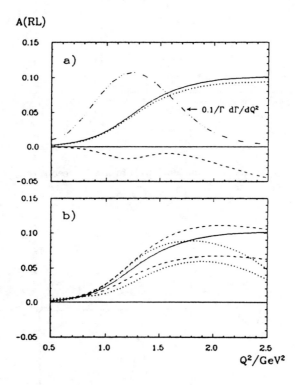

Figure 76: Expected asymmetry A_{RL} for different D/S-ratios

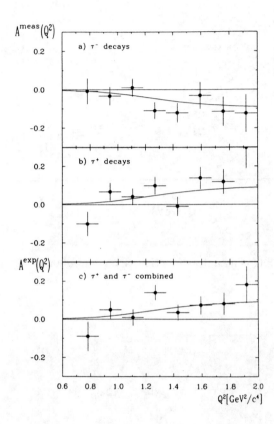

Figure 77: Measured asymmetry compared to model predictions assuming negative ν_τ helicity and $D/S = -0.11$

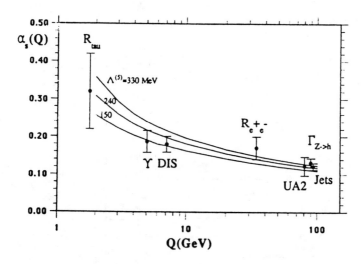

Figure 78: α_s-coupling constant as function of the four-momentum transfer Q^2

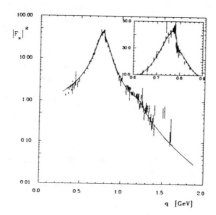

Figure 79: π-formfactor measured in $e^+e^- \to \pi^+\pi^-$ compared to different parametrizations

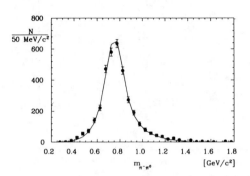

Figure 80: Comparison of the measured $\pi^-\pi^0$ invariant mass spectrum with the prediction of the CVC hypothesis using the parametrization of

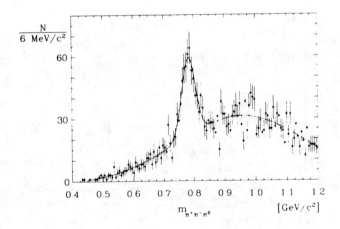

Figure 81: Measured invariant mass spectrum of the $\pi^-\pi^+\pi^0$-subsystem in $\tau^- \to \pi^-\pi^-\pi^+\pi^0\nu_\tau$ decay

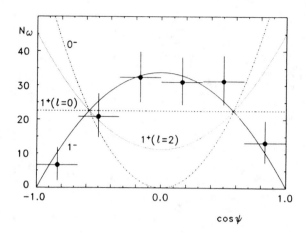

Figure 82: Angular distribution of the bachelor π^--meson in the $\omega\pi^-$ final state of the $\tau^- \to \pi^-\pi^-\pi^+\pi^0\nu_\tau$ decay

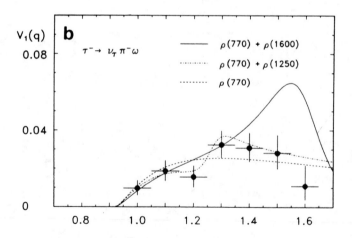

Figure 83: Measured spectral function of the $\pi^-\pi^-\pi^+\pi^0$-system compared to parametrizations of e^+e^--data using the CVC hypothesis

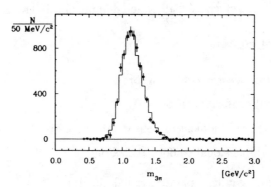

Figure 84: $\pi^-\pi^-\pi^+$ invariant mass spectrum of mesons from $\tau^- \to \pi^-\pi^-\pi^+\nu_\tau$ decay. The full line represents the fit of the ISGUR et al. model

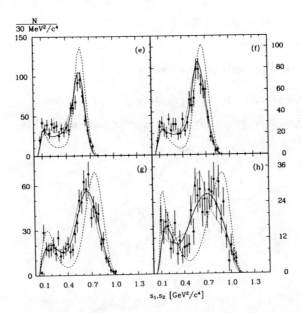

Figure 85: Dalitz plot of $\pi^-\pi^+$-subsystem for different invariant masses Q^2 of the $\pi^-\pi^-\pi^+$-system produced in the $\tau^- \to \pi^-\pi^-\pi^+\nu_\tau$ decay. Full line ISGUR model with $D/S = -0.11$, broken line pure D- and points pure S-wave

AN ELEMENTARY INTRODUCTION TO CONFORMAL FIELD THEORIES

J. Weyers
Institut de Physique Théorique,
Université Catholique de Louvain, Belgium

Table of contents

Lecture I §1 Conformal groups and algebras

§2 Global conformal theories

§3 The energy-momentum tensor and the basic Ward identities

§4 Two dimensional conformal theories

Lecture II §1 Strips, cylinders, planes : radial quantization

§2 Free bosons

§3 The central charge and the Virasoro algebra

§4 In- and -out-states

Lecture III §1 Highest weight states

§2 Kac determinants

§3 A few remarks on the Ising model

Lecture IV §1 Affine Kac-Moody algebras

§2 Enveloping Virasoro algebra or 2d current algebra

§3 The coset construction

References

Lecture I

§1 Conformal groups and algebras

Let R^n be a pseudo-euclidian space with flat metric $\eta_{\mu\nu} = diag\,(1,1,\cdots;-1,-1,\cdots)$ of signature $(p,q) - p+q = n-$ and $ds^2 = \eta_{\mu\nu}dx^\mu dx^\nu$ the square of the line element.

Under an arbitrary change of coordinates $x \to x'$, we have

$$ds^2 = g'_{\mu\nu}(x')dx'^\mu dx'^\nu = \eta_{\rho\sigma}dx^\rho dx^\sigma$$

and thus

$$\eta_{\mu\nu} \to g'_{\mu\nu}(x') = \frac{\partial x^\rho}{\partial x'^\mu}\frac{\partial x^\sigma}{\partial x'^\nu}\eta_{\rho\sigma} \tag{1}$$

By definition, the conformal group in n dimensions is the subgroup of these coordinate transformations which leaves the metric invariant up to a (local) scale change. In other words, the change of coordinates $x \to x'$ is an element of the conformal group iff

$$g'_{\mu\nu}(x') = W(x)\eta_{\mu\nu} \tag{2}$$

Obviously, for translations and Lorentz transformations

$$x \to x' = x + c$$
$$x \to x' = \Lambda x\,,\quad \Lambda \in SO(p,q)$$

we have $W(x) = 1$. Hence these coordinate changes belong to the conformal group. Similarly, for dilatations i.e.

$$x \to x' = kx$$

we find $W(x) = \frac{1}{k^2}$ and thus they too are conformal. Are there other conformal transformations ? To find out, consider an infinitesimal coordinate change

$$x^\mu \to x'^\mu = x^\mu + \varepsilon^\mu(x) \tag{3}$$

From Eq.(1) one obtains

$$g'_{\rho\sigma} = \eta_{\rho\sigma} - (\partial_\rho\varepsilon_\sigma + \partial_\sigma\varepsilon_\rho) \tag{4}$$

Hence Eq.(3) will be a conformal transformation provided

$$\partial_\rho\varepsilon_\sigma + \partial_\sigma\varepsilon_\rho = K(x)\eta_{\rho\sigma}$$

Tracing this relation with $\eta^{\rho\sigma}$ gives

$$\partial_\rho\varepsilon_\sigma + \partial_\sigma\varepsilon_\rho = \frac{2}{n}(\partial\cdot\varepsilon)\eta_{\rho\sigma} \tag{5}$$

as the condition for Eqs(3) to be conformal. When Eq.(5) can be solved, we have of course $W(x) = 1 - \frac{2}{n}(\partial \cdot \varepsilon)$.

For $n \geq 3$, a little algebra shows that in addition to the cases ε constant (tanslations) or linear in x (dilatations or Lorentz transformations), Eq.(5) admits solutions with ε quadratic in x but no others. Indeed $\varepsilon_\mu = a_\mu x^2 - 2x_\mu(a \cdot x)$ satisfies Eq.(5) and corresponds to a so-called "infinitesimal special conformal transformation".

In summary, we have thus learned that for $n \geq 3$, there are four types of infinitesimal conformal transformations:

(A) $\varepsilon^\mu = c^\mu$ (translation)

(B_1) $\varepsilon^\mu = \omega^\mu_\nu x^\nu$ (ωantisymmetric)   ("Lorentz" transformation)

(B_2) $\varepsilon^\mu = \lambda x^\mu$ (dilatations)

(C) $\varepsilon^\mu = a^\mu x^2 - 2x^\mu(a \cdot x)$ (special conformal transformations)

The algebra generated by these infinitesimal transformations (i.e. the algebra generated by $\varepsilon \cdot \partial$ for all ε^μ's given in A-C) has a total of $\frac{1}{2}(n+1)(n+2)$ generators and is locally isomorphic to the algebra of $SO(p+1, q+1)$. The counting of generators goes as follows: n translations, $n\frac{(n-1)}{2}$ Lorentz transformations, 1 dilatation and n special conformal transformations.

The corresponding finite conformal transformations read

$$x \to x' = x + c \tag{6a}$$

$$x \to x' = \Lambda x \qquad \text{with} \ \ W(x) = 1 \tag{6b}$$

$$x \to x' = kx \qquad \text{with} \ \ W(x) = \frac{1}{k^2} \tag{6c}$$

$$x \to x' = \frac{x + ax^2}{1 + 2a \cdot x + a^2 x^2} \quad \text{with} \ \ W(x) = (1 + 2a \cdot x + a^2 x^2)^2 \tag{6d}$$

Note that special conformal transformations can be written as $\frac{x'^\mu}{x'^2} = \frac{x^\mu}{x^2} + a^\mu$ i.e. as an "inversion" followed by a translation. Instead of the scale factor $W(x)$ defined in Eq.(2), we will find in the following that it is often more convenient to use the Jacobian, J, of the conformal transformations:

$$J = \frac{D(x')}{D(x)} = \det\left(\frac{\partial x'^\mu}{\partial x^\nu}\right) = [W(x)]^{-n/2} \tag{7}$$

For $n = 2$, on the other hand Eqs(5) become, in the euclidian case $\eta_{\mu\nu} = \delta_{\mu\nu} (\mu, \nu = 1, 2)$:

$$\partial_1 \varepsilon_1 = \partial_2 \varepsilon_2$$

$$\partial_1\varepsilon_2 = -\partial_2\varepsilon_1 \tag{8}$$

namely the Cauchy-Riemann equations which express the necessary and sufficient conditions for the complex function $\varepsilon(z) = \varepsilon_1 + i\varepsilon_2$ (resp. $\bar{\varepsilon}(\bar{z}) = \varepsilon_1 - i\varepsilon_2$) to be analytic in the complex variable $z = x_1 + ix_2$ (resp. $\bar{z} = x_1 - ix_2$). We have thus discovered that in two dimensions infinitesimal conformal transformations coïncide with analytical coordinate transformations. To determine the algebra of these infinitesimal transformations, we take for basis

$$z \to z' = z + \varepsilon_n(z)$$
$$\bar{z} \to \bar{z}' = \bar{z} + \bar{\varepsilon}_n(\bar{z})$$

where $\varepsilon_n(z) = -z^{n+1}$ and $\bar{\varepsilon}(\bar{z}) = -\bar{z}^{n+1}$

$$(n \in \mathbb{Z})$$

The infinitesimal generators are then, respectively,

$$\ell_n = -z^{n+1}\partial_z \quad \text{and} \quad \bar{\ell}_n = -\bar{z}^{n+1}\partial_{\bar{z}} \tag{9}$$

and one readily obtains

$$[\ell_m, \ell_n] = (m-n)\ell_{m+n} \quad [\bar{\ell}_m, \bar{\ell}_n] = (m-n)\bar{\ell}_{m+n} \quad [\ell_m, \bar{\ell}_n] = 0 \tag{10}$$

In other words, the local conformal algebra i.e. the algebra of infinitesimal conformal transformations is the direct sum $\mathcal{V} \oplus \bar{\mathcal{V}}$ of two isomorphic infinite dimensional algebras sometimes called the "classical Virasoro algebra" for reasons to be explained in Lecture 2.

However not every local holomorphic or antiholomorphic (i.e. analytical in z or $\bar{z}$) transformation can be "exponentiated" to give a "globally" well defined change of coordinates. To see this, the simplest is to go to the Riemann sphere $S^2 \equiv \mathbb{C} \cup \infty$ obtained by "compactifying" the complex plane i.e. adding the point at infinity.

To any point P on the sphere corresponds one and only one point of the complex plane and vice versa while the point corresponding to the north pole N of the sphere is the "point at infinity". Now, it is a theorem that the only transformations of the Riemann sphere which are everywhere well defined correspond to

$$z \to \frac{az+b}{cz+d} \qquad \bar{z} \to \frac{\bar{a}\bar{z}+\bar{b}}{\bar{c}\bar{z}+\bar{d}}$$

where $a, b, c, d \in \mathbb{C}$ and $ad - bc = 1$ i.e. the group $SL(2,\mathbb{C})/\mathbb{Z}_2 \approx SO(3,1)$ which is of course the "global conformal group" discussed at the beginning of this lecture but particulalized to the case $n = 2$. In this $SL(2, \mathbb{C})$ form, we find

—translations $\longleftrightarrow \begin{pmatrix} 1 & C \\ 0 & 1 \end{pmatrix}$ with $C = c_1 + ic_2$

—Lorentz transformations i.e. rotations $\longleftrightarrow \begin{pmatrix} e^{i\theta/2} & 0 \\ 0 & e^{-i\theta/2} \end{pmatrix}$

—dilatations $\longleftrightarrow \begin{pmatrix} k^{1/2} & 0 \\ 0 & k^{-1/2} \end{pmatrix}$

—special conformal transformations $\longleftrightarrow \begin{pmatrix} 1 & 0 \\ A & 1 \end{pmatrix}$ with $A = a_1 + ia_2$

Infinitesimally, this means that only those transformations which are generated by $\ell_{\mp 1}, \ell_0$, and their antiholomorphic partners $\bar{\ell}_{\pm 1}, \bar{\ell}_0$ are everywhere well defined on the Riemann sphere. The same result follows directly by requiring that the generator of an arbitrary holomorphic conformal transformation $g(z) = \sum_{n=-\infty}^{+\infty} a_n \ell_n = -\sum_{n=-\infty}^{+\infty} a_n z^{n+1} \partial_z$ be well defined at $z = 0$ and $z = \infty$:
at $z = 0$: $a_n \neq 0$ only for $n \geq -1$
at $z = \infty$: make the transformation $w = \frac{1}{z}$ to obtain $a_n \neq 0$ only for $n \leq 1$.
Of course similar results hold for the antiholomorphic generators as well.

In summary

(1) In any dimension, n, we have a *"global" conformal group* which is essentially $SO(p+1, q+1)$.

(2) For $n = 2$, the global conformal group can be written in termps of complex coordinates $(z = x_1 + ix_2, \bar{z} = x_1 - ix_2)$

$$z \to z' = \frac{az + b}{cz + d} \qquad \bar{z} \to \bar{z}' = \frac{\bar{a}\bar{z} + \bar{b}}{\bar{c}\bar{z} + \bar{d}'}$$

where $a, b, c, d \in \mathbb{C}$ and $ad - bc = 1$.

Under such a transformation,

$$ds^2 = dx_1^2 + dx_2^2 = dzd\bar{z} \to dz'd\bar{z}' = \left(\frac{1}{cz+d}\right)^2 \left(\frac{1}{\bar{c}\bar{z}+\bar{d}}\right)^2 dzd\bar{z}$$

(3) For $n = 2$ and *only* for $n = 2$, there is in addition to the global conformal group and its algebra, an infinite local conformal algebra generated by the ℓ_n's and $\bar{\ell}_n$'s of Eqs(9) and (10). (The "global algebra" is generated by $\ell_{\pm 1}, \ell_0$ and $\bar{\ell}_{\pm 1}, \bar{\ell}_0$).

§2 Global Conformal Theories

Whether in statistical models or in field theory, it is straightforward to define, for any dimen- sion n, a "global conformal theory". Roughly speaking, one requires all degrees of freedom (fields) to be expressible in terms of some fields with "simple" transformation properties under the global conformal group. N-point functions of these "simple" fields are then required to be covariant. More precisely

(1) the vacuum is invariant under global conformal transformations

(2) there are "quasi-primary" fields $A_k(x)$ which under conformal transformations $x \to x'$ transform as

$$A_k(x) \to A'_k(x) = J^{\Delta_k/n} A_k(x') \tag{11}$$

where J is the jacobian of the transformation given by Eq.(7) and Δ_k is the scale (or conformal) dimension of the field A_k. Infinitesimally, we have $x \to x' = x + \delta x$ with

$$\delta x_\mu = \delta c_\mu + [\delta\lambda \delta_{\mu\nu} + \delta\omega_{\mu\nu} + \delta a_\mu x_\nu - 2x_\mu \delta a_\nu] x_\nu \tag{12}$$

$(\mu, \nu = 1, \cdots n)$

$$A_k(x) \to A'_k(x) = A_k(x) + \delta A_k(x)$$

and

$$\delta A_k(x) = [\delta x \cdot \nabla + \frac{\Delta_k}{n}(\nabla \cdot \delta x)] A(x) \tag{13}$$

where $\nabla \cdot \delta x = n\delta\lambda - 2x \cdot \delta a$.

(3) N-point functions of quasi-primary fields are *covariant* under global conformal transformations, namely

$$< A_1(x_1) \cdots A_N(x_N) > = J_1^{\Delta_1/n} \cdots J_N^{\Delta_N/n} < A_1(x'_1) \cdots A_N(x'_N) > \tag{14}$$

Here, $J_\ell = |\frac{\partial x'}{\partial x}|_{x=x_\ell}$ ($\ell = 1, \cdots N$). The infinitesimal form of Eq.(14) is seen to be

$$\sum_{k=1}^{N} < A_1(x_1) \cdots \delta A_k(x_k) \cdots A_N(x_N) > = 0 \tag{15}$$

(4) Any other field in the theory can be expressed in terms of the quasi-primary fields.

Eqs (14) or (15) determine the 2 and 3-point functions of the quasi primary fields. For example, for a 2-point function, Eq.(14) reads

$$< A_1(x_1) A_2(x_2) = J_1^{\Delta_1/n} J_2^{\Delta_2/n} < A_1(x'_1) A_2(x'_2) > \equiv F^{(2)}(x_1, x_2)$$

Now, invariance under translations and rotations implies that

$$F^{(2)}(x_1, x_2) \equiv G(r_{12}) \quad \text{where} \quad r_{12} = |x_1 - x_2| = (g_{\mu\nu} x_1^\mu x_2^\nu)^{1/2}$$

Covariance under a pure scale transformation then requires $G(r_{12}) = \lambda^{\Delta_1 + \Delta_2} G(\lambda r_{12})$ i.e. $G(r_{12}) = \frac{C}{r_{12}^{\Delta_1 + \Delta_2}}$ with C an arbitrary constant. Finally covariance under special conformal transformations gives the constraint $G(r_{12}) = \frac{1}{D_1^{\Delta_1}} \cdot \frac{1}{D_2^{\Delta_2}} G\left(\frac{r_{12}}{D_1^{1/2} D_2^{1/2}}\right)$ where $D_i = (1 + 2a \cdot x_i + a^2 x_i^2)$ $i = 1, 2$.
Substituting the form of $G(r_{12})$ gives finally

$$F^{(2)}(x_1, x_2) \begin{cases} = \frac{C}{r_{12}^{2\Delta}} & \text{if } \Delta_1 = \Delta_2 = \Delta \\ = 0 & \text{if } \Delta_1 \neq \Delta_2 \end{cases} \tag{16}$$

Exactly the same reasoning leads to

$$F^{(3)}(x_1, x_2, x_3) = \frac{C_{123}}{r_{12}^{\Delta_1 + \Delta_2 - \Delta_3} r_{23}^{\Delta_2 + \Delta_3 - \Delta_1} r_{31}^{\Delta_3 + \Delta_1 - \Delta_2}} \tag{17}$$

Fortunately this is where it ends !! For 4-point (or higher n-point) functions, some arbitrariness creeps in (this is fortunate because otherwise there would be essentially one and only one conformal theory !). The origin of the arbitrariness comes from the so called "anharmonic ratio" of four points i.e.

$$\frac{r_{12} r_{34}}{r_{13} r_{24}} \tag{18}$$

It is straightforward to check that Eq.(18) is invariant under special conformal transformations as well as under translations, rotations and dilatations. Hence, a 4-point function can at best be determined up to an arbitrary function of the anharmonic ratio of the 4 points.

Before applying and extending these considerations to two dimensional theories let us push a bit further the analysis of the n dimensional case.

§3 The energy-momentum tensor and the basic Ward identities

Let us still imagine that we have an n dimensional conformal theory and let us ask the following question: what are the variations of N-point functions under an *arbitrary* infinitesimal local change of coordinates ? We will postulate the answer to be

$$\sum_{k=1}^{N} \langle A_1(x_1) \cdots \delta A_k(x_k) \cdots A_N(x_N) \rangle + \int d^n x \langle A_1(x_1) \cdots A_N(x_N) T_{\mu\nu}(x) \rangle \partial_\mu \delta x_\nu = 0 \tag{19}$$

where $T_{\mu\nu}(x)$ is the energy-momentum tensor. The second term in Eq.(19) expresses the fact that the energy-momentum tensor is the response of the theory to an infinitesimal change of metric. Its form is basically determined by the fact that is must be linear in δx_ν and that it must vanish when δx_ν = cte. A more precise justification of Eq.(19) is rather lenghty and will not be given here. Let me note that Drouffe and Itzykson, for example, take Eq.(19) as a definition of $T_{\mu\nu}(x)$ and I am rather sympathetic to that point of view !

Anyway let us assume Eq.(19) and examine the constraints which follow from global conformal transformations. As evident from Eq.(15), the second term in Eq.(19) must then vanish

- for translations, this is trivially the case

- for rotations, we must have $T_{\mu\nu} = T_{\nu\mu}$ (symmetric)

- for dilatations $\partial_\mu \delta x_\nu = \lambda \delta_{\mu\nu}$ and thus $T_{\mu\mu} = 0$ (traceless)

- finally for special conformal transformations

$$\partial_\mu \delta x_\nu = 2(x_\mu \delta a_\nu - x_\nu \delta a_\mu) - \delta_{\mu\nu} \delta a \cdot r$$

and hence there are no new constraints on $T_{\mu\nu}$.

If we assume furthermore that for an arbitrary variation of the coordinates (not necessarily conformal), the variations of the fields can still be written as in Eq.(13) - where of course δx_μ will *not* be given by Eq.(12) anymore - then we can integrate Eq.(19) by parts to obtain

$$\partial_\mu \langle A_1(x_1) \cdots A_N(x_N) T_{\mu\nu}(x) \rangle = \sum_{k=1}^{N} \left\{ \delta^{(n)}(x-x_k) \frac{\partial}{\partial (x_k)^\nu} - \frac{\Delta k}{n} \partial_\nu \delta^{(n)}(x-x_k) \right\}$$

$$\langle A_1(x_1) \cdots A_N(x_N) \rangle \qquad (20)$$

which expresses energy momentum conservation when "inserting" $T_{\mu\nu}(x)$ in an N-point function! One cannot go much further in the general case, so let us now look more closely at the case $n = 2$ and discuss what to make of the local conformal algebra !

§4 Two dimensional conformal theories

In terms of the euclidian variables (x_1, x_2) we have defined the complex variables

$$z = x_1 + ix_2 \qquad \bar{z} = x_1 - ix_2$$

and we have seen that $ds^2 = dx_1^2 + dx_2^2 = dz\, d\bar{z}$ i.e. the "complex" components of the metric are $g_{zz} = g_{\bar{z}\bar{z}} = 0$ and $g_{z\bar{z}} = g_{\bar{z}z} = 1/2$. Note also that $\partial_z =$

$\frac{1}{2}(\partial_{x_1} - i\partial_{x_2}) \equiv \partial$ and $\partial_{\bar{z}} = \frac{1}{2}(\partial_{x_1} + i\partial_{x_2}) \equiv \bar{\partial}$. Since $T_{\mu\nu}x^\mu x^\nu$ is a scalar, one easily obtains for a symmetric and traceless $T_{\mu\nu}$, its "complex" components:

$$T_{zz} = \frac{1}{2}(T_{11} - iT_{12}) \quad T_{\bar{z}\bar{z}} = \frac{1}{2}(T_{11} + iT_{12}) \quad T_{z\bar{z}} = T_{\bar{z}z} = 0 \qquad (21)$$

and energy momentum conservation now reads

$$\bar{\partial} T_{zz} = 0 = \partial T_{\bar{z}\bar{z}} \qquad (22)$$

Thus T_{zz} is a function of z only which we write $T_{zz} \equiv T(z)$ and for $T_{\bar{z}\bar{z}}$ which depends only on $\bar{z}$ we write $T_{\bar{z}\bar{z}} \equiv \bar{T}(\bar{z})$. Perhaps I should explain the notation to avoid misunderstandings: as we have already seen in Eq.(10) and now again in Eq.(22) we have "similar" structures arising in z and $\bar{z}$. It is then extremely convenient to regard z and $\bar{z}$ as independent variables (this corresponds to taking the euclidian coordinates (x_1, x_2) as complex numbers i.e. to go from $\mathbf{R}^2$ to $\mathbf{C}^2$): quantities depending on or related to z will be called the "holomorphic part" and the similar quantities which depend on or are related to $\bar{z}$ will be called the "antiholomorphic" part and will be denoted by the same symbol as their holomorphic partner but with a bar on top !

We have seen that a global conformal transformation reads in complex coordinates

$$z \to z' = f(z) = \frac{az+b}{cz+d}$$

$$\text{and } J = f'(z)\bar{f}'(\bar{z}) = \frac{1}{(cz+d)^2}\frac{1}{(\bar{c}\bar{z}+\bar{d})^2}$$

$$\bar{z} \to \bar{z}' = \bar{f}(\bar{z}) = \frac{\bar{a}\bar{z}+\bar{b}}{\bar{c}\bar{z}+\bar{d}}$$

A quasi-primary field has the transformation properties

$$A(z,\bar{z}) \to A'(z,\bar{z}) = (f'(z))^h (\bar{f}'(z))^{\bar{h}} A(f(z),\bar{f}(\bar{z})) \qquad (23)$$

where h and $\bar{h}$ are the conformal weights of the field A. I insist once more on the fact that h and $\bar{h}$ are *not* complex conjugates of each other. To make contact with the notation of §2, note that

$$\Delta_A = h + \bar{h} \qquad (24)$$

The quantity $s = h - \bar{h}$ is sometimes called the "spin" of the field $A(z,\bar{z})$. This misnomer comes from the fact that for a rotation

$$z \to e^{i\varphi}z$$

we have

$$A \to A' = e^{i(h-\bar{h})\varphi} \; A(e^{i\varphi}z, e^{-i\varphi}\bar{z})$$

The infinitesimal form of Eq.(23) reads

$$z \to f(z) = z + \varepsilon(z)$$
$$A \to (1+\partial\varepsilon)^h(1+\bar{\partial}\bar{\varepsilon})^{\bar{h}} A(z+\varepsilon(z), \bar{z}+\bar{\varepsilon}(\bar{z}))$$

i.e.
$$\delta A = \delta_z A + \delta_{\bar{z}} A$$

with
$$\delta_z A = \{\varepsilon(z)\partial + h\varepsilon'(z)\} A(z,\bar{z}) \tag{25a}$$
$$\delta_{\bar{z}} A = \{\bar{\varepsilon}(\bar{z})\bar{\partial} + \bar{h}\bar{\varepsilon}'(\bar{z})\} A(z,\bar{z}) \tag{25b}$$

We now have all the tools and the appropriate notation to define what we mean by a "local conformal theory". We are now allowing arbitrary analytic transformations of the coordinates z and $\bar{z}$.

Clearly we cannot ask for invariance of N-point functions anymore: that would be too much! (Check it !!). So what *do* we mean ?

First of all we want fields with "nice transformation properties": these are "primary fields" namely fields which are quasi-primary for arbitrary analytical transformations. Precisely, $A(z,\bar{z})$ is a primary field if for an arbitrary infinitesimal transformation

$$z \to z' = z + \varepsilon g(z) \qquad (\varepsilon = \text{infinitesimal real constant})$$
$$z \to \bar{z}' + \bar{z} + \varepsilon \bar{g}(\bar{z})$$

we have
$$\delta A(z,\bar{z}) = \varepsilon \left\{ g\partial + h\frac{\partial g}{\partial z} + \bar{g}\bar{\partial} + \bar{h}\frac{\partial \bar{g}}{\partial \bar{z}} \right\} A(z,\bar{z}) \tag{26}$$

Equivalently a primary field $A(z,\bar{z})$ of conformal weight $(h,\bar{h})$ is defined by

$$z \to f(z) \qquad \text{arbitrary !}$$
$$A(z,\bar{z}) \to (\partial f)^h (\bar{\partial}\bar{f})^{\bar{h}} A(f(z), \bar{f}(\bar{z})) \tag{27}$$

Next we plug Eq.(26) into Eq.(19), noting that

$$T_{\mu\nu}\partial_\mu x_\nu = 2\varepsilon\{T(z)\bar{\partial}g(z) + \bar{T}(\bar{z})\partial\bar{g}\} \tag{28}$$

From here, the quickest way to proceed is to take $g(z) = \frac{1}{w-z}$ and to use the identity

$$\bar{\partial}\frac{1}{z-w} = \delta^{(2)}(z-w)$$

358 Introduction to Conformal Field Theories

This leads at once to the fundamental identities of a local conformal theory

$$\sum_{k=1}^{N} \left[\frac{h_k}{(z-z_k)^2} + \frac{1}{z-z_k} \partial_{z_k} \right] \langle A_1(z_1, \bar{z}_1) \cdots A_N(z_N, \bar{z}_N) \rangle$$
$$= \langle T(z) A_1(z_1, \bar{z}_1) \cdots A_N(z_N, \bar{z}_N) \rangle \qquad (29a)$$

and

$$\sum_{k=1}^{N} \left[\frac{\bar{h}_k}{(\bar{z}-\bar{z}_k)^2} + \frac{1}{\bar{z}-\bar{z}_k} \partial_{\bar{z}_k} \right] \langle A_1(z_1, \bar{z}_1) \cdots A_N(z_N, \bar{z}_N) \rangle$$
$$= \langle \bar{T}(\bar{z}) A_1(z_1, \bar{z}_1) \cdots A_N(z_N, \bar{z}_N) \rangle \qquad (29b)$$

Alternatively, using Eq.(28) in Eq.(19), one can integrate the right hand side by parts and use the divergence theorem.

Eqs.(29a) can also be written in integral form i.e.

$$\sum_{k=1}^{N} [h_k g'(z_k) + g(z_k) \partial_{z_k}] \langle A_1(z_1, \bar{z}_1) \cdots A_N(z_N, \bar{z}_N) \rangle =$$
$$= \oint_C \frac{dz}{2\pi i} g(z) \langle T(z) A_1(z, \bar{z}) \cdots A_N(z_N, \bar{z}_N) \rangle \qquad (30a)$$

where the contour C encloses all z_i. One proceeds similarly with Eq.(29b). In the following, I will stop repeating that whatever manipulation is made on the holomorphic variables, it can be performed mutatis mutandis for the antiholomorphic ones.

Eq.(30a) can be read as follows: "inside N-point functions" the conformal transformation properties of a primary field $A(z\bar{z})$ can be expressed as

$$\delta_z A = \oint \frac{dw}{2\pi i} \varepsilon(w) T(w) A(z, \bar{z}) \qquad (31)$$

where the contour encloses the point z. In other words, the short distance expansion of primary fields with the energy momentum tensor encodes all their conformal properties !!

Let me summarize the main results obtained so far for 2 dimensional conformal theories:

(G) global conformal invariance leads to the following differential equations for N point functions of quasi primary fields

$$(G_1) \sum_{k=1}^{N} \partial_{z_k} \langle A_1 \cdots A_N \rangle = 0 \qquad \text{(translations)} \qquad (32)$$

$$(G_2) \sum_{k=1}^{N} (z_k \partial_{z_k} + h_k) \langle A_1 \cdots A_N \rangle = 0 \quad \text{(dilatations and rotations)} \qquad (33)$$

$$(G_3) \sum_{k=1}^{N} (z_k^2 \partial_{z_k} + 2h_k z_k) \langle A_1 \cdots A_N \rangle = 0 \qquad \text{(special conformal)} \qquad (34)$$

(L) for a local theory we have in addition

$$(L) \sum_{k=1}^{N} \left\{ \frac{h_k}{(z-z_k)^2} + \frac{1}{z-z_k} \partial_{z_k} \right\} \langle A_1 \cdots A_N \rangle = \langle T(z) A_1 \cdots A_N \rangle \quad (35)$$

In view of these extra constraints it is perhaps not too surprising that, at least in some cases, a local conformal theory can be completely solved i.e. all N-point functions are explicitly calculable!

Lecture II

§1 Strips, cylinders, planes : radial quantization

Let us start with Euclidian "space" and "time" coordinates σ^1 and σ^0. If the space were Minkowskian $\sigma^1 \pm \sigma^0$ would correspond to left and right movers; in Euclidian space we have instead $\zeta, \bar\zeta = \sigma^1 \pm i\sigma^0$ i.e. holomorphic or antiholomorphic dependence.

Compactifying the space coordinate, $\sigma^1 \equiv \sigma^1 + 2\pi$ (i.e. we identify σ^1 and $\sigma^1 + 2\pi$) gives us a cylinder, which we map (conformally) to the

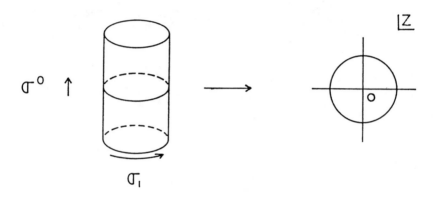

z plane

$$\zeta \to z = \exp\zeta$$

The infinite past (resp future) on the cylinder is mapped on the point $z = 0$ (resp ∞) of the z-plane while an equal time surface σ^0=cte goes over into a circle around the origin in the z-plane. Note also that time reversal $\sigma^0 \to -\sigma^0$ now reads $z \to \frac{1}{\bar z}$ and that the hamiltonian on the cylinder becomes the dilatation generator on the z plane. The analogue of canonical (equal time) commutator quantization on the cylinder is "radial quantization" on the z plane: this is an extremely convenient procedure for it allows optimal use of the full battery of complex function theory, contour integrals etc $\cdots$

To motivate a bit the procedure, recall that in the case of an exact (continuous) symmetry, we have, by Noether's theorem, a conserved current $j^\mu : \partial_\mu j^\mu = 0$. The corresponding charge $Q = \int d^{n-1}x j_0(x)$ is the generator for infinitesimal variations of any field $A(x)$ under (infinitesimal) symmetry operations

$$\delta_\epsilon A = \epsilon[Q, A]_{E.T.}$$

In two dimensions for local coordinate transformations which are all generated by the energy-momentum tensor, we thus expect for the charges (formal) expressions of the form

$$Q = \frac{1}{2\pi i} \oint dz T(z)\epsilon(z) + d\bar{z}\, \bar{T}(\bar{z})\, \bar{\epsilon}(\bar{z}) \tag{1}$$

and we want a "contour prescription" which corresponds to the cylinder equal time commutator

$$\left[\int d^{n-1}x j_0(x), A(y, y_0) \right]_{x_0=y_0}$$

The prescription is quite simple: define the radial ordering R on the z plane

$$\begin{aligned} R\, A(z)B(w) &= A(z)\, B(w) \text{ for } |z| > |w| \\ &= +B(w)\, A(z) \text{ for } |w| > |z| \end{aligned} \quad \text{(for "bosonic" fields)}$$

then,

$$\begin{aligned} \left[\int d^{n-1}x j_0(x), A \right]_{E\cdot T, CYL} &\to \frac{1}{2\pi i} \oint_w dz\, \epsilon(z) RT(z) A(w, \bar{w}) \\ &+ \frac{1}{2\pi i} \oint_{\bar{w}} d\bar{z}\, \bar{\epsilon}(\bar{z}) R\bar{T}(\bar{z}) A(w, \bar{w}) \end{aligned} \tag{2}$$

where the z contour is thightly woven around the point w. The above prescription corresponds graphically to

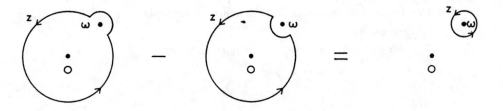

From Eqs(I-26) and (1) we have for primary fields

$$\delta_{\epsilon, \bar{\epsilon}} A(w, \bar{w}) = h\partial\epsilon(w) A(w, \bar{w}) + \epsilon(w) \partial A(w, \bar{w})$$

$$+\bar{h}\,\bar{\partial}\,\bar{\epsilon}(\bar{w})\,A(w,\bar{w}) + \bar{\epsilon}(\bar{w})\bar{\partial}A(w,\bar{w})$$

and $\delta_{\epsilon,\bar{\epsilon}}A(w,\bar{w}) = \frac{1}{2\pi i}\oint_w dz\, \epsilon(z)\, R(T(z)A(w,\bar{w}) +$ antiholomorphic part. Thus

$$R(T(z)A(w,\bar{w}) = \frac{h}{(z-w)^2}A(w,\bar{w}) + \frac{1}{z-w}\partial_w A(w,\bar{w}) + \cdots \quad (3)$$

and a similar formula holds for the antiholomorphic part.

Thus as was hinted in lecture I, the short distance operator product expansion of the energy momentum tensor with a primary field encodes in a particularly simple way the conformal transformation properties of this field.

To simplify the notation we will conform to the usual practise of dropping the symbol R and of giving only the holomorphic part of formulas. We thus write for a primary field

$$T(z)A(w) = \frac{h}{(z-w)^2}A(w) + \frac{1}{z-w}\mathcal{L}_{-1}A(w) + \sum_{n=0}^{\infty}(z-w)^n \mathcal{L}_{-n-2}A \quad (4)$$

The fields $\mathcal{L}_{-n}A (n>0)$ are called the *descendants* of A. Writing $\mathcal{L}_0 A = hA$ unifies the notation for all terms in the short distance expansion and the set of fields $\{\mathcal{L}_n A\}_{n=0,1\ldots}$ is called a *conformal family*. Note that $\mathcal{L}_{-1}A(w) = \partial A(w)$.

In the z-plane contour formalism, we have

$$\mathcal{L}_{-n}A(w) = \oint_w \frac{dz}{2\pi i}(z-w)^{-n+1}T(z)A(w) \quad (5)$$

(remember our conventions !).

What are the conformal transformation properties of descendant fields ? To answer the question we need to consider $T(w_1)T(w_2)$ for $w_1 \to w_2$ and the easiest way to proceed is to look first at a simple example.

§2 Free Bosons

The action reads, in a convenient normalization:

$$S = \int \frac{d^2x}{2\pi}\frac{1}{2}\sum_{\mu=1,2}(\partial_\mu \Phi)^2 \quad (6)$$

and solutions of the equations of motion correspond to left and right movers respectively. Writing, in the $z,\bar{z}$ coordinates

$$\Phi = \frac{1}{\sqrt{z}}(\phi(z) + \bar{\phi}(\bar{z})) \quad (7)$$

we easily find from $T_{\mu\nu} = -\partial_\mu\Phi\partial_\nu\Phi + \frac{1}{2}(\partial\Phi)^2\delta_{\mu\nu}$ that

$$T = T_{zz} =: -\frac{1}{2}(\partial\phi(z))^2 : \quad (8)$$

and
$$\bar{T} = T_{\bar{z}\bar{z}} =: -\frac{1}{2}(\bar{\partial}\bar{\phi})^2 :$$

and a little algebra gives for the holomorphic part of the propagator

$$\langle \phi(z)\phi(w) \rangle = -\ln(z-w) \tag{9}$$

Clearly $\phi(z)$ is NOT a conformal field. From Eq.(9) we infer that

$$\langle \partial\phi(z)\partial\phi(w) \rangle = -\frac{1}{(z-w)^2} \tag{10}$$

and there is thus the possibility that $\partial\phi(z)$ is a primary field of weight $(1,0)$. Let us check in detail that this is indeed the case. To perform the calculation let me remind you that the normal order prescription is defined by, e.g.

$$\begin{aligned} T(w) &= -\frac{1}{2} : (\partial\phi(w))^2 : \\ &= -\frac{1}{2}\lim_{z\to w}\left[\partial\phi(z)\partial\phi(w) + \frac{1}{(z-w)^2}\right] \end{aligned} \tag{11}$$

Using Wick's theorem we easily compute the singular part of $T(z)\partial\phi(w)$:

$$\begin{aligned} T(z)\partial\phi(w) &= -\frac{1}{2}: (\partial\phi(z))^2 : \partial\phi(w) = 2(-\frac{1}{2})\langle\partial\phi(z)\partial\phi(w)\rangle\partial\phi(z) + \cdots \\ &= \frac{1}{(z-w)^2}\partial\phi(z) + \cdots = \frac{1}{(z-w)^2}\{\partial\phi(w) + (z-w)\partial^2\phi(w)\cdots\} \\ &= \frac{1}{(z-w)^2}\partial\phi(w) + \frac{1}{z-w}\partial(\partial\phi) + \cdots \end{aligned} \tag{12}$$

Hence $\partial\phi(w)$ is indeed a primary field of weight $(h=1,0)$.
Clearly
$$\delta_\epsilon \partial\phi = \partial\epsilon\partial\phi + \epsilon\partial(\partial\phi) \tag{13}$$

and one checks from Eq.(12) that, as expected,

$$\begin{aligned} \oint_w \frac{dz}{2\pi i}\epsilon(z)T(z)\partial\phi(w) &= \oint_w \frac{dz}{2\pi i}\epsilon(z)\left\{\frac{\partial\phi(w)}{(z-w)^2} + \frac{1}{z-w}\partial^2\phi + \cdots\right\} \\ &= \partial\epsilon\partial\phi + \epsilon\partial^2\phi. \end{aligned}$$

With similar techniques it is also straightforward to check that

$$E(\alpha,\phi) =: \exp i\alpha\phi(w) : \tag{14}$$

is a primary field of weight $h = \frac{\alpha^2}{2}$, i.e.

$$T(z)E(\alpha,\phi) = \frac{\alpha^2}{2}\frac{1}{(z-w)^2}E(\alpha,\phi) + \frac{1}{z-w}\partial_w E(\alpha,\phi) + \cdots$$

It is very instructive to look at the short distance expansion of the energy momentum tensor with itself:

$$\begin{aligned}
T(z)T(w) &= \left(-\frac{1}{2}\right)^2 :\partial\phi(z)\partial\phi(z)::\partial\phi(w)\partial\phi(w): \\
&= \left(-\frac{1}{2}\right)^2 \{2(\langle\partial\phi(z)\partial\phi(w)\rangle)^2 + 4\partial\phi(z)\partial\phi(w)\langle\partial\phi(z)\partial\phi(w)\rangle \\
&\rightarrow \frac{1/2}{(z-w)^4} + \frac{2}{(z-w)^2}\left\{-\frac{1}{2}(\partial\phi(w))^2\right\} + \frac{1}{z-w}\partial\left[-\frac{1}{2}(\partial\phi(w))^2\right] + \cdots\} \\
&\rightarrow \frac{1/2}{(z-w)^4} + \frac{2}{(z-w)^2}T(w) + \frac{1}{z-w}\partial T(w) + \cdots \quad (15)
\end{aligned}$$

because of the first term (the so called anomaly or "central charge") we thus find out that $T(w)$ is *not* a primary field !

§3 The central charge and the Virasoro Algebra

From the simple example of the previous paragraph, you will now readily accept that, in general, one will find in any conformal theory that

$$T(z)T(w) = \frac{c/2}{(z-w)^4} + \frac{2}{(z-w)^2}T(w) + \frac{1}{z-w}\partial T(w) + \cdots \quad (16)$$

The coefficient c of the first term is known as the central charge of the theory: its value depends on the particular theory one is considering but it *never* vanishes in a quantum theory. Eq.(15) tells us that $c = 1$ in a free boson theory. One also finds that for a free fermion $c = 1/2$.

Anyway, the important point is that the first term of Eq.(16) is always there. Had this term been absent, Eq.(16) would simply have meant that $T(w)$ has conformal weight $h = 2$.

What is the effect of the central charge on the conformal transformation properties of $T(z)$ or, perhaps more to the point, *what* are the conformal transformation properties of $T(z)$? Let us first look at the vacuum expectation value of $T(z)$: taking, as usual,

$$\langle T(z)\rangle = 0 \quad (17)$$

what happens under an infinitesimal local conformal transformation ?

$$\begin{aligned} z &\rightarrow z' = z + \varepsilon g(z) \\ T &\rightarrow T' = T + \delta T \end{aligned}$$

From (Eq.I-30a) one easily gets

$$\langle \delta T(z)\rangle = \varepsilon \oint_z \frac{dz'}{2\pi i} g(z')\langle T(z')T(z)\rangle \quad (18)$$

Since we know from Eq.(16) that

$$\langle T(z)T(w)\rangle = \frac{c/2}{(z-w)^4}$$

we find

$$\langle \delta T(z)\rangle = \varepsilon c/2 \oint \frac{dz'}{2\pi i} \frac{g(z')}{(z'-z)^4} = \varepsilon \frac{c}{12} g'''(z) \ ! \qquad (19)$$

Remember that for *global* infinitesimal conformal transformations, $g(z)$ is at most quadratic in z. In this case the RHS of Eq.(19) does indeed vanish: the vacuum is invariant under global conformal transformations ! It is not invariant under local transformations but we never asked that it should be !

From Eq.(19), we infer the infinitesimal transformation properties of $T(z)$, namely

$$\delta_\varepsilon T(z) = \varepsilon[2g'(z) + g(z)\partial_z]T(z) + \varepsilon \frac{c}{12} g'''(z) \qquad (20)$$

It is by no means trivial to integrate this. But it can be done and one finds that under $z \to f(z)$

$$T(z) \to (\partial f)^2 T(f(z)) + \frac{c}{12}\{f,z\} \qquad (21)$$

where $\{f,z\}$ is known as the "Schwartzian derivative" and given by

$$\{f,z\} = \frac{f'''}{f'} - 3/2 \left(\frac{f''}{f'}\right)^2 \qquad (22)$$

I will not attempt to prove Eq.(21) here. Nevertheless, I feel that it is instructive to check the formula in the free boson case. Writing Eq.(11) as

$$T(z) = \lim_{z_1,z_2 \to z} -\frac{1}{2}\partial_{z_1}\phi(z_1)\partial_{z_2}\phi(z_2) - \frac{1}{2}\partial_{z_1}\partial_{z_2}\ln z_{12} \qquad (23)$$

we have, under a finite transformation

$$z \to u = f(z)$$
$$\phi(z) \to \tilde\phi(u)$$

$$T \to \lim_{z_1,z_2 \to z}\left\{f'(z_1)f'(z_2)\left[-\frac{1}{2}\partial_{u_1}\tilde\varphi\partial_{u_2}\tilde\varphi - \frac{1}{2}\partial_{u_1}\partial_{u_2}\ln u_{12}\right] + \frac{1}{2}\partial_{z_1}\partial_{z_2}\ln\frac{u_{12}}{z_{12}}\right\} \qquad (24)$$

where I have added and subtracted the term $\frac{1}{2}f'(z_1)f'(z_2)\partial_{u_1}\partial_{u_2}\ln u_{12} = \frac{1}{2}\partial_{z_1}\partial_{z_2}\ln u_{12}$. The last term of Eq.(24) gives the finite effect of the anomaly. Putting $z_1 = z + \delta$ and $z_2 = z$, a little algebra then confirms Eq.(21) for $c = 1$.

A rather nice intuitive picture of the anomaly follows from the "cylinder to z plane" conformal transformation $w \to z = e^w$. Clearly, in this case $\{z, w\} = -\frac{1}{2}$ and thus

$$T_{CYL}(w) = \left(\frac{\partial z}{\partial w}\right)^2 T(z) - \frac{c}{24} \qquad (25)$$

and the anomaly then appears as a Casimir effect i.e. a shift in the vacuum energy due to finite size effects.

Let me repeat once more the main point of this discussion: the central charge is a "quantum effect" i.e. it comes from the necessary ordering prescription in a quantum field theory and thus never vanishes (except of course for a "classical" theory).

It is now most convenient to expand $T(z)$ in a Laurent series

$$T(z) = \sum_{n \in \mathbb{Z}} z^{-n-2} L_n \qquad (26)$$

(note that under a scale transformation $z \to \frac{z}{k}, T \to k^2 T(\frac{z}{k})$ and $L_{-n} \to k^n L_{-n}$).

Equivalently we can formally define the modes of $T(z)$ by a contour integral

$$L_n = \oint \frac{dz}{2\pi i} z^{n+1} T(z)$$

and it is now straightforward to determine the algebra generated by these modes, remembering that

$$\left[\oint dz A(z), \oint dw B(w)\right] = \oint dw \oint_w dz R(A(z)B(w))$$

Thus,

$$[L_n, L_m] = \oint \frac{dw}{2\pi i} \oint_w \frac{dz}{2\pi i} z^{n+1} w^{m+1} T(z) T(w)$$
$$= \oint \frac{dw}{2\pi i} \oint_w \frac{dz}{2\pi i} z^{n+1} w^{m+1} \left\{ \frac{c/2}{(z-w)^4} + 2\frac{T(w)}{(z-w)^2} + \frac{\partial T(w)}{z-w} + \cdots \right\}$$

and since

$$\oint \frac{dz}{2\pi i} f(z) \frac{1}{(z-w)^k} = \frac{1}{(k-1)!} \partial_z^{k-1} f(z)|_{z=w}$$

we readily obtain

$$[L_n, L_m] = \oint \frac{dw}{2\pi i} \{\frac{c}{12} n(n^2 - 1) w^{n-2} w^{m+1} + 2(n+1) w^n w^{m+1} T(w)$$
$$+ w^{n+1} w^{m+1} \partial T(w) + \text{regular terms}\}$$

Integrating by parts
$$w^{m+n+2}\partial T = \partial\{w^{m+n+2}T\} - (m+n+2)w^{m+n+1}T(w),$$
we have, finally
$$\begin{aligned}[L_n, L_m] &= \oint \frac{dw}{2\pi i} \left\{\frac{c}{12}n(n^2-1)w^{n+m-1} + (n-m)w^{n+m+1}T(w)\right\} \\ &= \frac{c}{12}n(n^2-1)\delta_{n+m,0} + (n-m)L_{n-m} \end{aligned} \quad (27)$$
which is the famous Virasoro algebra, including the central charge term c. Note that the central charge does not affect the closed subalgebra consisting of L_0 and $L_{\pm 1}$.

Note also that repeating the whole procedure for the antiholomorphic part of the energy momentum tensor will lead to a second Virasoro algebra i.e.
$$\left[\bar{L}_n, \bar{L}_m\right] = \frac{\bar{c}}{12}n(n^2-1)\delta_{n+m;0} + (n-m)\,\bar{L}_{n+m} \quad (28)$$
with $\bar{c}$ the antiholomorphic central charge.

These results are truly fundamental: the algebraic structure underlying any (local) conformal theory consists of two commuting copies of an infinite dimensional algebra, the Virasoro algebra. The values of c and $\bar{c}$ distinguish one theory from another. Finally, in all cases, the global conformal group generated by $L_0, L_{\pm 1}$ and $\bar{L}_0, \bar{L}_{\pm 1}$ remains an exact symmetry irrespective of the values of c and $\bar{c}$.

§4 In- and Out-states

I now want to explore a little bit the structure of unitary representations of the infinite dimensional Virasoro algebra. As in the finite dimensional case it is much simpler to work with "states" than with operators. There are some subtleties in defining in and out states which I now describe briefly (and a bit sloppily). For any primary or quasi primary field $A(\sigma^0, \sigma^1)$ in a euclidian theory defined on a cylinder, there is of course no problem in defining the corresponding in-state:
$$|A_{\text{in}}> = \lim_{\sigma^0 \to -\infty} A(\sigma^0, \sigma^1)|0> \quad (29)$$
Since, as we have seen $\sigma^0 \to -\infty$ corresponds to $z, \bar{z} \to 0$ in the z plane we can write
$$|A_{\text{in}}> = \lim_{z,\bar{z}\to 0} A(z,\bar{z})|0> \quad (30)$$
It would appear naively that a similar procedure with $\sigma^0 \to +\infty$ or correspondingly for $z, \bar{z} \to \infty$ would define an out-state. This is not quite true however. The right answer follows from the requirement
$$\langle A_{\text{out}}|A_{\text{in}}\rangle = 1 \quad (31)$$

For a primary or quasi primary field, we have by definition that

$$\langle 0|A(z,\bar{z})A(w,\bar{w})|0\rangle = \frac{1}{(z-w)^{2h}}\frac{1}{(\bar{z}-\bar{w})^{2\bar{h}}}$$

Taking the limit $w, \bar{w} \to 0$, we get from Eq.(31)

$$\langle A_{\text{out}}| = \lim_{z,\bar{z}\to\infty} \langle 0|A(z,\bar{z})z^{2h}\bar{z}^{2\bar{h}} \tag{32}$$

Clearly, in a unitary theory we also want $\langle A_{\text{out}}| = (|A_{\text{in}}>)^\dagger$. Implementing this definition of "adjoint" on Eq.(32) leads to a somewhat peculiar looking relation for the "adjoint" of an holomorphic operator

$$O^\dagger(z) = O(\frac{1}{\bar{z}})\frac{1}{\bar{z}^{2h_0}} \tag{33}$$

For the energy momentum tensor,

$$T^\dagger(z) = \sum_{n\in \mathbf{Z}} \frac{L_n^\dagger}{\bar{z}^{n+2}} \quad \text{(by definition)}$$

and $T(\frac{1}{\bar{z}})\frac{1}{\bar{z}^4} = \sum L_n \bar{z}^{n-2} = \sum_{n\in\mathbf{Z}} \frac{L_n}{\bar{z}^{-n+2}}$. We then obtain from Eq.(33) that

$$L_n^\dagger = L_{-n} \tag{34}$$

which should be viewed as expressing the hermiticity of the energy momentum tensor in a unitary conformal field theory.

Note also that regularity of $T(z)|0>$ when $z\to 0$ leads to

$$L_m|0> \equiv 0 \quad m \geq -1 \tag{35}$$

Global conformal invariance requires, as we have seen repeatedly, that

$$L_{-1}|0> = L_0|0> = L_1|0> \equiv 0$$

In a local theory we get for "free", so to speak, an infinite set of extra conditions, namely $L_n|0>= 0$ for $n > 1$. Taking the adjoint of Eq.(35) gives

$$<0|L_m^\dagger \equiv 0 \quad m \geq -1$$

or, using Eq.(34),

$$<0|L_m = 0 \quad m \leq 1$$

It is perhaps intuitively useful to view the modes $L_n(n \in \mathbf{Z})$ of the energy momentum tensor as splitting into "creators", when n is negative, and "annihilators" (n positive). The Virasoro algebra then appears as some complicated coupled oscillator algebra.

Lecture III

§1 Highest weight states

Having done all the preliminary work in the last lecture, we are now ready to investigate representations of the Virasoro algebra. With a primary field $A(z,\bar{z})$ of conformal weight $(h,\bar{h})$ we construct the in-state

$$|h,\bar{h}> = A(0,0)|0>$$

From the short distance expansion of, say, the holomorphic part of the energy momentum tensor with $A(z,\bar{z})$ one easily finds

$$[L_n, A(z,\bar{z})] = \oint_z \frac{dw}{2\pi i} w^{n+1} T(w) A(z,\bar{z}) \qquad (1)$$

$$= h(n+1)z^n A(z,\bar{z}) + z^{n+1} \partial A(z,\bar{z})$$

Consequently, $\qquad [L_n, A(0,\bar{z})] = 0 \quad n > 0$

and
$$L_0|h,\bar{h}> = h|h,\bar{h}> \qquad (2)$$

$$L_n|h,\bar{h}> = 0 \quad n > 0 \qquad (3)$$

Similar equations are obtained from the antiholomorphic part $\bar{T}(\bar{z})$ for $\bar{L}_0$ and $\bar{L}_n$. Remember that $L_0 \pm \bar{L}_0$ are the generators of dilatations and rotations. Thus $h \pm \bar{h}$ are identified with the scaling dimension and the "spin" of the state. As done previously we focus on the holomorphic part of fields and define a "highest weight state" by the equations

$$L_0|h> = h|h> \qquad L_n|h> = 0 \quad n > 0 \qquad (4)$$

Equivalently, we have for the out state $<h|$,

$$<h|L_0 = h<h| \quad \text{and} \quad <h|L_n = 0 \quad n < 0 \qquad (5)$$

From the highest weight (in) state $|h>$, the only non trivial states are obtained by hitting it with a string of negative modes i.e.

$$L_{-n_1} L_{-n_2} \cdots L_{-n_k} |h> \qquad (6)$$

Collectively, such states are called the "descendants" of $|h>$ and, clearly, these are the states which we must "organize" to describe representations of the Virasoro algebra. But let met first make some comments on modes and descendants:

As for $T(z)$, we may expand the holomorphic part of a primary field in modes i.e.
$$A(z) = \sum A_n z^{-n-h} \tag{7}$$
and one easily obtain from Eq.(1)
$$[L_n, A_m] = \{n(h-1) - m\} A_{m+n} \tag{8}$$
and
$$|h> = A_{-h}|0> \tag{9}$$
From Eq.(II-4), one has for descendant fields
$$\mathcal{L}_{-n} A(w, \bar{w}) = \oint \frac{dz}{2\pi i} \frac{1}{(z-w)^{n-1}} T(z) A(w, \bar{w}) \tag{10}$$
Hence
$$L_{-n}|h> = L_{-n}(A(0)|0>) = (\mathcal{L}_{-n}A)(0)|0> \tag{11}$$
and thus descendant states are created by descendant fields. By the way Eq.(10), applied to the unit operator gives
$$\mathcal{L}_{-2}\mathbb{1} = \int \frac{dz}{2\pi i} \frac{1}{z-w} T(z) = T(w) \tag{12}$$
meaning that the energy-momentum tensor is a descendant of the unit operator.

Let us now come back to our highest weight state $|h>$. From $|h>$ we generate an infinite dimensional vector space by acting with arbitrary linear combinations of L_{-n_k}'s on $|h>$. This goes under the sweet name of a "Verma module" and clearly provides a representation of the Virasoro algebra.

Since
$$L_0 L_{-n}|h> = (n+h) L_{-n}|h> \tag{13}$$
we can organize the Verma module in subspaces corresponding to the eigenvalue $h + N$ of L_0. $N \geq 0$ is called the level. It is easily seen, from Eq.(13) that
$$L_0 L_{-n_1} L_{-n_2} \cdots L_{-n_k}|h> = (n_1 + n_2 + \cdots n_k + h) L_{-n_1} L_{-n_2} \cdots L_{-n_k}|h> \tag{14}$$

So at a given level N we have in the Verma module as many states as there are "partitions" $p(N)$ of N. We order the partitions $(n_1, n_2, \cdots n_k)$ according to $n_1 \geq n_2 \geq n_3 \cdots \geq n_k \geq 0$.

The "Verma module" then looks as follows

level	dim.	states
0	h	$\|h>$
1	$h+1$	$L_{-1}\|h>$
2	$h+2$	$L_{-2}\|h>$; $L_{-1}^2\|h>$
3	$h+2$	$L_{-3}\|h>$; $L_{-2}L_{-1}\|h>$; $L_{-1}^3\|h>$
⋮		
n	$h+n$	$p(n)$ states.

To make things as clear as possible, the "ordering" of partitions simply gets rid of states like $L_{-1}L_{-2}|h>$ (at level 3): this state is a linear combination of $L_{-3}|h>$ and $L_{-2}L_{-1}|h>$ which have been kept in the module. In case you are interested, there is a marvelous formula due to Euler for the partitions $p(n)$. It reads

$$\sum_{n=0}^{\infty} p(n) q^n = \frac{1}{\prod_{m=1}^{\infty}(1-q^m)} \tag{15}$$

where, conventionally, $p(0) = p(1) = 1$.

In general, a Verma module does not span an irreducible representation of the Virasoro algebra. The reason is that there can be *null vectors*. As we will see these null vectors appear at specific values of h and c (the central charge). To construct an irreducible representation one has to remove from the Verma module (quotientize is the technical term) all null states and their descendants (which, of course, are also null vectors).

The question is thus when are there null vectors ?
At level 1 ? If $L_{-1}|h> = 0$, applying L_1 and using the Virasoro algebra relations, this implies $h = 0$ and thus the highest weight state is the vacuum.
At level 2, the question becomes more interesting.
Let us set $\quad L_{-2}|h> + x L_{-1}^2 |h> = 0 \tag{16}$

Apply L_1 to Eq.(16) to obtain (remember Eq.(4))

$$[L_1, L_{-2}]|h> + x[L_1, L_{-1}^2]|h> = 0 \tag{17}$$

then use Eqs(II-27) to get

$$[3 + 2xh + 2x(h+1)]L_{-1}|h> = 0 \tag{18}$$

and since we assume $h \neq 0$, we can solve for x

$$x = \frac{-3}{2(2h+1)} \tag{19}$$

Now apply L_2 and follow the same procedure:

$$[L_2, L_{-2}]|h> + x[L_2, L_{-1}^2]|h>= 0 \tag{20}$$

$$(4h + \frac{c}{2} + 6hx)|h>= 0 \tag{21}$$

$$\text{or} \quad c = \frac{2h(5-8h)}{2h+1} \tag{22}$$

Higher L_n's ($n > 0$) give nothing new. Thus for a given h, at the particular values of c given by Eq.(22), we do have a null vector at level 2:

$$\left(L_{-2} - \frac{3}{2(2h+1)}L_{-1}^2\right)|h>= 0 \tag{23}$$

and obviously all descendants of this state will be null vectors as well.

A primary field which leads to a null vector at level N is said to be "degenerate at level N". For a primary field degenerate at level 2, Eq.(23) leads to a set of second order differential equations to be satisfied for any n-point functions involving this field !

More precisely, using Eqs(I-35) and (II-4) are obtain in a theory where c is fixed and the field A_1 has weight h, corresponding to Eq.(22) i.e.

$$h_1 = h_\pm = \frac{5 - c \pm \sqrt{(1-c)(25-c)}}{16} \tag{24}$$

$$\frac{3}{2(2h_1+1)} \frac{\partial^2}{\partial z_1^2} \langle A_1(z_1) \cdots A_n(z_n) \rangle = \sum_{p=2}^{n} \left\{ \frac{h_p}{(z_1 - z_p)^2} + \frac{1}{z_1 - z_p} \partial_{z_p} \right\} \langle A_1(z_1) \cdots A_n(z_n) \rangle \tag{25}$$

In other words, null states or degenerate primary fields are very desirable physically since they lead to Eqs analogous to Eq.(25) for n-point functions !

To conclude this paragraph, let me point out that Eq.(23) is frame independent ! More precisely let me define the "null fields"

$$\xi_\pm(z) = \frac{3}{2(2h_1+1)} \partial_z^2 A(z) - \mathcal{L}_{-2} A(z) \tag{26}$$

with h_1 given by Eq.(24) and $A(z)$ a primary field of weight h.

By definition, under a conformal transformation

$$z \to \zeta(z)$$

we have

$$A(z) = A'(\zeta) \left(\frac{d\zeta}{dz}\right)^h \tag{27}$$

and
$$T(z) = T'(\zeta)\left(\frac{d\zeta}{dz}\right)^2 + \frac{1}{12}c\{\zeta, z\} \qquad (28)$$

To find the transformation properties of, say, $\mathcal{L}_{-2}A$ one compares the short distance expansions of $T(z')A(z)$ and $T'(\zeta')A'(\zeta)$ - this is quite a lenghty calculation - while for $\partial_z^2 A(z)$ one has

$$\begin{aligned}\partial_z^2 A(z) &= \partial_\zeta^2 A'(\zeta)\left(\frac{d\zeta}{dz}\right)^{h+2} + \partial_\zeta A'(\zeta)(2h+1)\left(\frac{d\zeta}{dz}\right)^h \frac{d^2\zeta}{dz^2} \\ &+ hA'(\zeta)\left\{(h-1)\left(\frac{d\zeta}{dz}\right)^{h-2}\left(\frac{d^2\zeta}{dz^2}\right) + \left(\frac{d\zeta}{dz}\right)^{h-1}\frac{d^3\zeta}{dz^3}\right\}\end{aligned}$$

trivially obtained from Eq.(27). The net result can then be shown to be

$$\xi_\pm(z) = \xi'_\pm(\zeta)\left(\frac{d\zeta}{dz}\right)^{h_1+2}$$

and thus identifying $\xi_\pm(z)$ to 0 is indeed frame independent.

§2 Kac Determinants

To determine whether there is a non trivial linear relation between the vectors at a given level of the Verma module is in principle straightforward: one computes the determinant of the matrix obtained from the inner products of all vectors at that level. If this determinant vanishes, there is a linear combination of vectors which will have zero norm i.e. there is a null vector. Note also that if this determinant is negative, there is going to be a state with negative norm and hence the theory will not be unitary. In a remarkable "tour de force", Kac has given a simple closed formula for this determinant !

As a preliminary exercice, let us calculate the Kac-determinant at level 2

$$\det\begin{vmatrix} \langle h|L_2 L_{-2}|h\rangle & \langle h|L_1^2 L_{-2}|h\rangle \\ \langle h|L_2 L_{-1}^2|h\rangle & \langle h|L_1^2 L_{-1}^2|h\rangle \end{vmatrix} = \det\begin{vmatrix} 4h + \frac{c}{2} & 6h \\ 6h & 4h(1+2h) \end{vmatrix} \qquad (29)$$

The last explicit form follows directly from the Virasoro algebra, for example $\langle h|L_2 L_{-2}|h\rangle = \langle h|[L_2, L_{-2}]|h\rangle$ since $L_2|h> = 0$ then $[L_2, L_{-2}] = 4L_0 + \frac{c}{12} 2 \cdot 3 = 4L_0 + \frac{c}{2}$, hence $\langle h|L_2 L_{-2}|h\rangle = 4h + \frac{c}{2}$.

Similar manipulations give the explicit entries of Eq.(29). From this equation, we have

$$\begin{aligned}\det M_{N=2}(c,h) &= 2(16h^3 - 10h^2 + 2h^2 c + hc) \\ &= 32(h - h_{1,1}(c))(h - h_{1,2}(c))(h - h_{2,1}(c))\end{aligned}$$

with $h_{1,1}(c) = 0$ while $h_{1,2}, h_{2,1} = \frac{1}{16}(5-c) \mp \frac{1}{16}\sqrt{(1-c)(25-c)}$ just as we found out earlier.

Note that the 0 of the determinant corresponding to $h = 0$ comes from a "descendant" of the null state at level 1. Clearly if $L_{-1}|0\rangle = 0$ then $L_{-1}(L_{-1}|0\rangle) = 0$ too.

In general, it is quite obvious that if there is a null state at level n, it will generate, at level $N > n$, precisely $p(N - n)$ null states given by $L_{-n_1} \cdots L_{-n_k}|$ null state at level $n >= 0$ with $\sum_i n_i = N - n$.

Let me now give Kac's formula. It reads

$$\det M_N(c, h) = cte \prod_{\substack{r,s = 1 \\ 1 \leq rs \leq N}}^{N} (h - h_{r,s})^{p(N-rs)} \tag{30}$$

"cte" is a positive constant and the zeroes of the determinant, $h_{r,s}$, are easier to write in terms of a quantity m (in general complex) rather than the central charge c. The precise relation between m and c is

$$c = 1 - \frac{6}{m(m+1)} \tag{31a}$$

or

$$m = -\frac{1}{2} \pm \frac{1}{2}\sqrt{\frac{25 - c}{1 - c}} \tag{31b}$$

and the explicit form of $h_{r,s}(m)$ is as follows

$$h_{r,s}(m) = \frac{[(m+1)r - ms]^2 - 1}{4m(m+1)} \tag{32}$$

From these formulas, it is possible to give a complete analysis of unitary representations of the Virasoro algebra. Let me simply state the result of that analysis. There are three "regions" of c where one can have unitary representations:

· for $c > 1$ and $h \geq 0$ det M_N never vanishes and one can show that all eigenvectors of the matrix M_N have positive definite norms.

· for $c = 1$ det $M_N = 0$ for $h = \frac{(r-s)^2}{4}$ and there are no negative norm states.

· $0 < c < 1, h > 0$ is the delicate region which requires some subtlety in the analysis. It turns out that the whole region is excluded except for isolated values of c which are given by (31a) but with m a positive integer $m \geq 3$. This series of unitary representations of the Virasoro algebra is called the "minimal unitary series". Amazingly enough, to each of the values of m (or c) corresponds a solvable statistical model. In particular $m = 3$ corresponds to the (critical) ising model as will briefly be described in the next paragraph.

For each value of c (or m) in the minimal series, there are $\frac{m(m-1)}{2}$ allowed values of h which are precisely given by the zeroes of det M_N, namely $h_{r,s}$ but with $1 \leq r \leq m-1$ and $1 \leq s \leq r$. But since the $h_{r,s}$ are invariant under $r \to m-r$ and $s \to m+1-s$, it is convenient to present the allowed values of h in $m(m-1)$ "conformal grids" which you will find in learned books on the subject.
for $m=3$ one has $c=\frac{1}{2}$ and the conformal grid is given by

$s \uparrow$ $\vec{r}$

1/2	0
1/16	1/16
0	1/2

i.e. the only allowed weights are 0, $\frac{1}{2}$ and $\frac{1}{16}$. Note that we have in this case the following null vectors:
at level one : $L_{-1}|o> = 0$
at level two : $(L_{-2} - \frac{3}{4}L_{-1}^2)|1/2> = 0$
and : $(L_{-2} - \frac{4}{3}L_{-1}^2)|1/16> = 0$.
The conformal grid for $m=4$ (or $c=7/10$) looks like

$s \uparrow$ $\vec{r}$

3/2	7/16	0
3/5	3/80	1/10
1/10	3/80	3/5
0	7/16	3/2

and I leave it as an exercice to identify the null vectors at level one, two and three.

As a last comment, let me define the character of a highest weight representation of the Virasoro algebra: with $q = \exp 2\pi i\tau$, and $Im\tau > 0$.

$$\chi_{c,h}(\tau) = Tr\, q^{L_0-c/24} \sum_{n=0}^{\infty} \dim(h+n) q^{n+h-c/24} \tag{33}$$

$\chi_{c,h}(\tau)$ is thus a generating function for the number of linearly independent states at level n i.e. with eigenvalue $h+n$ of L_0. The meaning of the parameter τ only appears when one goes to the torus. Also there is an intimate relation between these characters and Jacobi's elliptic functions which, unfortunately. I have no time to discuss here.

§3 A few remarks on the ising model

To illustrate a little bit all the techniques which I have described in these lectures, let me sketch a rather beautiful calculation of a 4-point function in the

Ising model. In the previous paragraph, I have mentioned that the Ising model corresponds to the first theory in the minimal unitary series i.e. $m = 3$ or $c = 1/2$. This is only for the holomorphic part of the theory. In fact, the "full theory" has $c = \bar{c} = 1/2$ and the primary fields are $(0,0); (\frac{1}{16}, \frac{1}{16})$ and $(1/2, 1/2)$. (This precise content follows from modular invariance, but let us just accept it for the purpose of illustration).

$(0,0)$ is of course the identity operator while $(\frac{1}{16}, \frac{1}{16})$ and $(\frac{1}{2}, \frac{1}{2})$ are respectively the spin σ and energy density ε. This follows from the two-point correlation functions:

$$\langle \sigma_n \sigma_0 \rangle \sim \frac{1}{|n|^\eta} \text{ with } \eta = 1/4$$

$$\sim \frac{1}{r^{2h+2\bar{h}}} \text{ (see Lecture I) and thus } h_\sigma = \bar{h}_\sigma = \frac{1}{16}$$

Similarly
$$\langle \varepsilon_n \varepsilon_0 \rangle \sim \frac{1}{|n|^{2(2-\frac{1}{\nu})}} \text{ with } \nu = 1$$

$$\sim \frac{1}{r^{2\tilde{h}+2\tilde{\bar{h}}}} \text{ and thus } \tilde{h}_\varepsilon = \tilde{\bar{h}}_\varepsilon = \frac{1}{2}$$

The model is, as you know, exactly solvable and all correlators can be computed exactly. What I want to do here is simply outline the calculation of one particular four point correlator namely

$$\Sigma_4 = \langle \sigma(1)\, \sigma(2)\, \sigma(3)\, \sigma(4) \rangle$$

(a) when we take $\begin{array}{l} 1 = r_1 + R \\ 2 = r_2 + R \end{array}$ and let $R \to \infty$ $\Sigma_4 \to \langle \sigma(r_1)\sigma(r_2) \rangle \langle \sigma(r_3)\sigma(r_4) \rangle$
which fixes the normalization of Σ_4.

(b) next, from global conformal invariance we easily find that

$$\Sigma_4 = \left(\frac{z_{14} z_{32}}{z_{12} z_{34} z_{13} z_{42}} \right)^{1/8} \left(\frac{\bar{z}_{14} \bar{z}_{32}}{\bar{z}_{12} \bar{z}_{34} \bar{z}_{13} \bar{z}_{42}} \right)^{1/8} f(x, \bar{x}) \tag{34}$$

where $x = \frac{z_{12} z_{34}}{z_{14} z_{32}}$ is the anharmonic ratio of the four points

(c) the primary field σ is degenerate at level 2. Hence

$$\left\{ \frac{4}{3} \frac{\partial^2}{\partial z_1^2} - \frac{1}{16} \left(\frac{1}{z_{12}^2} + \frac{1}{z_{13}^2} + \frac{1}{z_{14}^2} \right) - \left(\frac{1}{z_{12}} \partial_{z_2} + \frac{1}{z_{13}} \partial_{z_3} + \frac{1}{z_{14}} \partial_{z_4} \right) \right\} \Sigma_4 = 0 \tag{35}$$

and simple manipulations transform this equation into

$$\left\{ x(1-x) \frac{\partial^2}{\partial x^2} + (\frac{1}{2} - x) \frac{\partial}{\partial x} + \frac{1}{16} \right\} f(x) = 0 \tag{36}$$

and a similar equation for the antiholomorphic part of f. With the regularity conditions which follow from (a) one finds

$$f = \sqrt{1 \pm \sqrt{1-x}} \;\; \sqrt{1 \pm \sqrt{1-\bar{x}}}$$

where the $\pm$'s are uncorrelated. There are thus four solutions from which "monodromy" (i.e. single valuedness) selects one ! The final answer is

$$\Sigma_4 = \frac{1}{\sqrt{2}} \left| \frac{z_{14} z_{32}}{z_{12} z_{34} z_{13} z_{42}} \right|^{1/4}$$

$$\left\{ \sqrt{1 + \sqrt{1-x}} \sqrt{1 + \sqrt{1-\bar{x}}} + \sqrt{1 - \sqrt{1-x}} \sqrt{1 - \sqrt{1-\bar{x}}} \right\} \quad (37)$$

Lecture IV

§1 Affine Kac-Moody algebras

Let me start with a set of (1,0) conformal fields, $J^a(z)$ sometimes called "simple currents". The index a simply labels the different currents. An elementary dimensional analysis restricts the singular parts in their short distance expansion to be something like

$$J^a(z)J^b(w) = \frac{\tilde{k}^{ab}}{(z-w)^2} + \frac{if^{abc}}{(z-w)}J^c(w) + \cdots \tag{1}$$

The f^{abc}, necessarily antisymmetric in a and b can be shown to satisfy a Jacobi identity (this follows from "associativity" of the short distance expansion): thus they are structure constants of some Lie algebra. For simplicity, I will restrict myself to the $SU(N)$ algebras. I will also choose the central extension "matrix" $\tilde{k}^{ab}$ to be a multiple of the identity matrix. Going to the mode expansion

$$J^a(z) = \sum_{n \in \mathbb{Z}} J_n^a z^{-n-1} \tag{2}$$

Eq.(1) then corresponds to the commutator algebra

$$\left[J_m^a, J_n^b\right] = i\sqrt{2}f^{abc}J_{m+n}^c + km\delta^{ab}\delta_{m+n,o} \tag{3}$$

which is known as a Kac-Moody algebra. f^{abc} are the "usual" structure constants (i.e. ε^{abc} for $SU(2)$, Gell-Mann's f^{abc} for $SU(3)$ etc $\cdots$). The normalization factor $\sqrt{2}$ has been chosen for later convenience (obviously it corresponds to a normalization of k, the so-called central extension). The indices a, b, c run from 1 to $N^2 - 1$ (for $SU(N)$).

Clearly the zero modes satisfy the ordinary Lie algebra commutation relations albeit in a funny normalization

$$\left[J_0^a, J_0^b\right] = i\sqrt{2}f^{abc}J_0^c \tag{4}$$

I want to discuss a little bit these "Kac-Moody" (KM) theories because it is widely believed that they "exhaust" all rational conformal theories (i.e. conformal theories with a finite number of primary fields). KM theories are also known as "2-dim. current algebra theories" and they provide a host of non trivial exactly solvable quantum field theories. They are also relevant to string theory but that is of no concern to us in these lectures.

The representation theory of the infinite dimensional KM algebra of Eq.(3) is quite similar to that of the Virasoro algebra. In particular one easily defines multiplets of primary fields and the corresponding "highest weight states".

Consider a multiplet (r) of fields, $A^\ell_{(r)}(z)$, corresponding to some representation (r) of $SU(n)$. It will be called a primary multiplet if

$$J^a(z)A^\ell_{(r)}(w) \sim \sum_m \frac{(T^a_{(r)})^\ell_m}{z-w} A^m_{(r)}(w) + \cdots \qquad (5)$$

where the $T^a_{(r)}$'s are matrix representations of (r). Such a primary multiplet creates a multiplet of highest weight states that spans a representation of the zero mode algebra i.e.

$$|(r),\ell> \equiv A^\ell_{(r)}(0)|0> \qquad (6)$$

and
$$J^a_0|(r),\ell> = \sum_m (T^a)^m_\ell |(r),m> \qquad (7)$$

From the regularity requirement of $J^a(z)|0>$ when $z \to 0$, one also gets that

$$J^a_n|0> = 0 \qquad (n>0) \qquad (8)$$

and
$$J^a_n|(r),> = 0 \qquad (n>0) \qquad (9)$$

The derivation of the Ward identities of "affine symmetry" is also completely analogous to the purely conformal (or Virasoro) case: let $\varepsilon^a(z)$ parametrize an infinitesimal local $SU(n)$ transformation; we then compute $\oint \frac{dz}{2\pi i} \varepsilon^a(z) J^a(z)$ inserted in an N-point function

$$\oint \frac{dz}{2\pi i} \varepsilon^a(z) \langle J^a(z) A_{(r_1)}(w_1,\bar{w}_1) \cdots \rangle = \sum_{k=1}^N \langle A_{(r_1)}(w;\bar{w}_1) \cdots \delta_\varepsilon A_{(r_k)}(w_k,\bar{w}_k) \cdots \rangle$$

where $\delta_\varepsilon A_{(r)}(w,\bar{w}) = \varepsilon^a T^a_{(r)} A_{(r)}$ to obtain

$$\langle J^a(z) A_{(r_1)}(w_1 \bar{w}_1) \cdots \rangle = \sum_{k=1}^N \frac{T^a_{(r_k)}}{z-w_k} \langle A_{(r_1)}(w_1,\bar{w}_1) \cdots A_{(r_k)}(w_k,\bar{w}_k) \cdots \rangle \qquad (10)$$

Let me also point out that the "affinization" of the $SU(n)$ algebra corresponding to Eq.(3) is often denoted $\widehat{SU(n)}$. Clearly a similar procedure can be implemented for other Lie groups.

§2 Enveloping Virasoro Algebra or 2d current algebra

From the (1,0) currents introduced earlier, it is tempting to try to build up an energy momentum tensor as a quadratic form. Historically this is called the "Sugawara construction". How does it work?

We start by assuming

$$T(z) = \alpha \sum_a :J^a(z)J^a(z): \qquad (11)$$

where the sum goes from 1 to $(n^2 - 1)$ for $SU(n)$ and the normal ordering prescription must now be consistent with Eq.(1) i.e.

$$\sum_a : J^a(z)J^a(z) := \lim_{z \to w} \sum_a J^a(z)J^a(w) - \frac{\tilde{k}(n^2 - 1)}{(z - w)^2} \qquad (12)$$

In Eq.(11) α is a constant which will be determined by consistency. Indeed, we want the $J^a(z)$'s to transform as (1,0) fields in a theory where $T(z)$ is given by Eq.(11) i.e.

$$T(z)J^a(w) = \frac{J^a(w)}{(z-w)^z} + \frac{\partial J^a(w)}{z-w} + \cdots \qquad (13)$$

Note that, in terms of modes, Eq.(11) implies that

$$L_n = \alpha \sum_{m=-\infty}^{+\infty} \sum_a = J^a_{m+n} J^a_m : \qquad (14)$$

while Eq.(13) is equivalent to the commutation relations

$$[L_m, J^a_n] = -n J^a_{m+n} \qquad (15)$$

We now apply L_{-1} to a highest weight state using Eqs (7), (9) and (14)

$$L_{-1}|(r) >= 2\alpha J^a_{-1} T^a_r |(r) > \qquad (16)$$

where, for simplicity I dropped summation signs and matrix element indices. (The factor of 2 comes from Eq.(14) $L_{-1} = \alpha : J^a_{-1} J^a_0 : +\alpha : J^a_0 J^a_{-1} : + \cdots$).

Next we apply J^b_1 to both sides of Eq.(16) usign Eqs (3), (7) and (15).
For the LHS, we have

$$\begin{aligned} J^b_1 L_{-1}|(r) > &= L_{-1} J^b_1 |(r) > -[L_{-1}, J^b_1]|(r) > \\ &= J^b_0 |(r) >= T^b_{(r)} |(r) > \end{aligned} \qquad (17a)$$

while the RHS gives

$$2\alpha J^b_1 J^a_{-1} T^{(a)}_r |(r) >= 2\alpha [J^b_1, J^a_{-1}] T^a_{(r)} |(r) >$$

$$\begin{aligned} &= 2\alpha [i\sqrt{2} f^{bac} J^c_0 + k\delta^{ab}] T^a_{(r)} |(r) > \\ &= 2\alpha i\sqrt{2} f^{bac} J^c_0 J^a_0 |(r) > +2\alpha k T^b_{(r)} |(r) > \\ &= 2\alpha i\sqrt{2} f^{bac} \frac{1}{2}[J^c_0, J^a_0]|(r) > +2\alpha k T^b_{(r)} |(r) > \\ &= -2\alpha f^{bac} f^{cad} J^d_0 |(r) > +2\alpha k T^b_{(r)} |(r) > \\ &= 2\alpha f^{acb} f^{acd} T^d_{(r)} |(r) > +2\alpha k T^b_{(r)} |(r) > \\ &= 2\alpha (n + k) T^b_{(r)} |(r) > \end{aligned}$$

$$\tag{17b}$$

where in the last line I have used $f^{acb}f^{acd} = n\delta^{bd}$ (which is consistent with the "funny" normalization used in Eq.(4)).

Comparing Eqs (17a) and (17b) gives us the final result

$$T(z) = \frac{1/2}{n+k}\sum_a :J^a(z)J^a(z): \tag{18}$$

from which one easily calculates the central charge

$$c_{SU(n)} = \frac{k(n^2-1)}{n+k} \tag{19}$$

(calculating the short distance expansion of $T(z)T(w)$ using Eq.(1) or, equivalently calculating $[L_2, L_{-2}]$ from Eq.(14)).

To continue, let me concentrate on $\widehat{SU(2)}$. I will furthermore assume that $|(r)>$ is an irreducible representation of the zero mode algebra i.e. of $SU(2)$. I want to show now that *k is an integer* (justifying a posteriori the "funny" normalization of the algebra).

Note first, that the "usual" $SU(2)$ generators are given by

$$I_{\pm} = \frac{1}{\sqrt{2}}(J_0^1 \pm J_0^2) \quad I_3 = \frac{1}{\sqrt{2}}J_0^3 \tag{20}$$

i.e.

$$[I_+, I_-] = 2I_3 \text{ etc} \cdots$$

In particular, the eigenvalues of I_3 are integers or half integers. However, from Eq.(3), specialized to the $SU(2)$ case, one readily checks that

$$\tilde{I}_+ = \frac{1}{\sqrt{2}}(J_{-1}^1 + iJ_{-1}^2) \quad \tilde{I}_- = \frac{1}{\sqrt{2}}(J_{+1}^1 - iJ_{+1}^2) \quad \tilde{I}_3 = I_3 - \frac{1}{2}k$$

also satisfy the usual $SU(2)$ algebra hence the eigenvalues of $2\tilde{I}_3$ i.e. of $2I_3 - k$ must be integers hence $k \in \mathbb{Z}$. k is called the *"level"*.

For unitary representations of $\widehat{SU(2)}$ - this means that $J_m^+ = J_{-m}$ by arguments similar to the Virasoro case - let us take $|(r)>$ to correspond to a spin j representation of $SU(2)$, then from

$$\begin{aligned}\|\tilde{I}_+|J,J>\|^2 &= \langle J,J|\tilde{I}_-\tilde{I}_+|J,J\rangle \\ &= \langle J,J|[\tilde{I}_-,\tilde{I}_+]|JJ\rangle = \langle JJ|-2I_3+k|JJ\rangle \\ &= -2J+k \geq 0 \end{aligned} \tag{21}$$

we see that the level must be *positive*.

From Eq.(21) we also learn that at a given level k, the only allowed primary representations must have $2J \leq k$ i.e. $J = 0, \frac{1}{2}, 1 \cdots \frac{k}{2}$ exhaust all possibilities.

Finally let me point out that using Eq.(18) or the definition of L_0 by Eq.(14), one finds

$$L_0|J,J> = \frac{J(J+1)}{k+2}|J,J> \qquad (22)$$

Also, I should at least mention that $\widehat{SU(2)}_k$ is elegantly realized in the Wess-Zumino-Witten lagrangians where the level k corresponds to the coefficient of the "topological term" in these lagrangians.

§3 The coset construction

We have seen in the previous paragraph that

$$c_{SU(2)} = \frac{3k}{k+2}$$

Obviously these central charges are ≥ 1. There is a very simple and very general construction, called the coset construction, which considerably enlarges the possible ranges of c. To conclude these lectures, I would like to outline this "coset construction" and show, in particular, how to recover the "minimal unitary series" of the Virasoro algebra from a Kac-Moody theory.

The essence of the "coset construction" is simply to consider a group G and a subgroup H. Let us denote the G currents by J_G^a and the H currents J_H^i (a runs from 1 to dim G and i from 1 to dim H and J_H^i is a subset of J_G^a). We can now construct two energy momentum tensors namely

$$T_G(z) = cte \sum_a : J_G^a(z) J_G^a(z) : \qquad (23a)$$

$$T_H(z) = (cte)' \sum : J_H^i(z) J_H^i(z) : \qquad (23b)$$

(the constants are in fact determined but this is irrelevant for the present discussion).

By definition we have that

$$T_G(z) J_H^i(w) = \frac{J_H^i(w)}{(z-w)^2} + \frac{\partial J_H^i(w)}{z-w} + \cdots \qquad (24)$$

as well as

$$T_H(z) J_H^i(w) = \frac{J_H^i(w)}{(z-w)^2} + \frac{\partial J_H^i(w)}{z-w} + \cdots \qquad (25)$$

Eqs (24) and (25) simply express that $J_H^i(z)$ has conformal weight (1,0) in both theories given by Eqs (21).

From Eq.(24) and (25), it follows that $T_G - T_H$ has a *regular* short distance expansion with J_H^i and thus with T_H. This means that
$$T_G = (T_G - T_H) + T_H \equiv T_{G/H} + T_H \tag{26}$$
gives an orthogonal decomposition of the Virasoro algebra generated by T_G into two commuting (since their short distance expansion is regular) subalgebras.

Now, since $T_{G/H}$ and T_H have no singular part in their short distance expansion, clearly
$$\begin{aligned} T_G(z)T_G(w) &= \frac{1/2 c_G}{(z-w)^4} + \cdots \\ &= T_{G/H}T_{G/H} + T_H T_H = \frac{1/2 c_{G/H}}{(z-w)^4} + \frac{1/2 c_H}{(z-w)^4} \end{aligned} \tag{27}$$
hence
$$c_{G/H} = c_G - c_h \tag{28}$$

Consider now the specific case of diagonal embeddings i.e. coset spaces of the type $G \times G/G$. It is trivial to verify that the level in the denominator is fixed to be the sum of the levels of the two groups in the numerator. Thus a simple example of a coset construction is given by
$$G/H = \frac{SU(2)_k \times SU(2)_1}{SU(2)_{k+1}} \tag{29}$$
and, from Eq.(27) we have
$$c_{G/H} = \frac{3k}{k+2} + 1 - \frac{3(k+1)}{k+3} = 1 - \frac{6}{(k+2)(k+3)} \tag{30}$$
namely, as promised, the c values for the minimal unitary series when $k = 1, 2, \cdots$.

In a bit less than four hours of lecturing, I have hardly begun to describe the amazingly rich structure of 2 dim. conformal theories. Much to my regret I had to completely leave out very important topics such as modular invariance for example. I can only hope that the elementary introduction given in these lectures will motivate some of you to learn more about the subject !

ACKNOWLEDGEMENTS

Parts of these notes were written up during a visit at SLAC. I am grateful to the SLAC Theory group and to Prof. R. Blankenbeckler for their warm hospitality.

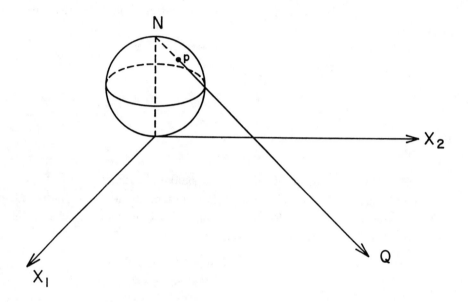

References

Since these lectures were meant as an elementary introduction to conformal field theories, I do not have the slightest pretense of originality. In fact, I remorselessly plundered various sets of lectures already published on the subject and, in particular

- P. Ginsparg : Applied Conformal Field Theory in Fields, Strings and Critical Phenomena, Les Houches, Session XLIX, 1988. Ed. by E. Brezin and J. Zinn-Justin, Elsevier (1989).

- C. Itzykson, J-M. Drouffe : Théorie statistique de champs, Vol.2, Intereditions et Editions du CNRS (1989).

- Y. St Aubin : Phénomènes critiques en deux dimensions et invariance conforme (Louvain, 1987).

- J. Bagger : CFT à la BPZ. "III Mexican School of Particles and Fields". Mexico 1988, J.L. Lucio and A. Zepeda, eds (World Scientific).

Extensive portions of my lectures have been taken almost verbatim from these references.

The fundamental paper on 2d conformal theories is

- A.A. Belavin, A.M. Polyakov and A.B. Zamolodchikov : Nucl. Phys. B241 (1984) 333.

The Virasoro algebra - including the central charge - was introduced in M. Virasoro, Phys. Rev. D1 (1970) 2933.

A readable book on $SL(2,C)$ and some of its subgroups is the following:

- J. Lehner : "A short course in Automorphic Functions". Holt, Rinehart and Winston 1966.

A discussion of energy momentum tensors can be found in Y. St Aubin's lectures.

For a particularly insightful discussion of quantization procedures see

- P.A.M. Dirac, Rev. Mod. Phys. 21 (1949) 392.

- S. Fubini, A.J. Hansen and R. Jackiw, Phys. Rev. **D7** (1973) 1732.

The "short distance" or operator product expansion is due to

- K. Wilson, Phys. Rev. **D10** (1974) 2445.

The triviality of representations of the Virasoro algebra when $c = 0$ is shown in

- J.F. Gomes, Phys. Lett. 171B (1986) 75.

The relation between central charge and Casinir effect is discussed in

- H.W.J. Blöte, J.L. Cardy, M.P Nightingale, Phys. Rev. Lett. **56** 742 (1982).

- I. Affleck, Phys. Rev. Lett. **56** 746 (1986).

The litterature on representations of the Virasoro algebra is immense. A few "classic" papers are

- B.L. Feigin and D.B. Fuchs : "Verma modules over the Virasoro algebra" Lecture Notes in Mathematics, Vol. 1060, Springer, New York (1985).

- P. Goddard, A. Kent and D. Olive, Comm. Math. Phys. 103 (1986) 105.

- D. Friedan, Z. Qiu and S. Shenker, Phys. Rev. Lett. **52** (1984) 1575; Comm. Math. Phys. 107 (1986) 535.

Transformation properties of descendant fields are discussed in C. Itzykson, J.M. Drouffe op.cit.

The formula for the Kac determinant is proven explicitly e.g. in

- C. Thorn, Nucl. Phys. B 248 (1984) 551.

Concerning the minimal unitary series, see the earlier references to Friedan et al, Goddard et al as well as

- R.J. Baxter, J. Phys. A13 L61 (1980).

- G.E. Andrews, R.J. Baxter and J.P. Forrester, J. Stat. Phys. 35 (1984) 193.

A good review of the Ising model as a conformal theory can be found in

- J. Cardy "Conformal Invariance" in Domb and Lebowitz, Phase Transitions, Vol. 11, Academic Press (1987).

see also Itzykson and Drouffe op.cit. and references quoted therein.

The calculation of the 4-point function sketched here is due to

- Vl. S. Dotsenko and V.A. Fateev, Nucl. Phys. B251 (1985) 691.

A readable introduction to Kac-Moody algebras and the coset construction is given by

- P. Goddard and D. Olive, Int. J. Mod. Phys. A1 (1986) 303.

   A much more difficult treatise on the subject is

- V. Kac : "Infinite Dimensional Lie Algebras", Cambridge University Press (1985)

   The original Sugawara construction is found in

- H. Sugawara, Phys. Rev. 170, 1659 (1968).

See also

- P. Goddard, W. Nahm and D. Olive, Phys. Lett. 160B (1985) 111.

   The "Wess-Zumino-Witten" models are discussed in Ginsparg, Itzykson, Drouffe, Bagger op. cit. The original litterature is

- J. Wess and B. Zumino, Phys. Lett. **37B** (1971) 95.

- E. Witten, Comm. Math. Phys. 92 (1984) 455.

QUANTUM MECHANICS AND DECOHERENT HISTORIES

J. Weyers

Institut de Physique Théorique,
Université Catholique de Louvain, Belgium

Do you know the difference between a physicist, a philosopher and a theologian ? As I learned from my colleague, but friend nevertheless, Jean Pestieau: a physicist is someone who looks for a black cat in a black room; a philosopher is someone who looks for a black cat in a black room where there is no black cat and a theologian is someone who looks for a black cat in a black room where there is no black cat but $\cdots$ finds one !

This seminar is about some physics issues in quantum mechanics (QM) - of course within the limits of my understanding of the subject ! I will not be concerned at all with the philosophical questions or implications which the rather weird conceptual framework of QM usually leads to. In other words, I will just try to look for the black cat hoping it is there !

In standard QM courses one is usually taught what has become known as the "Copenhagen Interpretation of QM"[1]. Why on earth do we need an "interpretation" of a physical theory ? Well, this is a philosophical question, isn't it ? Hence, I will $\cdots$ not answer it !

The dominant figures in the development of this Quantum Orthodoxy were Bohr, Schrödinger, Heisenberg and Dirac. Significant contributions were also made by Born, Wigner, etc $\cdots$ [1].

Among the "big four", I must admit having a lot of sympathy for Schrödinger. He had but one passion in his life [2] : women ! As far as Bohr is concerned, I find it a bit difficult, in 1992, to understand what precisely made him THE central figure in the development of QM. From what he has written [3] on the subject - at least the parts which I have read - he appears rather allergic to any kind of mathematical formalism but extremely fond of "verbalizations" in the description of the behaviour of quantum systems ! "Particle-wave duality", the "principle of complementarity", a "quantum phenomenon" are but of few of his linguistic inventions. These philosophical inclinations notwithstanding, Bohr did indeed play a proeminent role in the development of QM. In particular, it was he who stood up to Einstein and countered his various attacks on quantum theory. Perhaps the best illustration of Bohr's overpowering personality is given by $\cdots$ Heisenberg. In the paper [4] where he announces the discovery of the uncertainty relations - a landmark in QM -, Heisenberg apologizes (sic !), in a note added in proof, for the incomplete and superficial aspects of his analysis and announces a forthcoming paper by Bohr where a "serious" analysis will be presented ! Amazing !! I have added Dirac's name to the list of "founding fathers" for obvious reasons: Dirac invented the "canonical quantization" method [5], developed a quantization procedure for constrained systems [6] and, at the very least, initiated the path integral formulation [7] of QM later to be developed by Feynman [8]. Not a bad record ! So even if Dirac did not, stricto sensu, contribute to the elaboration of the Copenhagen interpretation, he deserves a special place among the developers of the orthodoxy for the sheer importance of his work on the subject. To give credit where it belongs, I must also remind you that, historically, the probabilistic interpretation of the wave function is due to Born [9].

Anyway, QM and its orthodox interpretation were quickly taken up and "axiomatized" by two of the leading mathematicians of the time: J. van Neumann [10] and H. Weyl [11]. One of them even gave a (wrong [12]) proof that there was no alternative to the orthodox view ! Nevertheless, the Copenhagen interpretation was not, and still is not, without rather eminent "heretics": de Broglie, Schrödinger himself - who played some yo-yo between heresy and orthodoxy - and, of course Einstein, as I already mentioned. To make a long story short, let me simply remind you that Einstein's final assault on QM came with the EPR [13] "paradox" in 1935. When this too failed, it appeared for a while that the day of the heretics was over despite some continuing "unhappiness" of quite a few people with the orthodox view.

The comeback of the heretics will be discussed a bit later in this talk. First, let me describe, à la hussarde, the main points of the Copenhagen Interpretation:

(I) Somewhere, there is a classical world !

(II) ALL information obtainable on a quantum system, is given by its wave function.

(III) Time evolution: when you don't watch a quantum system, the evolution of its wave function is causal, deterministic and given by the Schrödinger equation:

$$i\hbar \frac{\partial \psi}{\partial t} = H\Psi$$

(IV) Measurement "theory": when you do watch i.e. measure some properties of a quantum system,

    a) you cannot measure everything you might want to (uncertainty relations)

b) $|\langle m|\Psi\rangle|^2$ gives the probability of the result m for a measurement

c) if the result m is obtained, the wave function "reduces" to $|m>$.

It is perhaps useful to illustrate these various points with a few citations from Heisenberg [14]: "The Copenhagen interpretation of Quantum Theory starts from a paradox. Any experiment in physics $\cdots$ is to be described in the terms of classical physics. We cannot and should not replace the concepts of classical physics by any others".

"The probability function - Heisenberg's name for the wave function - represents a mixture of two things, partly a fact and partly our knowledge of a fact $\cdots$ It represents a tendency for events and our knowledge of events. The probability function can be connected with reality only if one essential condition is fulfilled: if a new measurement is made to determine a certain property of the system $\cdots$ the result of the measurement will be stated in terms of classical physics".

"The theoretical interpretation of an experiment requires three distinct steps:

(1) the translation of the initial experimental situation into a probability function;

(2) the following up of this function in the course of time;

(3) the statement of a new measurement to be made on the system, the result of which can then be calculated from the probability function.

For the first step, the fulfillment of the uncertainty relations is a necessary condition. The second step cannot be described in terms of the classical concepts: there is no description of what happens to the systems between the initial observation and the next measurement. It is only in the third step that we change over again form the "possible" to the "actual" $\cdots$ A real difficulty in the understanding of this interpretation arises, however, when one asks the famous question: but what happens "really" in an atomic event".

"$\cdots$ it is important to remember that in natural science we are not interested in the universe as a whole but $\cdots$ it is important that a large part of the universe does not belong to the object $\cdots$"

Except for the last quotation whose relevance will appear a bit later, it is difficult to give a more lucid description of what QM is all about !

Undoubtedly the weirdest aspect of QM is that it does not pretend to describe "reality" anymore but rather what we can know about reality: in some sense the only contact between reality and QM occurs in the "measurement process". But there are other pecularities !

Even the staunchest supporter of Copenhagen must admit that point (I) makes this interpretation of QM questionable ! The dichotomy quantum world ⟷ classical world (or micro ⟷ macro or probabilistic ⟷ deterministic) implied by (I) and which Heisenberg calls a paradox, is in fact no paradox at all ! It just doesn't make (logical) sense. Note in particular that it precludes the existence of an isolated quantum system. We will see later on how elegantly this trap of the Copenhagen interpretation is avoided in the decoherent histories approach.

Alternatives where point (II) does not hold are known as "hidden variable scenarios". Shortly I will describe a bit more one of these scenarios, the last (?) great heresy, which is due to Bohm [15]. The sharpest onslaught on hidden variable scenarios came from "Bell's inequalities" [16]. However they do not invalidate Bohm's interpretation of QM.

Point (III) embodies the linear superpostion principle and all the problems that go with it, namely why are superpositions of classically different states not observed ? For example, why can't I be in a superposition of two position eigenstates: one in my wife's bed and the other in the bed of the lady next door ? When I suggested this possibility to my beloved wife, she $\cdots$ didn't believe QM ! Shrödinger's cat [17] - was it black ? - is a better known example of an absurd superposition state for a classical object.

How to get rid of these superpositions is not an academic question. Wigner [18], for example, suggested the "conscious mind" as the destroyer of superpositions !! If nothing else, the desperate nature of this solution reflects the importance of the problem. A more obvious way of getting rid of superpositions is to make the equation $\cdots$ non linear ! This has been done time and again [19]. Decoherent histories will lead to a much more appealing solution to the problem.

Finally, point (IV) has also led to several objections concerning the "nature of reality", the "role of the observer", the "acausal reduction of the wave function" etc, etc $\cdots$ For example, one of Bell's last papers [20] is entitled "Against measurement". Here too, I believe that the decoherent histories approach provides a satisfactory answer to most of the problems.

The preceeding discussion is of course extremely superficial but it is only meant as an illustration of some of the interpretation problems of QM. Let me also mention "objections" to QM which were raised concerning the "unphysical" nature of the wave function (de Broglie, Schrödinger) or which were based on physical realism (Einstein). But as I said at the beginning of this talk, I do not want to discuss issues concerning reality, causality, locality or other - lities.

Rather, as promised, let me spend a little bit of my time on Bohm's approach to QM [15]. This theory puts some "substance" in the rather vague ideas of de Broglie and Schrödinger which I just mentioned. It also goes quite some way in meeting Einstein's requirements of physical realism. The starting point is that a non relativistic quantum particle <u>is</u> a particle, namely it has, at all times, a well defined <u>position</u> in space (let me take it as one dimensional for simplicity). A "Bohmian particle" has thus a position x <u>and</u> a wave function Ψ. Hence we are in a hidden variable scenario where the hidden variable is the position of the particle ! What a ludicrous nomenclature ! The evolution of (x, Ψ) is as simple as it possibly can be and is given by

$$\dot{x} = \frac{\hbar}{m} Im \frac{\frac{\partial \Psi}{\partial X}}{\Psi}$$
$$i\hbar \frac{\partial \Psi}{\partial t} = H\Psi$$

Clearly Ψ has become a kind of "pilot wave" which guides the motion of the particle; Ψ is now a field which is endowed with physical properties (e.g. it can carry enery).

Bohm's theory is very clever indeed and cannot easily be dismissed. It does give a more "physical" picture of e.g. the double-slit "experiment" : the electron does go through one hole <u>or</u> the other and still there are interferences because of the pilot wave. Also there are no "macroscopic" superpositions in this theory. Bohm's theory is not QM as we know or practise it (e.g. there is no uncertainty relation). It is some alternative theory to QM and, as such, it should simply be confronted with experiment ! Unfortunately, as some recent panegyrists [21] of Bohm's have argued, you can always "massage" the pilot wave's initial conditions (unknowable in the sense of classical statistical mechanics) to agree with all predictions of ordinary non relativistic QM ! Beuh !

I have tried to present Bohm's point of view as fairly as possible but let me hasten to say that I very strongly dislike it. Luckily there is a flaw in Bohm's approach (temporary or terminal, I do not know for sure): the theory cannot be made relativistic. I suspect that even if it could, it would violate, say, the CPT theorem since it is a "non-local" theory (by the way it is this last property which allows it to avoid the ax of Bell's inequalities). Honestly, I would very much prefer a clean experimental kill to vague arguments based on taste or even on theorems!

Finally let me mention yet another interpretation of QM, namely Everett's many worlds [22]. This one ⋯ I love, but as a science fiction fan not as a physicist.

So where do we stand with all these interpretation "problems" ?

Well, Copenhagen QM works fantastically well - it is high time to remind ourselves of this fact! QM plus relativity predicted antimatter which was found;

QED gives the $g-2$ of electrons and muons to an incredible precision. The standard model, at least its electroweak part, is getting more and more precise. Of course, I do not mean to imply that there are no conceptual problems in a relativistic quantum theory, but it certainly works beautifully.

Nevertheless the fact remains that Copenhagen QM is an incomplete interpretation of the theory, if only because of point (I).

Since 1980 but especially in the last few years, very significant progress has been made in completing and sharpening the interpretation of QM. This progress is the result of work by Griffiths [23], Zurek [24], Omnes [25] and Gell-Mann and Hartle [26] and rests on the concepts of "history" and "decoherence". These concepts do not require a drastic revision of quantum orthodoxy but they "refocus" the issues, so to speak.

Let me try, in the last part of this seminar, to briefly sketch what "decoherent histories" are and how they are used to solve most of the interpetation problems raised earlier. My presentation will closely follow the approach of Gell-Mann and Hartle (GMH) but I have no time for anything more than a gross outline of some of their ideas. I can only advise you to read their papers.

The starting point of GMH is that they are interested in the universe as a whole ! (remember the last quotation form Heisenberg ?). Assuming that QM underlies the laws of physics, the universe and everything it contains must then be described in terms of QM. Thus no "outside observer" anymore and no classical world to start with.

Consequently, for a closed quantum system such as the universe, point (I) of the Copenhagen interpretation is replaced by the initial conditions on the system as given, say, by a density matrix $\rho(0)$.

Suppose we know $\rho(0)$ and the hamiltonian (or action) of our system. Suppose also that we have some information about its past : what can we predict about the future ? GMH insist on the fact that there is "no certainty in this world". Thus by predictions one means an assignment of probabilities but with some errors. Probabilities must of course obey the usual rules of probability theory (positivity, additivity on disjoint sets etc $\cdots$). Accepting some error means that the corresponding rules will only be approximately obeyed.

By predictions about the future one thus means (approximate) probabilities for events or sequences of events. What is an event ? To motivate the answer to this question, consider some observable 0. With this observable one can associate projection operators P and in the Heisenberg picture $P(t) = \exp i\frac{Ht}{\hbar} \, P(0) \exp -i\frac{Ht}{\hbar}$.

An exhaustive set of exclusive alternatives is given by a particular set,

k, of projectors $P^k_{\alpha_k}(t)$ - where α_k's are the alternatives. The set is exhaustive if $\sum_{\alpha_k} P^k_{\alpha_k}(t) = 1$ and the alternatives are exclusive if $P^k_{\alpha_k} P^k_{\beta_k} = \delta_{\alpha\beta} P^k_{\beta_k}$. If the $P^k_{\alpha_k}(t)$ are one dimensional, the set is complete.

Simple examples of such sets are $P_\uparrow = |\uparrow\rangle\langle\uparrow|$ and $P_\downarrow = |\downarrow\rangle\langle\downarrow|$, which are a complete exhaustive set of exclusive alternatives for spin projections of a spin 1/2 particle. Similarly, $P_{\Delta_i} = \int_{\Delta_i} |x> dx <x|$ with disjoint Δ_i's whose union covers the whole domain of x is another exhaustive set of exclusive alternatives (but this time incomplete). We identify an event α_k (or equivalently a "property α_k") with the projector $P^k_{\alpha_k}(t)$. An event α_k is thus one particular element of an exhaustive set of exclusive alternatives !

The probability of an event α_k, given the initial density matrix is given as usual by

$$p(\alpha_k) = Tr(P_{\alpha_k}(t)\rho(0)P_{\alpha_k}(t)) = Tr(P_{\alpha_k}(t)\rho(0))$$

We are now ready to define the concept of history: a history is a particular time ordered sequence of events. Thus history $\iff \{P^1_{\alpha_1}(t_1), P^2_{\alpha_2}(t_2) \cdots P^n_{\alpha_n}(t_n)\}$ with $\alpha_1, \cdots \alpha_n$ fixed and $t_1 < t_2 < t_3 \cdots < t_n$.

Let me get a few more definitions out of the way. A set of alternative histories is a set of histories where the alternatives α_k run over all their possible values. A completely fine grained history is a history where all $P^k_{\alpha_k}(t_k)$ are one dimensional projectors. One history is a coarse graining of another history when the projectors defining the first history (at one or more than one instant in time) are sums of the corresponding projectors of the second one. The inverse operation is called fine graining.

As indicated above, the aim of the theory is to determine the probabilities or certain events or sequences of such events. In other words we want to assign a probability to a specific history or probabilities to elements of a set of alternative histories. However, in general, this will not be possible: because of interferences, the probabilities that one might try to assign to sets of completely fine grained histories will not satisfy the sum rules of probability theory and hence will not be probabilities ! In other words a certain amount of appropriate coarse graining will always be necessary in sets of alternative histories to which probabilities can be consistently assigned.

As an example let me consider, following Hartle [26], a two-event history such as the double-slit: at time $t = t_0 = 0$, an electron beam, say, is emitted from a source; at time t_1 it passes through one of the two slits and at time t_2 it reaches some spot on the screen. At time t_1, we have the events u or d - the electron passing trough the upper or lower slit - with corresponding projectors $P_u(t_1)$ and

$P_d(t_1)$, while at time t_2, $P_{\Delta_i}(t_2)$ are the projectors associated with an electron hitting the (one dimensional screen) in the interval Δ_i.

Let the system be described at $t = 0$ by $|\Psi>$. It evolves to time t_1 i.e.

$$|\Psi> \longrightarrow e^{-iHt_1}|\Psi> \quad (\hbar = 1)$$

at which time we ask about the events u or d. The two states

$$P_u e^{-iHt_1}|\Psi> \quad \text{and} \quad P_d e^{-iHt}|\Psi>$$

then evolve until time t_2 where the events Δ_i occur.

$$|\Delta_i, t_2; u, t_1; \Psi> = P_{\Delta_i} e^{-iH(t_2-t_1)} P_u e^{-iHt_1}|\Psi>$$
$$\text{and} \quad |\Delta_i, t_2; d, t_1; \Psi> = P_{\Delta_i} e^{-iH(t_2-t_1)} P_d e^{-iHt_1}|\Psi>$$

are thus the states corresponding to the two histories

$$(\Psi, t=0) \rightarrow (u, t_1) \rightarrow (\Delta_i, t_2) \quad \text{or} \quad \{P_u(t_1), P_{\Delta_i}(t_2)\}$$
$$(\Psi, t=0) \rightarrow (d, t_1) \rightarrow (\Delta_i, t_2) \quad \text{or} \quad \{P_d(t_1), P_{\Delta_i}(t_2)\}$$

Can one assign probabilities to these histories, or, equivalently, are

$$p_u(\Delta_i) = p(\Delta_i, t_2; u, t_1) = <\Delta_i, t_2; u, t_1; \Psi|\Delta_i, t_2; u, t_1; \Psi>$$

and

$$p_d(\Delta) = p(\Delta_i, t_2; d, t_1) = <\Delta_i, t_2; d, t_1; \Psi|\Delta_i, t_2; d, t_1; \Psi>$$

honest to God probabilities ?

If they were, the history

$$(\Psi, t_0 = 0) \longrightarrow (\Delta_i, t_2) \quad \text{or} \quad \{\mathbb{1}(t_1), P_{\Delta_i}(t_2)\}$$

(which is a coarse graining of the two event histories) should have probability $p(\Delta) = p_u(\Delta) + p_d(\Delta)$.

We know very well that this will not be the case because of interference i.e. $<\Delta_i, t_2; u, t_1; \Psi|\Delta_i, t_2; d, t_1; \Psi> \neq 0$.
Only when the interferences between alternative histories in a given set disappear, will it be possible to assign probabilities to these histories. Such histories are said to <u>decohere</u>.

According to GMH, the fundamental formula of QM for two event histories then reads in a general notation

$$\text{Tr}\left[P_{\alpha_2}^2 e^{-iH(t_2-t_1)} P_{\alpha_1}^1 e^{-iHt_1} |\Psi><\Psi| e^{iHt_1} P_{\alpha_1'}^1 e^{iH(t_2-t_1)} P_{\alpha_2}^2 \right]$$
$$= \text{Tr}\, P_{\alpha_2}^2(t_2) P_{\alpha_1}^1(t_1) \rho P_{\alpha_1'}^1(t_1) P_{\alpha_2}^2(t_2) \cong p(\alpha_2, t_2; \alpha_1, t_1; \rho)\delta_{\alpha_1,\alpha_1'}$$

This formula allows indeed to <u>identify</u> (at least in principle) the sets of alternative decoherent histories i.e. those sets of alternative histories for which the left hand side of the last equation (approximately) vanishes for different events at time $t_1(\alpha_1 \neq \alpha_1')$ and then to <u>calculate</u> their probabilities.

More generally, one defines the decoherence functional of a pair of histories.

$$D_{[\alpha],[\alpha']} = \text{Tr}\left[P^n_{\alpha_n}(t_n)\cdots P^1_{\alpha_1}(t_1)\rho P^1_{\alpha_1'}(t_1)\cdots P^n_{\alpha_n'}(t_n)\right] \text{ with } [\alpha] = (\alpha_1,\cdots\alpha_n)$$

and the fundamental formula then reads

$$D_{[\alpha],[\alpha']} = p([\alpha])\delta_{[\alpha],[\alpha']}$$

Note that D is always diagonal in the last index and that decoherence is automatic for single event histories. Note also that decoherence not only depends on the histories but also on ρ !

Let me now end with some broad indications on some directions on the program of GMH for making this decoherent history approach a complete interpretation of QM. Preliminary work [27] [28] along these lines does justify a certain dose of optimism !

First, one should <u>derive</u> the existence of a classical (or quasiclassical) domain from QM. Clearly, coarse graining of sets of alternative decoherent histories does not spoil decoherence. Inversely, by fine graining, one will eventually obtain among all sets of histories one or several "maximal sets of decoherent histories" : any further fine graining of these destroys the decoherence. These maximal sets are the natural candidates for (quasi) classical domains: their histories are expected to follow very closely all classical correlation patterns. Examples are given in ref [28].

A typical "measurement" of a quantum variable is then simply viewed as a correlation of this variable with variables in a quasi classical domain and one immediately recovers the Copenhagen rules for assigning probabilities to the results of a measurement. Decoherence is really at the heart of the matter here: it replaces the notion of measurement in the Copenhagen interpretation but it is more precise, more objective and it does not need anything like an observer.

What is the physical origin of decoherence ? Roughly speaking it comes essentially from the extremely large number of degrees of freedom of the "environment" and decoherence is thus very much like "dissipation".

As a final remark, let me point out that decoherence also allows to get rid of superposition states for macroscopic systems [29]: different positions, say, will be correlated with orthogonal sets of the environment variables and decoherence will occur very quickly indeed ! Did I see a flicker of disappointment in the eyes of the lady next door ?

References

[1] See the beautiful collections of reprints in J.A. Wheeler and W.H. Zurek, Eds., "Quantum Theory and Measurement" (Princeton University, Princeton, NJ, 1983).

[2] See e.g. Moore "Schrödinger", Cambridge University Press (1989).

[3] e.g. N. Bohr : "Atomic Physics and Human Knowledge" (Wiley, New York, 1958).

[4] W. Heisenberg, Zeit. für Physik **43** (1927) 172-198, translated in Ref. 1.

[5] P.A.M. Dirac, The Principles of Quantum Mechanics (Clarendon, Oxford, 1930).

[6] P.A.M. Dirac, Proc. Roy. Soc. (London) A246, 326 (1958).

[7] P.A.M. Dirac, The Lagrangian in Quantum Mechanics Physikalische Zeitschrift der Sowjet Union, Band 3, Heft 1 (1933).

[8] R.P. Feynman, Rev. Mod. Phys. **20**, 367 (1948).

[9] M. Born, Zeit. für Physik **37** (1926) 863-867, translated in Ref. 1.

[10] J. von Neumann : "Mathematische Grundlagen der Quantummechanik" (Springer, Berlin, 1932).

[11] H. Weyl, Gruppentheorie und Quantummechanik (1928).

[12] See e.g. J.S. Bell, Speakable and unspeakable in quantum mechanics (Cambridge, England, 1987).

[13] A. Einstein, B. Podolsky and N. Rosen, Phys. Rev. **47**, 777 (1935) reprinted in Ref. 1.

[14] W. Heisenberg, "Physics and Philosophy" (Harper, New York, 1958).

[15] D. Bohm, Phys. Rev. **85** (1952) 166-193.

[16] J.S. Bell, Physics 1, 195 (1964), reprinted in Refs 1. and 12.

[17] E. Schrödinger, Natuurwissenschaften bf 23 (807-812, 823-828, 844-849) translated in Ref. 1.

[18] E.P. Wigner in The Scientist speculates, I.J. Good, ed. reprinted in Ref. 1.

[19] For a recent attempt see e.g. S. Weinberg. Phys. Rev. Lett. **62**, 485 (1989).

[20] J.S. Bell, "Against Measurement" Physics World, August 90, 33-40.

[21] D. Dürr, S. Goldstein and N. Zanghi. J. Stat. Phys. **67**, 843 (1992).

[22] H. Everett III, Rev. Mod. Phys. **29**, 454 (1957).

[23] R. Griffiths, J. Stat. Phys. **36**, 219 (1984).

[24] W.H. Zurek, Phys. Rev. **D 24**, 1516 (1981), Phys. Rev. **D 26**, 1862 (1982).

[25] R. Omnes, Rev. Mod. Phys. **64**, 339 (1992) and references quoted therein.

[26] M. Gell-Mann and J.B. Hartle in "Complexity, Entropy and the Physics of Information", W. Zurek, Editor, Santa Fe Institute Studies in the science of complexity, N° (Addison-Wesley).
M. Gell-Mann and J.B. Hartle, "Alternative Decohering Histories in Quantum Mechanics", Caltech preprint 68-1694, UCSBTH-90-56.
J.B. Hartle "The Quantum Mechanics of Closed Systems" preprint UCSBTH-92-16.

[27] M. Gell-Mann and J.B. Hartle, "Classical Equations for Quantum Systems" preprint UCSBTH-91-15.

[28] H.F. Dowker and J.J. Halliwell "The Quantum Mechanics of History : the Decoherence Functional in Quantum Mechanics" preprint FERMILAB-PUB 92-44-A.

[29] E. Joos and H.D. Zeh, Z. Phys. **B 59**, 223 (1985).

LIST OF PARTICIPANTS

G. Acosta Avalos
(CIVESTAV-IPN.)

J. Appel
(Fermi National Accelerator Lab.)

J. I. Aranda
(CINVESTAV-IPN.)

M. Avila Aoki
(CINVESTAV-IPN.)

A. P. Balachandran
(Univ. Syracuse, EUA.)

Enrique Barradas
(Universidad Autónoma, Puebla.)

G. Branco
(Lisboa)

Alejandro Cabo
(Academia de Ciencias de Cuba.)

Pedro Campos
(CINVESTAV-IPN.)

E. Cantoral
(Univ. Autónoma, Puebla.)

Salvador Carrillo
(CINVESTAV-IPN.)

Heriberto Castilla
(CINVESTAV-IPN.)

Jorge Castiñeiras
(Academia de Ciencias de Cuba.)

Victoria Cerón Angeles
(Universidad Autónoma, Puebla.)

Esther Cervantes Robledo
(Univ. Autónoma, Puebla)

H. G. Compeán
(CIVESTAV-IPN)

J. G. Contreras
(CINVESTAV-IPN)

Ruben Cordero E.
(CINVESTAV-IPN)

Fco. J. Delgado Cepeda
(CINVESTAV-IPN)

J. L. Díaz Cruz
(CINVESTAV-IPN.)

J. Socorro Díaz
(Instituto de Física de la Univ.
de Guanajuato)

Luz María Díaz Rivera
(Univ. Autónoma, Puebla)

K. Eggert
(CERN.)

Joaquín Escalona
(Universidad de Morelos.)

Arturo Fernández
(Univ. Autónoma, Puebla)

Juan B. Flores
(CINVESTAV-IPN.)

Ricardo Gaytán
(CINVESTAV-IPN.)

A. García
(CINVESTAV-IPN.)

A. García Zenteno
(Instituto de Física, UNAM.)

Gabriel Germán
(Instituto de Física, UNAM)

Mario Guerrero Rangel
(Instituto de Física, UNAM)

J.J. Godina
(CINVESTAV-IPN)

José Luis González
(CINVESTAV-IPN)

Gerardo Gutiérrez
(CINVESTAV-IPN)

J.F. Gunion
(Univ. California, Davies)

A. Hernández
(CINVESTAV-IPN)

Angeles Hernández Ruiz
(Univ. Autónoma, Puebla)

Gerardo Herrera
(CINVESTAV-IPN)

Javier Miguel Hernández
(CINVESTAV-IPN)

R. Huerta
(CINVESTAV-IPN)

Rebeca Juárez
(Escuela Superior de Ciencias
Físico-Matemáticas)

Guadalupe López Bueno
(Univ. Autónoma, Puebla)

G. López Castro
(CINVESTAV-IPN)

Juan Carlos López Vieyra
(Universidad Autónoma
Metropolitana-Iztapalapa.)

J.L. Lucio Martínez.
(Instituto de Física de
la Univ. de Guanajuato)

Leonel Magaña
(CINVESTAV-IPN)

Gerardo Martínez
(Univ. Nacional de Colombia)

Rodolfo Martínez
(Facultad de Ciencias, UNAM)

Mario Maya
(Univ. Autónoma, Puebla)

Pablo Medina
(CINVESTAV-IPN)

Rafael Alberto Méndez Sánchez
(CINVESTAV-IPN)

Omar G. Miranda
(CINVESTAV-IPN)

Gerardo Mora Hernández
(Universidad Autónoma
Metropolitana-Iztapalapa

Matías Moreno
(Instituto de Física, UNAM)

List of Participants

P. Kileanowski
(CINVESTAV-IPN)

J. Dionicio Morales
(CINVESTAV-IPN)

Marcos Moshinski
(Instituto de Física, UNAM)

Gregorio Mota Porras
(Instituto de Física, UNAM)

Juan A. Nieto García
(Instituto de Física de
la Univ. de Guananjuato)

Ulises Necamendi
(CINVESTAV-IPN)

Octavio Obregón
(Instituto de Física de
la Univ. de Guanajuato)

Abdel Pérez
(CINVESTAV-IPN)

M. A. Pérez
(CINVESTAV-IPN)

Aurora Pérez
(Academia de Ciencias de Cuba)

A. Pich
(Academia de Ciencias de Cuba)

Luis O. Pimentel
(Universidad Autónoma)
Metropolitana-Iztapalapa

Merced Montesinos
(CINVESTAV-IPN)

Alejandro Noé Morales D.
(Instituto de Estudios Nucleares, UNAM

Alfonso Rosado
(Univ. Autónoma, Puebla)

José O. Rosas Ortíz
(CINVESTAV-IPN)

O. A. Sampayo
(CINVESTAV-IPN)

G. Sánchez Colón
(CINVESTAV-IPN)

Sully A. Sánchez Hdez.
(Univ. Autónoma, Puebla)

Rodolfo Sassot
(Universidad de la
Plata, Argentina)

Antonio del Sol
(Instituto de Física, UNAM)

Modesto Sosa
(Instituto de Física, de la
Univ. de Guanajuato)

Manuel Torres
(Instituto de Física, UNAM)

Eduardo Salvador Tútuti
(Instituto de Física, UNAM

List of Participants

J. Pullin
(Univ. Utah, EUA)

Marco A. Reyes
(CINVESTAV-IPN)

Alfonso Rosado
(Univ. Autónoma, Puebla)

S.A. Tomás Velázquez
(CINVESTAV-IPN)

Luis F. Urrutia
(Instituto Estudios Nucleares, UNAM)

Miguel Vargas
(Instituto de Física de la
Univ. de Guanajuato)

Elsa F. Vázquez Valencia
(CINVESTA-IPN)

D. Weggener
(Univ. Dortmund, Alemania)

J. Weyers
(Univ. Louvain, Bélgica)

José Wudka
(Univ. California, Riverside)

Ricardo Zayas
(Universidad Autónoma, Puebla)

A. Zepeda
(CINVESTAV-IPN)

Author Index

B

Balachandran, A. P., 1
Branco, G. C., 82

P

Pich, A., 95
Pullin, J., 141

R

Ruiz-Altaba, M., 191

W

Wegener, D., 228
Weyers, J., 348, 388

AIP Conference Proceedings

		L.C. Number	ISBN
No. 247	Global Warming: Physics and Facts (Washington, DC, 1991)	91-78423	0-88318-932-1
No. 248	Computer-Aided Statistical Physics (Taipei, Taiwan, 1991)	91-78378	0-88318-942-9
No. 249	The Physics of Particle Accelerators (Upton, NY, 1989, 1990)	92-52843	0-88318-789-2
No. 250	Towards a Unified Picture of Nuclear Dynamics (Nikko, Japan, 1991)	92-70143	0-88318-951-8
No. 251	Superconductivity and its Applications (Buffalo, NY, 1991)	92-52726	1-56396-016-8
No. 252	Accelerator Instrumentation (Newport News, VA, 1991)	92-70356	0-88318-934-8
No. 253	High-Brightness Beams for Advanced Accelerator Applications (College Park, MD, 1991)	92-52705	0-88318-947-X
No. 254	Testing the AGN Paradigm (College Park, MD, 1991)	92-52780	1-56396-009-5
No. 255	Advanced Beam Dynamics Workshop on Effects of Errors in Accelerators, Their Diagnosis and Corrections (Corpus Christi, TX, 1991)	92-52842	1-56396-006-0
No. 256	Slow Dynamics in Condensed Matter (Fukuoka, Japan, 1991)	92-53120	0-88318-938-0
No. 257	Atomic Processes in Plasmas (Portland, ME, 1991)	91-08105	0-88318-939-9
No. 258	Synchrotron Radiation and Dynamic Phenomena (Grenoble, France, 1991)	92-53790	1-56396-008-7
No. 259	Future Directions in Nuclear Physics with 4π Gamma Detection Systems of the New Generation (Strasbourg, France, 1991)	92-53222	0-88318-952-6
No. 260	Computational Quantum Physics (Nashville, TN, 1991)	92-71777	0-88318-933-X
No. 261	Rare and Exclusive B&K Decays and Novel Flavor Factories (Santa Monica, CA, 1991)	92-71873	1-56396-055-9
No. 262	Molecular Electronics—Science and Technology (St. Thomas, Virgin Islands, 1991)	92-72210	1-56396-041-9
No. 263	Stress-Induced Phenomena in Metallization: First International Workshop (Ithaca, NY, 1991)	92-72292	1-56396-082-6

No. 264	Particle Acceleration in Cosmic Plasmas (Newark, DE, 1991)	92-73316	0-88318-948-8
No. 265	Gamma-Ray Bursts (Huntsville, AL, 1991)	92-73456	1-56396-018-4
No. 266	Group Theory in Physics (Cocoyoc, Morelos, Mexico, 1991)	92-73457	1-56396-101-6
No. 267	Electromechanical Coupling of the Solar Atmosphere (Capri, Italy, 1991)	92-82717	1-56396-110-5
No. 268	Photovoltaic Advanced Research & Development Project (Denver, CO, 1992)	92-74159	1-56396-056-7
No. 269	CEBAF 1992 Summer Workshop (Newport News, VA, 1992)	92-75403	1-56396-067-2
No. 270	Time Reversal—The Arthur Rich Memorial Symposium (Ann Arbor, MI, 1991)	92-83852	1-56396-105-9
No. 271	Tenth Symposium Space Nuclear Power and Propulsion (Vols. I–III) (Albuquerque, NM, 1993)	92-75162	1-56396-137-7 (set)
No. 272	Proceedings of the XXVI International Conference on High Energy Physics (Vols. I and II) (Dallas, TX, 1992)	93-70412	1-56396-127-X (set)
No. 273	Superconductivity and Its Applications (Buffalo, NY, 1992)	93-70502	1-56396-189-X
No. 274	VIth International Conference on the Physics of Highly Charged Ions (Manhattan, KS, 1992)	93-70577	1-56396-102-4
No. 275	Atomic Physics 13 (Munich, Germany, 1992)	93-70826	1-56396-057-5
No. 276	Very High Energy Cosmic-Ray Interactions: VIIth International Symposium (Ann Arbor, MI, 1992)	93-71342	1-56396-038-9
No. 277	The World at Risk: Natural Hazards and Climate Change (Cambridge, MA, 1992)	93-71333	1-56396-066-4
No. 278	Back to the Galaxy (College Park, MD, 1992)	93-71543	1-56396-227-6
No. 279	Advanced Accelerator Concepts (Port Jefferson, NY, 1992)	93-71773	1-56396-191-1

No. 280	Compton Gamma-Ray Observatory (St. Louis, MO, 1992)	93-71830	1-56396-104-0
No. 281	Accelerator Instrumentation Fourth Annual Workshop (Berkeley, CA, 1992)	93-072110	1-56396-190-3
No. 282	Quantum 1/f Noise & Other Low Frequency Fluctuations in Electronic Devices (St. Louis, MO, 1992)	93-072366	1-56396-252-7
No. 283	Earth and Space Science Information Systems (Pasadena, CA, 1992)	93-072360	1-56396-094-X
No. 284	US-Japan Workshop on Ion Temperature Gradient-Driven Turbulent Transport (Austin, TX, 1993)	93-72460	1-56396-221-7
No. 285	Noise in Physical Systems and 1/f Fluctuations (St. Louis, MO, 1993)	93-72575	1-56396-270-5
No. 286	Ordering Disorder: Prospect and Retrospect in Condensed Matter Physics: Proceedings of the Indo-U.S. Workshop (Hyderabad, India, 1993)	93-072549	1-56396-255-1
No. 287	Production and Neutralization of Negative Ions and Beams: Sixth International Symposium (Upton, NY, 1992)	93-72821	1-56396-103-2
No. 288	Laser Ablation: Mechanismas and Applications-II: Second International Conference (Knoxville, TN, 1993)	93-73040	1-56396-226-8
No. 289	Radio Frequency Power in Plasmas: Tenth Topical Conference (Boston, MA, 1993)	93-72964	1-56396-264-0
No. 290	Laser Spectroscopy: XIth International Conference (Hot Springs, VA, 1993)	93-73050	1-56396-262-4
No. 291	Prairie View Summer Science Academy (Prairie View, TX, 1992)	93-73081	1-56396-133-4
No. 292	Stability of Particle Motion in Storage Rings (Upton, NY, 1992)	93-73534	1-56396-225-X
No. 293	Polarized Ion Sources and Polarized Gas Targets (Madison, WI, 1993)	93-74102	1-56396-220-9
No. 294	High-Energy Solar Phenomena A New Era of Spacecraft Measurements (Waterville Valley, NH, 1993)	93-74147	1-56396-291-8
No. 295	The Physics of Electronic and Atomic Collisions: XVIII International Conference (Aarhus, Denmark, 1993)	93-74103	1-56396-290-X

No. 296	The Chaos Paradigm: Developments an Applications in Engineering and Science (Mystic, CT, 1993)	93-74146	1-56396-254-3
No. 297	Computational Accelerator Physics (Los Alamos, NM, 1993)	93-74205	1-56396-222-5
No. 298	Ultrafast Reaction Dynamics and Solvent Effects (Royaumont, France, 1993)	93-074354	1-56396-280-2
No. 299	Dense Z-Pinches: Third International Conference (London, 1993)	93-074569	1-56396-297-7
No. 300	Discovery of Weak Neutral Currents: The Weak Interaction Before and After (Santa Monica, CA, 1993)	94-70515	1-56396-306-X
No. 301	Eleventh Symposium Space Nuclear Power and Propulsion (3 Vols.) (Albuquerque, NM, 1994)	92-75162	1-56396-305-1 (Set) 156396-301-9 (pbk. set)
No. 302	Lepton and Photon Interactions/ XVI International Symposium (Ithaca, NY, 1993)	94-70079	1-56396-106-7
No. 303	Slow Positron Beam Techniques for Solids and Surfaces Fifth International Workshop (Jackson Hole, WY 1992)	94-71036	1-56396-267-5
No. 304	The Second Compton Symposium (College Park, MD, 1993)	94-70742	1-56396-261-6
No. 305	Stress-Induced Phenomena in Metallization Second International Workshop (Austin, TX, 1993)	94-70650	1-56396-251-9
No. 306	12th NREL Photovoltaic Program Review (Denver, CO, 1993)	94-70748	1-56396-315-9
No. 307	Gamma-Ray Bursts Second Workshop (Huntsville, AL 1993)	94-71317	1-56396-336-1
No. 308	The Evolution of X-Ray Binaries (College Park, MD 1993)	94-76853	1-56396-329-9
No. 309	High-Pressure Science and Technology—1993 (Colorado Springs, CO 1993)	93-72821	1-56396-219-5 (Set)
No. 310	Analysis of Interplanetary Dust (Houston, TX 1993)	94-71292	1-56396-341-8